Animal Welfare in Animal Agriculture

Husbandry, Stewardship, and Sustainability in Animal Production

EDITED BY

Wilson G. Pond,
Fuller W. Bazer, and
Bernard E. Rollin

CRC Press
Taylor & Francis Group
Boca Raton London New York

CRC Press is an imprint of the
Taylor & Francis Group, an **informa** business

CRC Press
Taylor & Francis Group
6000 Broken Sound Parkway NW, Suite 300
Boca Raton, FL 33487-2742

© 2012 by Taylor & Francis Group, LLC
CRC Press is an imprint of Taylor & Francis Group, an Informa business

No claim to original U.S. Government works

Version Date: 20110621

International Standard Book Number: 978-1-4398-4842-5 (Hardback)

Library of Congress Cataloging-in-Publication Data
Animal welfare in animal agriculture : husbandry, stewardship, and sustainability in animal production / [editors] Wilson G. Pond, Fuller W. Bazer, Bernard E. Rollin.
p. cm.
Includes bibliographical references and index.
ISBN 978-1-4398-4842-5
1. Animal welfare. 2. Livestock--Social aspects. 3. Animal industry--Moral and ethical aspects. I. Pond, Wilson G., 1930- II. Bazer, Fuller Warren, 1938- III. Rollin, Bernard E.
HV4757.A55 2011
179'.3--dc23   2011022090

Visit the Taylor & Francis Web site at
http://www.taylorandfrancis.com

and the CRC Press Web site at
http://www.crcpress.com

Dedication

This book is dedicated to the memory of Dr. Stanley E. Curtis, whose seminal contributions to the advancement of the welfare and well-being of farm animals are legendary. The following excerpt from an article* written by Dr. Curtis in 2007 provides insight into the impact of his long-time contributions to the improvements of farm animal welfare. His lifetime efforts are an inspiration to all who seek to ensure animal well-being everywhere.

An important issue in animal agriculture nowadays is the public demand for evidence that animals on farms and ranches are being treated humanely, that animal state of being (ASB) is high most of the time. But, right now, how should ASB be assessed in production settings?

Important as this question is, scientists have yet to reach consensus as to how to accomplish that task. It is an unsettled area of knowledge that is seriously in need of more concerted attention. Animal-welfare scientists represent several disciplines, and therefore approaches, guiding principles, and vocabularies differ among them. These differences have led to confusion and misunderstanding among interested stakeholders.

Many animal-welfare scientists, following the classic, pioneering contributions of observations and thought by I.J.H. Duncan (Duncan and Wood-Gush, 1971; Duncan, 1996, 2001), have concluded that assessing ASB should be based mostly on animal feelings (Dawkins, 1980; McMillan, 2005). This ultimately may be the ideal methodology. But unfortunately, right now we are unable for certain to measure animal feelings (e.g., anxiety, fear, frustration, and pain) directly, objectively, and scientifically in the laboratory, let alone is it possible to do so in a production setting. ("Measure" herein is used in the sense of "to ascertain the extent or quantity of by comparison with a standard.") As Duncan (2002) has pointed out, the measurement of the behavior patterns postulated to be correlated with negative conscious feelings in animals can itself be objective and scientific. It is at the step of the interpretation of such observations of behavior in terms of any associated ill feelings where the feelings approach is still scientifically uninformed and wanting with respect to the practical usefulness of that approach on farms and ranches today.

So, until such time as we do know how to interpret putative behavioral indicators of reduced animal feelings, and how to quantitatively transform those indicators into valid measures of animal feelings, some are instead advocating the use of objectively measurable animal-performance traits as indicators. The bases of this performance-based approach include 1) the principle that what cannot be measured cannot be managed; 2) the fact that we now can objectively measure productive and reproductive performance traits but not animal feelings; and 3) the fact that reductions in performance traits are early, sensitive indicators that ASB is being deleteriously affected.

Much of the impediment to answering the big question of how to assess ASB may reside in the fact that many — probably most — animal-welfare scientists have virtuously dismissed an approach based on animal functions and performance, favoring instead an approach based mostly or totally

* We are deeply grateful to Dr. Wayne Kellogg, Editor in-Chief, *Professional Animal Scientist,* for his efforts in granting permission to publish in this college textbook the "Introduction" to a manuscript titled "Commentary: Performance Indicates Animal State of Being: A Cinderella Axiom," written by Dr. Curtis and published in *The Professional Animal Scientist* 23(2007): 573–583.

on animal feelings and mind. Some hold that "animal welfare is about how the animal feels" (e.g., Duncan, 1996) and others that "animal welfare is characterized by the absence of behavioral problems" (e.g., Ladewig, 2003). However, still others think that animal functions and performance also are extremely relevant.

Mench (1998a) noted a "growing sense that animal-welfare science has reached an impasse," and this probably owes largely to disagreement over what constitutes farm-animal welfare. This dichotomy epitomizes the spirit of scientific dialogue.

Wilson Pond, Fuller Bazer, and Bernard Rollin, Editors

LITERATURE CITED

Dawkins, M. S. 1980. *Animal Suffering: The Science of Animal Welfare*. Chapman & Hall, London, UK.

Duncan, I. J. H. 1996. Animal welfare defined in terms of feelings. *Acta Agric. Scand. A. Anim. Sci.* (Suppl. 27):29.

Duncan, I. J. H. 2001. Can we understand and use feelings of animals as a concept of animal welfare? In *Food Chain 2001. Proceedings of the European Union Conference*. Uppsala, Sweden. Page 131.

Duncan, I. J. H. 2002. Poultry welfare: Science or subjectivity? *Br. Poult. Sci.* 43:643.

Duncan, I. J. H., and D. G. M. Wood-Gush. 1971. Frustration and aggression in the domestic fowl. *Anim. Behav.* 19:500.

Ladewig, J. L. 2003. Of mice and men: Improved welfare through clinical ethology. In *Proceedings of the 37th International Congress of the International Society of Applied Ethology*, Abano Term, Italy. Page 29.

McMillan, F. D., Ed. 2005. *Mental Health and Well-Being in Animals*. Blackwell Publishing Professional, Ames, IA.

Mench, J. A. 1998a. Thirty years after Brambell: Whither animal welfare science? *J. Appl. Anim. Welfare Sci.* 1:91.

Contents

SECTION One   Roles of Animals in Society

SECTION Two   Treatment of Animals and Societal Concerns

SECTION Three   *Sustainable Plant and Animal Agriculture for Animal Welfare*

Forewords

ACADEMIC

Students in the twenty-first century are learning in an environment where science and technology advance at a rate that encourages rapid dissemination and implementation of ideas. Deliberations on the morality and ethics of resulting changes occur at a much slower rate, and generally not in the same courses that teach the science. Hence, many individuals have perspectives on animal welfare that are largely influenced by public debate in the mass media, particularly electronic media.

An understanding of animal welfare within the food system and of how and where changes in production systems might need to be made requires the integration of knowledge from many fields, the antithesis of the "pigeon-holing" that occurs so easily in academic programs and student minds. Science and economics play critical roles, but must be seen within the context of how modern production systems evolved. The "sound bite" approach to ethics leads to misconceptions such as large-scale production systems being invariably careless of animal welfare and the idea that humanity "enslaved" other species for strictly selfish purposes.

Students need to be given opportunities to look at animal welfare in a context that includes the historical development of animal domestication and of modern animal production systems. Animal behavior played a critical role in the determination of which species were amenable to domestication, yet it has not had a recognized status alongside genetics, nutrition, and physiology in most animal science curricula. Behavior is an essential monitor of animal welfare, especially in intensive systems, and, as such, needs to be better integrated into curricula. The growth in size and intensity of production units is in response to the explosive growth in the human population and its food demands, with animal and human behavior and welfare intimately connected.

Academia changes with glacial speed in comparison to the world of applied science and technology. Tradition and fiscal constraints mean that offered courses often fail to give students the training and encouragement to integrate concepts across disciplines. The volume of factual material expands constantly and conscious efforts must be made to offer course time that requires and encourages students to think through the issues related to animal welfare. This ability to reflect and integrate is essential if humans are to be capable of evaluating and improving animal welfare in modern production systems.

Elizabeth Oltenacu, PhD
Emerita, Department of Animal Science
Cornell University, Ithaca, NY

Public policy regarding the welfare of livestock and other animal species must be based on science and reason, not emotion. There is more need now for objective research and an informed public than ever before. Academia has been described as being largely preoccupied with lofty, remote, or intellectual pursuits, rather than those of practical application. In reality, academia is highly responsive to changing public attitudes and concerns, and the public is becoming increasingly interested in animal welfare. Academic institutions must compete for funding from public and private sources. Competition is also keen for the best students and for the reputation of being cutting-edge and relevant.

Colleges and universities originally taught the art of animal husbandry, but when public interest in science increased after World War II, the term *husbandry* was dropped in favor of the word *science*. Recently, there has been renewed interest in changing the names of the courses that teach husbandry back to *husbandry*, but as those courses are now based on the latest advances in science, a name change is not likely.

At most colleges and universities, courses and extracurricular opportunities are reviewed regularly by faculty peer groups and administrators. Input from students, alumni, and employers of graduates are often solicited and may be directly incorporated into the review process. Although many academic departments may wish to start new courses on farm animal welfare and related issues, new courses and faculty have been difficult to add during periods of tight budgets. Many programs, however, are responding by updating their existing courses. For example, many species-oriented production (husbandry) courses, meats courses, applied ethology, ethics, and capstone courses are adding modules on the audit process. Audits (Chapter 6) are a system to ensure that good husbandry practices are being followed, so they are a natural fit into classes that already teach the latest husbandry practices. These courses may also devote more time to the latest events affecting animal welfare issues.

Extracurricular programs that provide additional opportunities for students to get involved in animal welfare-related activities have greatly increased. In addition to the traditional judging teams, students on many campuses have organized clubs that assist local shelters, or are otherwise involved in animal rescue or similar projects. Quiz bowls in which students compete based on their knowledge of animal husbandry have been popular for many decades. A particularly innovative program is the annual Intercollegiate Animal Welfare and Assessment Judging Contest pioneered at Michigan State University. Colleges and universities from Canada and the United States are invited to send teams to two days of seminars and competition.

Interest in the field of animal welfare science has grown so much over the past 30 years that there is a shortage of professionals with graduate training in the United States. For example, the USDA's Food and Agricultural Sciences National Needs Graduate and Postgraduate Fellowship Grants Program for 2010 listed "animal well-being (ethologists; bioethicists)" as their highest priority-targeted expertise shortage area.

One of the main goals of academia is to stimulate people to think critically and seek out alternative viewpoints. Most agricultural animal well-being issues are not simple, although special interest groups on both sides of the issue often promote a simplistic version. With many electronic, print, and other sources of information readily available, people can easily pick the news sound bites and entertainment that come closest to their personal biases and avoid exposure to the other sides of many issues.

Funding is the biggest single problem facing researchers in farm animal welfare science. Producer and commodity groups have and continue to make significant contributions to animal welfare research, although their resources are very limited. The USDA's competitive grants programs have been the largest source of funding in the United States, although the funds need to be greatly increased and the success rates of receiving funding for proposals submitted to the program are generally 20% or less. People often ask animal welfare and activist groups for assistance in funding research projects, but the answer is almost always *no*. One problem is what is known in the business as "the vegan police," the more radical members who do not support any research.

Extension programs have been at the forefront of creating quality assurance and auditing programs that have had an industry-wide impact. Most major meetings of state and national producer organizations include demonstrations of low-stress handling, and those demonstrations attract the largest crowds. Educational programs on proper animal handling, best practices, auditing, and emergency euthanasia of livestock are not only in demand at extension meetings with farmers and ranchers, but are also requested by auction barns, slaughter plants, and livestock transport companies.

In conclusion, academia is needed more than ever to help policymakers and the public make rational decisions regarding animal welfare, environmental, and ethical issues.

Ted H. Friend, PhD
Department of Animal Science
Texas A & M University, College Station, TX

In the past, welfare research has concentrated on prevention of negative welfare aspects such as hunger, thirst, inadequate feed, injuries, disease, and fear or chronic stress. The current research is more focused on stimulation of positive welfare aspects. Welfare is more than prevention of suffering. It also includes the satisfaction of desires and needs of animals.

Current modern housing systems are poorly designed when considering the behavioral and adaptive needs of animals. Systems are often simple in design and boring to live in with no distraction material other than the group mates of the animal. Routine treatments such as tail docking and beak trimming have to be used to allow animals to survive and produce well in these systems. This is part of the reason that welfare of farm animals is often so poorly perceived in public opinion.

Animals like pigs and poultry prefer a rich environment because of their behavioral needs to play (which is important to develop their social skills) and to root (to find feed).

Several recent developments in animal science and related disciplines show that environmental enrichment can have significant effects on prevention of maladaptive behavior such as tail biting in pigs and has stress-reducing effects, improves feed intake, and prevents diarrhea in piglets around weaning. The enrichment material (e.g., long straw, wood branches, or peat) should be ingestible, odorous, chewable, deformable, and destructible and should be replenished regularly.

Such enrichment measures result in satisfaction of desires and needs and therefore contribute to positive welfare. Moreover, the animals also seem more robust when going though transitions like weaning in piglets, suggesting that improved welfare and improved production go hand in hand. From a welfare and production point of view, it is therefore important that experts in the field of behavioral sciences join forces with system designers to design systems that are built based on behavioral and adaptive needs of animals instead of breeding animals that will fit the current systems. The latter route will bring us to ethical discussion on whether animals' intrinsic values may be changed to fit our current systems. In addition, systems built on behavioral and adaptive needs of animals must be realistic, ecologically sound, and economically viable to be successful.

Implementation of welfare in practice has become an interdisciplinary challenge where animal scientists, system designers, ecologists, and economists must join forces. Is it realistic to think that such systems will get a place in a world where low-cost prices for meat are so important? The public concern about animal welfare is increasing and retailers and governments are well aware of this. In Western Europe, cage housing for layer hens soon will be forbidden by law and retailers demand pregnant sows to be non-tethered. A recently developed welfare-friendly system for laying hens was supported by welfare organizations, and eggs from this system are sold by retailers. Animal products from those new systems, which are perceived better by the public, may get a bigger share of the market, thereby helping the producers of those products. Therefore, we think that the time is here to meet the challenges by research using a multidisciplinary approach. This multidisciplinary approach should also have a place in our teaching of undergraduate and graduate courses at universities and in training of students at other schools. First, students must gain knowledge of different aspects of animal welfare, and then integrate this knowledge using system design and analyses.

Bas Kemp, PhD, and Martin Verstegen, PhD
Department of Animal Science
Wageningen University, Wageningen, the Netherlands

COMMERCIAL OPERATIONS

Scientists studying animal behavior, pain perception, and other issues relevant to animal welfare provide information that can be used to determine the effects of different production systems and practices on animal welfare. Science provides information that can be used to make ethical decisions, but it cannot provide all the answers. For example, a scientific experiment can provide data indicating that a certain procedure causes pain, but it cannot provide an ethical judgment on how much pain is acceptable. Furthermore, there may be differences of opinion on what is ethical. This is one of the reasons there are so many different animal agricultural practices all over the world. Economics is also a big factor. Practices detrimental to animal welfare may be used to lower costs. For example, the productivity of each individual laying hen is decreased when too many hens are jammed into a small cage. However, the overall cost for the eggs may be lower because fewer expensive buildings are required. The individual hen may suffer in the process of lowering the cost of eggs. Some of the main factors that compromise animal welfare include the following:

INADEQUATE MANAGEMENT AND LACK OF EMPLOYEE SUPERVISION

Some of the worst abusive treatment of animals occurs when overworked, poorly supervised employees commit acts of abuse and cruelty. Some examples are beating animals, dragging a crippled animal, throwing small animals, or jabbing them with sharp objects. Abusive practices can occur on both large and small farms. Many people assume that big farms have more abuse problems, but size is not a determining factor. The most effective way to prevent abuse is through good management.

NEGLECT

Starvation or inadequate diets are examples of neglect. Allowing manure to build up in an animal's stall until the animal is covered in filth is also neglect. Neglect can happen on both large and small farms.

ANIMAL BEHAVIORAL PREFERENCES IN INTENSIVE SYSTEMS VERSUS EXTENSIVE SYSTEMS

Almost everyone who cares about animal welfare can agree that deliberate abuse of animals and neglect are very detrimental to animal welfare. However, there is a much greater controversy and disagreement on an animal's behavioral needs. Scientists can measure, in an objective manner, an animal's motivation for an environmental enrichment such as straw for pigs to chew on or a secluded nest box for a laying hen. Research shows very clearly that animals prefer specific amenities. Therefore, to provide an acceptable level of animal welfare in an intensive animal production system, environmental enrichments are needed to satisfy what the animals "want" most.

Examples of extensive systems of animal production are grass-fed beef and free-range chickens. Producers in this extensive segment will sell to high-end markets of affluent, concerned consumers. Intensive segments of animal production will remain large-scale commercial producers who will sell animal products at more affordable prices. This sector will need to eliminate some of the most objectionable practices such as sow gestation stalls and small, cramped chicken cages. To provide affordable animal products, these systems will have to be intensive, but must also provide for the most highly motivated behavioral needs. One example that is already being implemented is colony housing for hens that provides nest boxes, perches, and a place to scratch.

BIOLOGICAL SYSTEM OVERLOAD

I predict that biological system overload will become one of the most serious animal welfare problems in the future. Animals have been pushed to produce more and more milk, meat, or eggs, and problems with lameness and weakness have already increased since the 1980s and may get worse.

Lameness in dairy cows has greatly increased and some pigs with heavy muscles are too weak to walk through the stockyard at a meat plant. There is a point where animal productivity should no longer be increased because the animal has difficulty functioning. Managers should strive for optimal productivity rather than maximum productivity. A dairy cow that lasts for three or four years of milking would probably be a good tradeoff between productivity, cost, and welfare compared to a cow that lasts for only two years of milking.

ECONOMIC FACTORS

Economic pressures can cause producers to cut corners and compromise animal welfare, but economic factors can also be forces to improve animal welfare. The treatment of animals at slaughter plants greatly improved after McDonald's Corporation and other restaurant companies started auditing slaughter plants. Large buyers are in a position to drive positive change. Handling and transport practices will improve when people are held financially accountable for death losses and injuries. When I worked with the restaurant companies to implement animal welfare audits, I saw huge improvements. Large buyers have the economic clout to enforce standards. This is why I spend large portions of my time working with large buyers of animal products to develop standards and conduct audits. The need for grocery stores and restaurants to audit animal welfare is equally important for both conventional agriculture and the organic/ natural sectors.

MEASURING WELFARE IS ESSENTIAL

People are able to manage the things that they can measure. To maintain high standards, managers need to measure welfare indicators such as the percentage of lame animals, skinny animals, animals with sores, animals with abnormal behavior, or dirty animals. In organic operations, coat condition should also be evaluated because lice treatments are often not used and bald spots on untreated cattle are not acceptable. Measuring is essential to prevent "bad from becoming normal." If a producer gets used to seeing a high percentage of lame cows, he or she may start to think that is normal. Animal handling should also be measured to prevent handling practices from reverting to being rough and inappropriate. Variables such as the percentage of immobile animals falling down or the percentage of those vocalizing during handling can be measured. Measurement enables a producer to determine if welfare is getting better or getting worse. Productivity is routinely measured. Welfare indicators should also be measured.

Temple Grandin, PhD
Department of Animal Science
Colorado State University, Fort Collins, CO
Grandin Livestock Handling Systems, Inc.

Preface

Animal welfare is a topic of great interest and importance to society. Animals are used for companionship, service, research, food, fiber, and by-products. Ongoing efforts to ensure the well-being and comfort of food animals are imperative for fulfillment of sustainable agriculture. Animal source foods provide important nutrients in the diets of humans and animals. A major challenge for society is the maintenance of a stable environment to support human and animal needs. Our intent is to link the societal challenge of sustaining animal and human welfare with a strong and viable food system ensured by stewardship of land, crops, animals, and natural resources.

The book is presented in three parts: Section 1: Roles of Animals in Society, Chapters 1–3; Section 2: Treatment of Animals and Societal Concerns, Chapters 4–8; and Section 3: Sustainable Plant and Animal Agriculture for Animal Welfare, Chapters 9–14. The Forewords, written by individuals representing academia and industry, underscore the need for the animal welfare discussion in this textbook. Increases in food production have occurred because of scientific, technological, and global marketing advances. New knowledge in soil, water, crop, and animal science has increased concurrently with advances in transportation and communication. This industrialization of agriculture has created urban societies in which the vast majority have little awareness and understanding of agriculture and food production. For example, during the 1950s, approximately 20% of the U.S. workforce was in farming; in 2011, the figure is approximately 1%.

A major challenge for society in the coming decades is to provide sufficient global food to meet the needs of an increasing human population. Demand for animal source foods is growing, especially in developing countries, to counter widespread malnutrition that continues to be a major insult to infants and children.

During the past 40 years, economics improved and per capita consumption of milk, meat, and eggs in developing countries has increased. In contrast, during the same period in the developed countries, average per capita animal source food consumption has declined slightly.

The care and welfare of all animals is a high priority for society. A prominent milestone in this movement began with the exposure a century ago of questionable practices used in animal slaughter plants. Progress in animal welfare reforms and oversight is an ongoing effort by those engaged in food animal production and laboratory animal care.

Concurrent with these ongoing efforts in animal welfare reform, several small but well-funded organizations are active in promoting efforts to curtail or eradicate food animal production and the use of laboratory animals in biomedical research. Such efforts may affect animal source food production and the use of animals in biomedical and agricultural research. Consequently, the nutritional and physiological well-being of infants, children, and other vulnerable humans is at risk, particularly in developing countries. However, it is important to distinguish between abolitionists, who accept no legitimate animal use, and those who seek to improve the treatment and well-being of food animals as well as animals used in biomedical and agricultural research.

This book is intended to provide a framework for open discussions related to those issues that embrace the concepts of nutrition, animal welfare, and freedom of food choices. Chapter authors are highly qualified and recognized experts in their respective fields of teaching, research, and public service. The book is primarily written for undergraduate college students in varying fields of study: animal sciences, animal behavior, animal welfare, plant sciences, environmental sustainability, sociology, economics, and nutrition. The subject of animal welfare reaches across society in general, both urban and rural, and has a significant impact on consumer attitudes and choices.

ACKNOWLEDGMENTS

The editors are grateful to many people for their efforts, individually and collectively, in producing this textbook. We are indebted to John Sulzycki and Jill Jurgensen of Taylor & Francis/CRC Press for their support, chapter authors for their leadership and team spirit in approaching a complex book structure requiring cooperation and collaboration among 40 authors across academic disciplines, interests, and perspectives, and Marsha Pond for her dedicated editorial assistance.

Contributor List

Aaron Alejandro, BS
Texas FFA Foundation
Austin, Texas

Kenneth Anderson, PhD
North Carolina State University
Raleigh, North Carolina

David Andrews, JD
Food & Water Watch
Washington, DC

Charles Arnot, APR
CMA Consulting and the Center for Food
    Integrity
Kansas City, Missouri

Joseph L. Baumert, PhD
University of Nebraska
Lincoln, Nebraska

Fuller W. Bazer, PhD
Texas A & M University
College Station, Texas
and
Seoul National University
Seoul, Republic of Korea

Christopher Boleman, PhD
Texas 4-H and Youth Development
    Texas AgriLife Extension Service
College Station, Texas

Donald M. Broom, PhD
University of Cambridge
Cambridge, UK

Judith L. Capper, PhD
Washington State University
Pullman, Washington

Cherie Carrabba, BS
Texas Junior Livestock Association
Crockett, Texas

J. S. Caton, PhD
North Dakota State University
Fargo, North Dakota

Terry Engle, PhD
Colorado State University
Fort Collins, Colorado

David Fraser, CM, PhD
University of British Columbia
Vancouver, Canada

Ted H. Friend, PhD
Texas A&M University
College Station, Texas

**Gail C. Golab, PhD, DVM, MACVSc
(Animal Welfare)**
Director, Animal Welfare Division
American Veterinary Medical Association
Shaumberg, Illinois

Temple Grandin, PhD
Grandin Livestock Handling Systems, Inc.
Colorado State University
Fort Collins, Colorado

Christa Hanson, PhD
Iowa State University Extension
Des Moines, Iowa

Bas Kemp, PhD
Wageningen University
Wageningen, The Netherlands

Frederick Kirschenmann, PhD
Iowa State University
Ames, Iowa

Duane C. Kraemer, DVM, PhD
Texas A & M University
College Station, Texas

Gregory Lardy, PhD
North Dakota State University
Fargo, North Dakota

Robert P. Martin
Pew Environment Group
Washington, DC

John J. McGlone, PhD
Texas Tech University
Lubbock, Texas

Alan McHughen, DPhil
University of California
Riverside, California

Christian E. Newcomer, VMD
Association for Assessment of Accreditation of
    Laboratory Animal Care International
Frederick, Maryland

Elizabeth Oltenacu, PhD
Cornell University
Ithaca, New York

Fred Owens, PhD
Pioneer Hi-Bred Int'l, a DuPont Business
Johnston, Iowa

R. Anne Pearson, PhD
University of Edinburgh
Roslin Midlothian, UK

Kevin R. Pond, PhD
Texas Tech University
Lubbock, Texas

Wilson G. Pond, PhD
Cornell University
Ithaca, New York

Bernard E. Rollin, PhD
Colorado State University
Fort Collins, Colorado

Paul Shapiro
Humane Society of the United States
Washington, DC

John N. Sofos, PhD
Colorado State University
Fort Collins, Colorado

Jarret D. Stopforth, PhD
Purac America, Inc.
Lincolnshire, Illinois

Steve L. Taylor, PhD
University of Nebraska
Lincoln, Nebraska

Paul B. Thompson, PhD
Michigan State University
East Lansing, Michigan

Duane Ullrey, PhD
Michigan State University
East Lansing, Michigan

Vincent Varel, PhD
USDA, ARS, U.S. Meat Animal
    Research Center
Clay Center, Nebraska

Martin Verstegen, PhD
Wagenigen University
Wageningen, The Netherlands

James Wells, PhD
USDA, ARS, U.S. Meat Animal
    Research Center
Clay Center, Nebraska

Section One

Roles of Animals in Society

1 Perspectives on Emergence of Contemporary Animal Agriculture in the Mid-twentieth Century

The Decline of Husbandry and the Rise of the Industrial Model

Bernard E. Rollin and Paul B. Thompson

CONTENTS

The domestication of animals occurred some 10,000 years ago and represented a milestone for the history of human civilization. The origin and sequence of domestication is a hotly debated topic among anthropologists and historians. Richard Bulliet (2005) argues that animals were probably first kept in captivity for use in sacrificial rites. This practice allowed ancient civilizations to observe which species were tame enough for use as work animals. Animals, notably cattle, provided labor and locomotion when they were harnessed to plows, sledges, and wagons beginning in about 4000 BC. Thus, animal agriculture was indispensable to accelerating the development of crop agriculture. The flesh and hides of sacrificial animals were routinely consumed by those in the royal house or the priesthood. Eventually, the habit of having the animals under human control at all times provided a constant and consistent food supply ready at hand. It also thereby created the leisure time necessary to societal progress.

However domestication actually occurred, humans selected among animals congenial to human management, and further shaped them in terms of temperament and production traits by breeding and artificial selection. These animals included cattle—dubbed by Calvin Schwabe the "mother of the human race"—sheep, goats, horses, dogs, poultry and other birds, swine, ungulates, and other animals capable of domestication. The animals provided food and fiber (meat, milk, wool, and leather); power to haul and plow; transportation; and served as weaponry (horses and elephants). As people grew more effective at breeding and managing the animals, productivity increased. As humans benefited, so arguably did the animals. They were provided with the necessities of life in a predictable way. Thus was born the concept of husbandry—the remarkable practice and articulation of the symbiotic contract humans made with farm animals.

"Husbandry" is derived from the Old Norse words "hus" and "bond"; the animals were bonded to one's household. The essence of husbandry was *care*. Humans put animals into the most ideal environment possible for the animals to survive and thrive, the environment for which they had evolved and been selected. In addition, humans provided them with sustenance, water, shelter, protection from predation, medical attention (as was available), help in birthing, food during famine, water during drought, safe surroundings, and comfortable appointments. Eventually, what was born of necessity and common sense became articulated in terms of a moral obligation inextricably bound up with self-interest. In the biblical story of Noah, we learn that even as God preserves humans, humans preserve animals. The ethic of husbandry is, in fact, taught throughout the Bible—animals must rest on the Sabbath even as we do; one is not to seethe a calf in its mother's milk (so we do not grow insensitive to animals needs and natures); and we can violate the Sabbath to save an animal. Proverbs tells us "the wise man cares for his animals." The Old Testament is replete with injunctions against inflicting unnecessary pain and suffering on animals, as exemplified in the strange story of Balaam who beats his ass, and is reprimanded by the animal's speaking through the grace of God.

The true power of the husbandry ethic is best expressed in the 23rd Psalm. There, in searching for an apt metaphor for God's ideal relationship to humans, the Psalmist invokes the good shepherd:

> The Lord is My shepherd; I shall not want.
> He maketh me to lie down in green pastures:
> He leadeth me beside the still waters.
> He restoreth my soul.

We want no more from God than what the good shepherd provides to his animals. Indeed, consider a lamb in ancient Judaea. Without a shepherd, the animal would not easily find forage or water, would not survive the multitude of predators the Bible tells us prowled the land—lions, jackals, hyenas, birds of prey, and wild dogs. Under the aegis of the shepherd, the lamb lives well and safely. In return, the animals provide their products and sometimes their lives, but while they live, they live well. Even slaughter, the taking of the animal's life, must be as painless as possible, performed with a sharp knife by a trained person to avoid unnecessary pain. Ritual slaughter was, in antiquity, a far kinder death than bludgeoning; most importantly, it was the most humane modality available at the time (despite its questionable status today).

The metaphor of the good shepherd is emblazoned in the Western mind. Jesus is depicted as both shepherd and lamb from the origin of Christianity until the present in paintings, literature, song, statuary, and poetry as well as in sermons. To this day, ministers are called shepherds of their congregation, and the word "pastor" is derived from "pastoral." In addition, when Plato discusses the ideal political ruler in the *Republic*, he deploys the shepherd–sheep metaphor: The ruler is to his people as the shepherd is to his flock. Qua shepherd, the shepherd exists to protect, preserve, and improve the sheep; any payment tendered to him is in his capacity as wage earner. So too the ruler again illustrates the power of the concept of husbandry on our psyches. Because of its close connection to God's putative relation to humans, husbandry has traditionally been a favored topic for sermons and homilies in the Judeo-Christian tradition. The concept of husbandry was regularly emphasized in the education of the young, both as a foundation for agriculture and as an exemplary value to reflect upon. Viewed from the perspective of agricultural ethics, the singular beauty of husbandry is that it was both an ethical and prudential doctrine. It was prudential in that failure to observe husbandry inexorably led to ruination of the person keeping animals. Not feeding, not watering, not protecting from predators, not respecting the animals' physical, biological, and physiological needs and natures, what Aristotle called their *telos*—the "cowness of the cow," the "sheepness of the sheep"—meant your animals did not survive and thrive, and thus neither did you. Failure to know and respect the animal's needs and natures had the same effect. Indeed, even Aristotle, whose worldview was fully hierarchical with humans at the top, implicitly recognized the contractual nature of husbandry when he off-handedly affirmed that although the natural role of animals is to serve man, domestic animals are "preserved" through so doing. The ultimate sanction of failing at husbandry—erosion of self-interest—obviated the need for any detailed ethical exposition of moral rules for husbandry. Anyone unmoved by self-interest is unlikely to be moved by moral or legal injunctions! Yet although one finds little written about animal ethics and little codification of that ethic in law before the twentieth century, there is no reason to suppose that husbandry was not also conceived in ethical terms. Indeed, the religious tradition discussed previously suggests just the opposite. If the shepherd did not tend his flock from a perspective of ethical compassion (along with self-interest), how could the metaphor of God as "my shepherd" have attained the resonance and meaning that it evidently has?

Given the overlap between ethics and self-interest in traditional husbandry, the bulk of what was articulated in animal ethics aimed at identifying overt, deliberate, sadistic cruelty, hurting an animal for no purpose or for perverse pleasure, or not providing food or water. The biblical prohibition against animal cruelty was continued and augmented in the rabbinical tradition as *Tsaar Baalei Chaim*—the suffering of living things. The prohibition against yoking an ox and an ass to the same plow arises out of concern of stress on the weaker animal. At the same time, of course, the Bible is replete with commandments that encourage good husbandry. Concern for cruelty to animals arises in the Catholic tradition in the writings of St. Thomas Aquinas. Despite the fact that animals enjoy no moral status in Catholic theology, Aquinas strictly forbids cruelty on the grounds (buttressed by modern psychology) that cruelty to animals leads inexorably to cruelty to humans.

Despite the sound and Solomonic basis for husbandry and its long history, this simple ethic was dealt a serious blow in the twentieth century. It is essential to stress that the widespread loss of husbandry among some producers was not the result of malice or thoughtlessness. It occurred through the eventual maturation of change processes that had long been at work in agricultural systems of European origin, ushered along by a series of technological innovations that were themselves accelerated in the years following World War II. By the closing decades of the twentieth century in some environments, these change processes had supplanted the ideas that had supported a relatively benign on-farm relationship between livestock and their human caregivers over the preceding centuries. By 1980, the philosophical vision of farming that held sway throughout the United States and other nations of European settlement had been swept away by a new understanding. In this new way of seeing things, agriculture is just another sector in the industrial economy. Like the energy or manufacturing sectors, the role of agriculture is to bring forth commodities for consumption in the marketplace, and to do so at the least possible cost. These changes were not brought about by a lack of concern for animals. The forces that created this philosophical revolution in the way that scientists, policymakers, and opinion leaders thought of agriculture are not uniquely or even primarily focused on the livestock sector.

Industrial agriculture is the inevitable result of unconstrained technological innovation on the one hand, combined with a singular neglect of the food system's unique contributions to quality of life on the other. The technology piece of the change process gave us industrial agriculture as a simple result of agricultural economics. Farm productivity is the ratio of farm output over input. Inputs include land, labor, and purchased goods such as seed, feed, fertilizer, and equipment. Outputs include salable farm products: in the animal sector, meat, milk, eggs, and animal by-products such as hides. A change in technology increases productivity when the new tools or techniques being used increase the outputs in the form of salable products while keeping the inputs in the form of land, labor, and other purchased goods constant. For an individual farm, an increase in productivity means that the farmer has more to sell. This is a good thing for the farmer as long as the price received for those commodity goods stays the same. With more to sell, the farmer has more income. The hitch is that as the new technology is widely adopted by other farmers, the entire farm sector has more to sell, and this creates a problem in agriculture that fuels the process of industrialization.

According to Economics 101, when supply goes up, prices must come down. Thus, as farm productivity grows, the total supply of farm commodities grows with it and prices fall. Eventually the farmer is back where he started. The ultimate benefit of an increase in productivity is passed on to consumers, who enjoy lower prices for food. However, something important has gone on in the meantime. Those farmers that adopted the new tools and techniques early made windfall profits before prices fell, while farmers who were late to adopt them were stuck with the problem of having to sell their meat, milk, and eggs for less than it cost to produce them. This, as any student of economics knows, leads to bankruptcy. When the bankrupt farms go up for auction, the early adopters are sitting there with windfall profits in their pockets, anxious to buy up the bankrupt farms. Agricultural economists call this the "technology treadmill." An individual farmer is running harder (producing more) to stay in the same place (maintain the same income). At the same time, less productive (and usually smaller) producers are constantly going bankrupt and leaving farming, while the ones still on the treadmill are getting bigger and bigger. When still newer tools and techniques come along, this process repeats itself all over again.

There are several ethical points to learn from the technology treadmill. The first point is that no farmer can afford *not* to adopt the most productive, state-of-the-art tools and techniques, and the smart ones are always the first to do so. If other farmers are producing for less, market prices will eventually adjust to reflect that fact, and the "laggard" (this is actually the term that rural sociologists once used to describe late adopters) will be forced to go out of business. From the individual farmer's perspective, there is no ethical choice to be made. Either you use the most productive technology or you are not a farmer at all. There is no point in trying to blame producers for this as a matter of ethics. They literally have no choice. The second point is if this were all that there was to say about the economics of farming, then there would be strong ethical arguments for thinking that the technology treadmill is a good thing. It is obviously *not* a good thing for the smaller, less productive farmers who are losing their farms, but it is important to remember that the cost of food is constantly coming down with every turn of the treadmill. This decline in the cost of food is a good thing for people who buy food. It is an especially good thing for people who spend a comparatively large portion of their income on food (i.e., the poor). Several generations of agricultural economists and policymakers were so impressed by this logic during the twentieth century that urging farmers to "get big or get out" was official U.S. government policy (Thompson, 2010).

However, there is more to the story.

Between the two World Wars, agricultural scientists and government officials became extremely concerned about supplying the U.S. public with enough cheap and plentiful food. First, after the Dust Bowl and the Great Depression, many people in agriculture had soured on farming. Agriculture was always subject to the vagaries of weather and economics, but never in U.S. history to the staggering extremes experienced in the unpredictable and incomprehensible events over which the individual was powerless. Second, reasonable predictions of urban and suburban encroachment on agricultural land were being made, with a resultant loss of land for food production. This tendency has in fact continued through the present. Today, rural property that was formerly used for dryland farming of winter wheat now can sell for $60,000 per acre for development use. Moreover, as farmland is developed into housing, homeowners do not wish to live next to animal production units that create odor and dust. Third, many farm people had been sent to both foreign and domestic urban centers as military personnel during both World Wars, thereby creating in them a reluctance to return to rural areas lacking in excitement and amenities. This problem is well illustrated by the post-World War I song, "How 'Ya Gonna Keep 'Em Down on the Farm (After They've Seen Paree)?" Fourth, having experienced the specter of literal starvation during the Great Depression, the American consumer was, for the first time in our history, fearful of an insufficient food supply. Fifth, projection of major population increases (that in fact happened) further fueled concern. Sixth, promises of better jobs in cities, for example in the automotive industry in Detroit, lured farm workers out of agricultural areas into urban areas by the promise of higher income than could be made on farms.

When the considerations of loss of land and diminution of agricultural labor are coupled with the rapid development of a variety of technological modalities relevant to agriculture during and after World War II and with the burgeoning belief in technologically based economics of scale, it was probably inevitable that animal agriculture would become subject to industrialization. This was a major departure from traditional agriculture and a fundamental change in agricultural core values—industrial values of efficiency and productivity replaced and eclipsed the traditional values of "way of life" and husbandry. Husbandry-based animal agriculture was about putting square pegs in square holes, round pegs in round holes, and creating as little friction as possible doing so. Animal welfare was linked conceptually to productivity—harming the animal's welfare diminished its productivity. To be sure, people did not always pursue their own interest and could be sloppy or abrasive in animal care despite the concomitant loss of productivity. However, the key point was that the two were closely tied together. As industrial agriculture began to take hold, academic departments of animal husbandry changed their names to departments of animal science, symbolically betokening a move to industry. Animal science, in fact, is defined in textbooks as the application of industrial methods to the production of animals. No husbandry person would ever dream of keeping animals evolved for extensive grazing confined in small cages. No husbandry person would ever dream of feeding blood and bone meal, poultry waste, or cement dust to farm animals, but such "innovations" are entailed by industrial/efficiency mindset and applied research.

With the industrialization of agriculture, people no longer needed to put square pegs in square holes, round pegs in round holes, but by using "technological sanders," could force square pegs into round holes and round pegs into square holes. In other words, animals could be placed into environments and housing systems that violated their biological and psychological natures without harming their productivity. Antibiotics, vaccines, bacterins, hormones, air-handling systems, and other technological innovations allowed us to put animals where their needs and natures were not met, where suffering in fact occurred. In a traditional husbandry system, these practices could have reduced farm productivity, but in the industrial system, they increased farm productivity from the economic standpoint. Using technology, productivity was severed from animal welfare. For example, the economically most efficient way to produce eggs maximizes the number of eggs produced per barn, rather than per bird. A modern poultry barn costs hundreds of thousands of dollars, while a chicken costs only a few cents. Stocking densities that maximize productivity sacrifice animal health in order to get the best return on the total investment.* Whereas, in husbandry agriculture, productivity and animal welfare went hand-in-hand, they were disconnected under an industrial approach, with animals suffering, but in ways irrelevant to productivity. However, small husbandry farms, operating on smaller profit margins, still exist today in the United States and worldwide.

By the last quarter of the twentieth century, a significant portion of animal agriculture had been channeled into industrialized confinement in the United States, Europe, Latin America, and Asia. Machines replaced human skilled labor, and industrialized agriculturalists boasted that agricultural intelligence was in the systems, not in husbandry-trained workers. Husbandry was often supplanted by industry in many areas of animal agriculture except for extensive sheep and cattle ranching. In these cases, not only was animal welfare adversely affected, but also new problems for agriculture arose. One issue was sustainability: in extensive cattle ranching, environmental sustainability was assured because if a cattle rancher overgrazed his pasture land, he essentially lost his livelihood. Industrial agriculture, on the other hand, did not represent a self-sustaining balanced equilibrium. A detailed account of the problems created by the industrialization of animal agriculture is presented in Chapter 4, but they are worth a brief summary here.

1. Environmental—Inexpensive fossil fuels are one of the main drivers for industrialization in all of agriculture, including animal production. Furthermore, such operations generate enormous amounts of manure. Unlike the valuable role of manure in pastoral agriculture, where it nourishes the soil, in confinement manure becomes a potential pollutant. Excess manure leaches into ground water and pours into surface water under conditions of high rain, as famously occurred in North Carolina. The wastes in turn produce significant odor, and eutrophication of streams, rivers, and lakes, that is, growth of undesirable algae and bacteria. In the central valley of California between San Francisco and Los Angeles, many

---

* In 2000, the Producer Committee for the United Egg Producers acknowledged this, increasing recommended space allocations from an industry average of 48 sq. in. per bird to 72 sq. in. per bird.

giant dairies have generated unprecedented air pollution consisting of organic volatile compounds, nitrous oxide, ammonia, and methane, eliciting unprecedented environmental regulations. Industrial operations also consume vast amounts of precious water.

2. Human health issues—Closely connected to environmental contamination are human health issues. Two-thirds of human infectious diseases are zoonotic, and close confinement allows infectious microorganisms to burn through populations, much like a cold in a dormitory. In addition, crowded conditions may be conducive to rapid mutation and development of new pathogens. When antibiotics or other drugs are used as a technological sander to compensate for unhealthy conditions or as a growth promotant at low levels, surface water from runoff of industrial animal production facilities can become polluted with pharmaceuticals. Many scientists believe that feeding antibiotics to livestock for growth promotion encourages resistance to antibiotic agents in important human pathogens and thus an end to such use of antibiotics in agriculture should be legislated. Others (De Haven, 2010) deny this claim. Worker health may also become a problem, both because of pathogens and because of bad air. In some swine barns, workers must wear respirators, although the animals do not! The air pollution mentioned earlier in the central valley of California is responsible for marked increased incidence of respiratory disease, cardiovascular problems, and pre-natal and neonatal health problems, as California health authorities told the Pew Commission on which one of us (BR) served.

3. Loss of small agriculture and destruction of rural communities—As mentioned, in some 26 years the United States had lost 87.8% of the swine producers operating in 1980 (Vansickle, 2002) with the hogs now produced by large companies. From over one million producers in the 1960s, by 2005 the number had fallen to 67,000 (USDA/NASS, 2005). As the small hog farmers have gone out of business, the once thriving communities they nurtured have become ghost towns. This in turn kills the communities. Moreover, in rural areas where large operators have become established, major cultural conflicts occur between traditional inhabitants and the migratory workers. In the face of these considerations, we must again recall Jefferson's admonition that small farms and farmers are the backbone of democracy; no one wishes to see major corporations monopolizing the food supply.

4. "Externalized costs"—What helped drive industrialized agriculture's evolution is the desire for "cheap food." Americans spend only 9% of their income on food, as opposed to the 20% spent by Europeans. However, it should be clear from our discussion that what one pays in the supermarket does not represent the true cost of animal products created by industrial methods. The Pew Commission was told by California state health officials that human health costs (in addition to the suffering associated with illness), for example, from pollution from dairies in the central valley of California cost every man, woman, and child in that area an estimated $3 billion, or $1000 per year in direct medical costs. The costs of environmental pollution and the cleanup it will eventually require are inestimable, and how does one cost-account the animals' suffering?

It has often been asked if those who developed industrial animal production methods were callous or oblivious to animal welfare. Most certainly not! They are, however, guilty of a major conceptual error. Since most of the developers come from experience and training in husbandry agriculture, they may have assumed that the same logic that governed husbandry would remain in industrial systems. That is, they thought that the new agriculture would preserve the close connection between productivity and animal welfare that one found in traditional agriculture. Hence, as we shall see in Chapter 5, industrial agriculturalists were disposed to treat productivity as definitive of welfare, forgetting the role of what we have called "technological sanders" in preserving productivity even while welfare is severely compromised.

Industrial agriculture created major welfare problems for farm animals that did not arise, or were insignificant, under husbandry agriculture.

In general, all animals in confinement agriculture (with the exception of beef cattle who live most of their lives on pasture, and are "finished" on grain in dirt feed lots, where they can actualize much of their nature) suffer from the same generic set of affronts to their welfare absent in husbandry agriculture.

1. Production diseases—By definition, a production disease is a disease that would not exist or would not be of serious epidemic import were it not for the method of production. Examples are liver and rumenal abscesses resulting from feeding cattle too much grain, rather than roughage. The animals that get sick are more than balanced out economically by the remaining animals' weight gain. Other examples are confinement-induced environmental mastitis in dairy cattle and "shipping fever." There are textbooks of production diseases, and well over 90% of what farm animal veterinarians treat is production diseases (Rollin, 2009).

2. Loss of workers who are "animal smart"—In large industrial operations such as swine factories, the workers are minimum wage, sometimes illegal, often migratory, with little animal knowledge. Confinement agriculturalists will boast that "the intelligence is in the system" and thus the historically collective wisdom of husbandry is lost, as is the concept of the historical shepherd, now transmuted into rote, cheap labor.

3. Lack of individual attention—Under husbandry systems, each animal is valuable. In intensive swine operations, the individuals are worth little. When this is coupled with the fact that workers are no longer caretakers, the result is obvious.

4. The lack of attention to animal needs determined by their physiological and psychological natures—As mentioned earlier, "technological sanders" allow us to keep animals under conditions violative of their natures, thus severing productivity from assured well being.

THE EGG INDUSTRY

Let us briefly examine some representative industrial systems to understand in specific terms the problems of animal welfare generated by industrialization of animal agriculture. Consider, for example, the egg industry, one of the first areas of agriculture to experience industrialization. On a typical nineteenth-century American farm, chickens ran free in barnyards, able to express their natural behaviors of moving freely, nest-building, dust-bathing, escaping from more aggressive animals, defecating away from their nests and, in general, fulfilling their natures as chickens. They fed on a combination of natural forage and waste products (table scraps, generally) from the farm household. Chickens were typically kept near the house and tended by women and children, who were not paid for their labor. "Egg money" is a phrase that refers to the income that a household would make by selling a few excess eggs off the farm. During this era, eggs were typically available only seasonally, as these free-ranging hens would turn their energies elsewhere as spring gave way to summer. This farmstead practice was first supplemented and then eventually often displaced by operations in which hundreds and eventually thousands of egg-laying hens were kept on litter in low buildings. Eggs were still gathered by hand, although now increasingly by low-wage workers, who also distributed milled feeds, collected dead birds, and were responsible for hygiene. The key technologies in this transition were in breeding, on the one hand, as the genetically diverse but broody flocks of yesteryear were displaced by leghorns that would lay eggs constantly, and electric lights, on the other, which regularized light cycles and broke the seasonal nature of egg production. Although still free ranging, birds in these systems were also beak trimmed to minimize cannibalism (Friedberg, 2008). This middle system, already well in place by the 1930s, was supplanted by the caged layer systems of the 1960s and 1970s in which hens were kept on wire and methods of egg collection and manure removal were completely automated. In its most economically efficient configuration, hens were stocked so densely in small cages so that some must stand on others. The trade association for the shell egg industry (i.e., eggs sold in shells) no longer recommends these stocking densities, although many producers who sell liquefied eggs to the food industry, as well as a minority of shell egg producers, still use them. Putting chickens in cages and putting the cages in environmentally controlled buildings requires large amounts of capital, energy, and technological "fixes." For example, it is necessary to run exhaust fans to prevent lethal build-up of ammonia. The value of each chicken is negligible so more chickens are needed; chickens are cheap, cages are expensive so as many chickens as is physically possible are crowded into cages. The vast concentration of chickens requires antibiotics, vaccines, and other drugs to prevent wildfire spread of disease in crowded conditions. Breeding of animals is oriented solely toward productivity; genetic diversity—a safety net allowing response to unforeseen change— is lost.

THE DAIRY INDUSTRY

Consider another example, the dairy industry, once viewed as the paradigm case of bucolic, sustainable animal agriculture, with grazing animals giving milk and fertilizing the soil with their manure for continued pasture. Although the industry wishes consumers to believe that this situation still exists—the California dairy industry ran advertisements proclaiming that California cheese comes from "happy cows," showing the cows in pastures—the truth is radically different. The vast majority of California dairy cattle spend their lives on dirt and concrete, and in fact never see a blade of pasture grass, let alone consume it.

Ubiquitous across contemporary agriculture, animals have been single-mindedly bred for productivity—in the case of dairy cattle, for milk production. Today's dairy cow produces three to four times more milk than 60 years ago. In 1957, the average dairy cow produced between 500 and 600 pounds of milk per lactation. Fifty years later, it is close to 20,000 pounds (The Colorado Dairy Industry, 2005; USDA/NASS, 2006). From 1995 to 2004 alone, milk production per cow increased 16%. A high percentage of the U.S. dairy herd is chronically lame (Nordlund, 2004; some estimates range as high as 30%), and these cows suffer serious reproductive problems. Whereas in traditional agriculture, a milk cow could remain productive for 10 or even 15 years, today's cow lasts slightly longer than two lactations, a result of metabolic burnout and the quest for ever-increasingly productive animals, hastened in the United States by the use of bovine somatotropin (BST) to further increase production. Such unnaturally productive animals naturally suffer from mastitis, and the industry's response to mastitis in portions of the United States has created a new welfare problem by docking of cow tails without anesthesia in a futile effort to minimize teat contamination by manure. (No husbandry person would so mutilate a cow, leaving her with an open wound and no way to chase flies.) Still practiced, this procedure has been definitively demonstrated not to be relevant to mastitis control (see Bagley, 2003). Arguably, the stress and pain of tail amputation coupled with the concomitant inability to chase away flies may well dispose cows to more mastitis. In a dairy, calves are removed from mothers shortly after birth, before receiving colostrum, creating significant distress in both mothers and infants. Bull calves may be shipped to slaughter or a feedlot immediately after birth, generating stress and fear. (Under husbandry, these animals would have been eaten as veal or sold locally.)

THE SWINE INDUSTRY

The intensive swine industry, which through a handful of companies is responsible for 85% of the pork produced in the United States, is also responsible for significant suffering that did not affect husbandry-reared swine. Certainly the most egregious practice in the confinement swine industry and possibly, given the intelligence of pigs, in all of animal agriculture is the housing of pregnant sows in gestation crates or stalls—essentially small cages. The recommended size for such stalls, in which the sow spends her entire productive life of about four years, with a brief exception we will detail shortly, according to the industry is 3 feet high ×2 feet wide ×7 feet long—this for an animal that may weigh 600 pounds or more. (In reality, many stalls are smaller.) The sow cannot stand up, turn around, walk, or even scratch her rump. In the case of large sows, they cannot even lie flat, but must remain arched. The exception alluded to is the period of farrowing—approximately three weeks—when the sow is transferred to a "farrowing crate" to give birth and nurse her piglets. The space for her is no greater, but there is a "creep rail" surrounding her so the piglets can nurse without being crushed by her postural adjustments.

Under extensive conditions, a sow will build a nest on a hillside so excrement runs off; forage an area covering a mile a day; and take turns with other sows watching piglets and allowing all sows to forage (Rollin, 1995). With the animal's nature thus aborted, she may exhibit bizarre and deviant behavior such as compulsively chewing on the bars of the cage, and endure foot and leg problems and lesions from lying on concrete in her own excrement. Keeping the sow confined is seen as more efficient, as she uses less feed and less labor is required to manage the animals.

Jim and Pamela Braun (1998), now activists opposing industrial pork production, explain how such changes seemed entirely rational to them when they were involved in installing a confinement system on their own farm. Their family-farm system of raising pigs outdoors in a barnyard began to fail in the late 1960s when they encountered difficulties in managing a porcine disease called MMA.

The only treatment was a series of shots strategically timed immediately after farrowing. If the sequence was missed, the piglets died. Even the tamest sows became very leery after receiving the first shot, and thousands of field-farrowed piglets died.

> In order to solve this and other problems in hog production, ...[a] concrete pit was built, and concrete slats were installed to service a 144 foot by 44 foot farrowing house that was totally enclosed. ... Each stall was its own self-contained sow hotel, with an automatic feeder, waterer, and manure removal system. We farrowed year round and the sows could not run from their shots, thereby helping to ensure the health and safety of the piglets. By the fall of 1974, six more buildings were added, and all of my father's hogs were on slatted floors and under aluminum roofs. ... Confinement solved many problems associated with hog production. The pigs were protected from the elements, which increased their feed efficiency and their rate of gain. Sow productivity was increased because they could be weaned and rebred to farrow no matter the season or weather. Also, left on their own outside, hogs develop a social structure and a pecking order that is rigidly enforced. Only those at the top of the hierarchy thrive. They receive the larger portions of feed by bullying the smaller and weaker hogs. Stronger and more dominant pigs mutilate and often kill weaker and smaller pigs. Grouping hogs into smaller, protected numbers inside helped to reduce the "Boss Hog" syndrome. (Braun and Braun, 1998, pp. 40–41)

They go on to acknowledge weaknesses in these systems (such as antibiotic use), but the main thrust of their indictment of industrial pig production emphasizes unfair and illegal pricing structures, unfair credit practices, and state and federal tax credits that corporations (seeking to integrate pig production) use to put the squeeze on independent producers (Braun and Braun, 1998, p. 50).

Two striking anecdotes tellingly underscore the difference between husbandry agriculture and its practitioners and industrial agriculture and its practitioners with regard to animal welfare. A few years ago, we observed some sharply contrasting incidents that dramatically highlight the moral difference between intensive and extensive agriculture. That particular year, Colorado cattle ranches, paradigmatic exemplars of husbandry, were afflicted by a significant amount of scours. Over two months, I (BR) talked to a half dozen rancher friends of mine. Every single one had experienced trouble with scours, and every one had spent more on treating the disease than was economically justified by the calves' monetary value. When these men were asked why they were being what an economist would term "economically irrational," they were quite adamant in their response: "It's part of my bargain with the animal; part of caring for them," one of them said. It is, of course, the same ethical outlook that leads ranch wives to sit up all night with sick marginal calves, sometimes for days in a row. If the issues were strictly economic, these people would hardly be valuing their time at 50 cents per hour—including their sleep time!

Now, in contrast to these uplifting moral attitudes, consider the following: One animal science colleague related that his son-in-law, who was raised on a ranch, was an employee in a large, total confinement swine operation. As a young man, he had raised and shown pigs, keeping them semi-extensively. One day he detected a disease among the feeder pigs in the confinement facility where he works, which would necessitate killing them because this operation did not treat individual animals, their profit margin being allegedly too low. Out of his long established husbandry ethic, he came in on his own time with his own medicine to treat the animals. He cured them. Management's response was to fire him on the spot for violating company policy! He kept his job and escaped with a reprimand only when he was able to prove that he had expended his own—not the company's—resources. He continued to work for them, but felt that his health had suffered by virtue of what I (BR) have called the "moral stress" he experienced every day; the stress growing out of the conflict between what he was told to do and how he morally believed he should be treating the animals. Eventually, he left agriculture altogether. These contrasting incidents, better than anything else we know, eloquently illustrate the large gap between the ethics of husbandry and industry.

This chapter has detailed the historical/conceptual basis for recent societal demands regarding farm animal welfare. Chapter 5 will interpret what form the social demand is currently taking. Viewpoints and approaches from a multidisciplinary group of educators and scientists are offered.

REFERENCES

Bagley, C.V. 2003. Tail docking of dairy cattle, http://extension.usu.edu/files/dairy/uploads/htms/taildock.htm, accessed 7/20/2011.

Braun, J. and Braun, P. 1998. Inside the industry from a family hog farmer. In: *Pigs, Profits and Rural Communities*, Thu, K. M. and Durrenberger, E. P., Eds. Albany, NY: State University of New York Press, pp. 39–56.

Bulliet, R. 2005. *Hunters, Herders and Hamburgers: The Past and Future of Human-Animal Relationships*. New York: Columbia University Press.

The Colorado Dairy Industry. 2005. *Quick Facts Based on 2005 Production,* provided by Bill Waites, CSV animal sciences chair.

De Haven, W.R. 2010. www.FeedstuffsFoodlink.com, March 8, 2010, p. 17.

Friedberg, S.E. 2008. The triumph of the egg. *Comparative Studies in Society and History* 50:400–423.

Nordlund, K., Cook, N.B., and Octzel, G.R. 2004. Investigation strategies for laminitis problem herds. *Journal of Dairy Science* 87 (E Suppl): E27–E35.

Rollin, B. 2009. Veterinary ethics and production diseases. *Cambridge Animal Health Research Reviews,* 10(2): 125–130.

Rollin, B.E. 1995. *Farm Animal Welfare*. Ames, IA: Iowa State University Press.

Thompson, P.B. 2010. *The Agrarian Vision: Sustainability and Environmental Ethics*. Lexington, KY: University of Kentucky Press.

USDA/NASS. 2005.

USDA/NASS. 2006. Milk Production and Milk Cows. http://www.nass.usda.gov/statisticsbystate

Vansickle, J. 2002. Profits slow decline in hog farm numbers, *Natural Hog Farmer*. http://www.nationalhog-farmer.com/mag/farming_profits_slow_decline/index.html

2 Contributions of Farm and Laboratory Animals to Society

Wilson G. Pond, R. Anne Pearson, Kevin R. Pond,
Christian E. Newcomer, Christopher Boleman,
Aaron Alejandro, and Cherie Carrabba

CONTENTS

INTRODUCTION

Wilson G. Pond

Animals in agriculture live on farms of all sizes ranging from a few animals per farm to several thousand. Farm animal welfare is of concern in enterprises varying widely in size and in environmental conditions. In this chapter, we describe the many contributions of farm animals in a global society representing the economic spectrum from d eveloped countries to developing countries.

The dominant role of farm animals in the global economy is centered on animal source food production. Foods of animal origin (fish, meat, milk, and eggs) provide an array of required nutrients that are not always present in adequate amounts in plant source foods. Consumption of animal products helps ensure sufficient intake of essential nutrients, including essential amino acids (particularly lysine, tryptophan, and threonine), essential fatty acids (omega-3 and omega 6), as well as numerous vitamins and essential mineral elements (Carnagey and Beitz, 2011; Knight and Beitz, 2011). In addition to these conventional food nutrients, a group of foods known as functional foods has been identified, most of which are unique to animal source foods. Several bioactive components of proteins and lipids in milk, fish, meat, and eggs (Austic, Hsu, and Larrtey, 2011) from animals have unique properties that provide enhanced physiological benefits to humans. An example of a functional food component from ruminant animals is conjugated linoleic acid (CLA). Research indicates it is anti-carcinogenic and may reduce cardiovascular disease (Santos, O'Donnell, and Bauman, 2011).

Also, evidence shows that some amino acids have functional roles in regulating key metabolic pathways in non-ruminant animals, for example, swine (Wu and Kim, 2011). The improved nutritional status of human populations is also associated with improved animal well-being as nutritional status of food animals improves.

In developing countries, demand for animal source foods is increasing as income rises. This increase in availability of animal source foods improves human nutrition, particularly in infants.

In addition, other important economic and cultural contributions of farm animals to society worldwide include production of animal fibers, leather, and pharmaceutical and biomedical products, as well as draft power and utilization of food processing wastes. Additional benefits include the enrichment of youth development through programs that enhance appreciation of the importance of animal care and well-being in food animal production, and an array of service functions, including companionship between humans and animals (addressed in Chapter 3).

REFERENCES

Austic, R.E., Hsu, K.-N., and Larrtey, F.M. 2011. Functional food components in animal source foods: Eggs from chickens. In: *Encyclopedia of Animal Science*, Pond, W.G. and Bell, A.W., Eds. Boca Raton, FL: CRC Press, pp. 466–469.

Carnagey, K.M. and Beitz, D.C. 2011. Animal source foods (ASFs) nutritional value. In: *Encyclopedia of Animal Science*, Pond, W.G. and Bell, A.W., Eds. Boca Raton, FL: CRC Press, pp. 27–29.

Knight, T.J. and Beitz, D.C. 2011. Animal source foods (ASFs): Improvements. In: *Encyclopedia of Animal Science*, Pond, W.G. and Bell, A.W., Eds. Boca Raton, FL: CRC Press, pp. 20–22.

Santos G.B., O'Donnell, A.M., and Bauman, D.E. 2011. Functional food components: Ruminant-derived foods. In: *Encyclopedia of Animal Science*, Pond, W.G. and Bell, A.W., Eds. Boca Raton, FL: CRC Press, pp. 470–472.

Wu, G. and Kim, S.W. 2011. Functional amino acids. In: *Encyclopedia of Animal Science*, Pond, W.G. and Bell, A.W., Eds. Boca Raton, FL: CRC Press, pp. 462–465.

FARM ANIMALS IN DRAUGHT AND TRANSPORT

R. Anne Pearson

INTRODUCTION

Animals have been used for agricultural work throughout the centuries, starting soon after cultivation began. They have been used to carry loads, cultivate fields, and pull carts as well as more specific tasks in harvesting and processing crops and trees and in water lifting and irrigation. As such, they make significant, but often ignored contributions to society. Despite the increase in mechanization and use of motorized forms of power throughout the world during the twentieth and twenty-first centuries, many people today continue to rely on animal power to complement human labor in agriculture and transport.

USE OF ANIMALS FOR WORK

Cattle are the most commonly used animals for work throughout the world. Water buffalo are also used in the humid tropics, and donkeys, horses, mules, and camels in the drier and temperate areas. Camels, yaks, llamas, dogs, and elephants are used in specific tasks in specific environments and even small ruminants have been used to transport agricultural goods in mountainous areas where flocks move locations with the seasons. Hence, working animals are maintained over a wide range of agro-ecological zones, but are particularly common on small mixe d farms where rain-fed crops are grown mainly for food production. On 70% of farms in developing countries, draught animals and humans provide the only power input. This is largely because on farms where size and scale of enterprise rule out mechanical power, animal power is the only means the farmers have of cultivating land, other than use of family labor.

Although draught animals make their greatest contribution in agriculture, they also have an important role in transport. It has been estimated that about 20% of the population of the world relies largely on animal transport of goods. Animal carts and sledges are used to transport goods and people in rural areas, especially where roads are unsuitable for motor vehicles. Animal power reduces the drudgery of many of the household activities such as water and fuel collection. Where wheeled vehicles cannot be used, such as in mountainous areas where roads are absent or poorly developed, pack animals may be used to transport goods. Working animals, particularly in North Africa and Asia, make a considerable and important contribution to the urban economy, being used to transport produce within urban areas. Many of the people owning and using these animals are landless people for whom the animal represents the main way of earning a living (see Pritchard, 2010).

Draught animals are also used in the timber industry and to power stationary equipment such as water pumps, sugar cane crushers, and grinding mills. Less widespread is their use in the movement of materials in small-scale building projects and road, dam, and reservoir construction. Working animals can also be found in certain niche operations in industrial enterprises—transporting fruits and sugar cane to road heads in plantations and moving bricks in brick factories, for example.

NUMBERS OF ANIMALS USED FOR WORK

It is impossible to obtain precise information on the number of animals used for work purposes in the world. Most countries maintain statistics on livestock numbers, but for ruminants, they do not identify use for work separately from use for beef or milk. In many places, large ruminants are multipurpose, being used for work, calf production, and ultimately beef as farmers try to make the best use of the feed resources available on their farms. Most donkeys and mules kept in developing countries can be assumed to be kept mainly for work. At least 60% of the horses kept in the tropics are kept for draught work. In recent years, mules have become more popular—farmers in Latin America are tending to replace their work oxen with mules and horses, and in North Africa, mules are increasingly being favored over donkeys and horses where available. Speed, stamina, longevity, and an ability to maintain body condition on low protein, high fiber diets have always made mules popular but expensive to purchase. A review commissioned by the Food and Agriculture Organization (FAO) gives details of recent trends in the use of livestock for work around the world (Starkey, 2010).

SKILLS IN SOCIETIES USING ANIMAL POWER

In some areas of the world, draught animals are part of the traditional way of cultivating the land. For instance, in Asia, North Africa, Ethiopia, Somalia, and in most of Latin America, people are accustomed to training and managing their work animals. Implements are readily available locally, usually made from local materials, with a local system to repair and replace them.

In other areas of the world, draught animal power is a more recent technology in cultivation and crop production. For instance, until recently in West Africa and much of Sub-Saharan Africa, animal diseases prevented the keeping of animals in many areas, and the traditional methods of cultivating the land used manual labor only. It is only within the twentieth century that many people have made use of draught animals on their farms in these areas. This follows the reduction in disease vector habitat and increased availability of veterinary treatments for the diseases. Because of the relative newness of the animal power technology in these areas, the support infrastructure is not always available locally. As a result, the animals and implements for purchase are expensive, and they involve considerable investment by the farmers before the farmers can see the benefits and the drawbacks for themselves. Often, implements are imported or manufactured by companies selling a range of agricultural equipment. Although spares may be available, the manufacturers or retailers can be some distance from the farm, and so repairs cannot be done *in situ* in the fields, as they often can be in systems that are more traditional.

A lack of skill can often be seen where farm animals are used in transport enterprises in more urban areas. In these operations, while some users have a long experience of working with animals, others have little experience in livestock keeping. Equids tend to be favored over ruminants for their greater speed in transport. The horse, mule, or donkey is used to provide a daily income, rather as a vehicle would be used, and may be regarded as an expendable item by some, with little care given to working practices or to the animal's management and health. Cattle, buffalo, and camels generally fare better, largely due to their resale value for meat. Thus, it is not surprising that the nongovernmental organizations (NGOs) and animal charities operating to improve working animal welfare and health more often voice welfare concerns for the working horse and donkey than for the ruminant.

PRODUCTION FROM WORKING ANIMALS

The output from work animals as a contribution to the community is more difficult to assess than that from beef or dairy animals. Draught force, speed, work, and power have all been used to assess output of working animals. Area ploughed or cultivated and distance traveled or load carried in transport are outputs that can be measured easily. Less immediate, perhaps, but important to the farmer, is the yield of the crop their working animals have helped to produce. Manure is an important by-product and one many small-scale farmers rely on to help maintain soil fertility, particularly as the costs of chemical fertilizers continue to rise, putting them out of reach of many small-scale farmers.

The amount of work an animal can do depends on the speed at which it works and the draught force generated. For a particular draught force, the speed determines the power output of the animal, that is, the rate at which the animal does the work. Therefore, these parameters are all closely related. Various aspects of the animal, the implement, the environment, and the operator all interact to determine the amount of work done in a day.

NUTRIENT REQUIREMENTS OF WORKING ANIMALS

Researchers have determined the nutrient requirements of working animals. Ruminants have received the most attention (Lawrence and Pearson, 1991). However, interest in the performance of working horses and donkeys has increased in recent years and their requirements are now more fully understood (Perez, Valenzuela, and Merino, 1996; Pearson, 2005). The main requirement for work is energy. Extra requirements for protein, minerals, and vitamins for work are not as large and can usually be met by the increase in food given to meet the additional energy requirements. Energy requirement during a working day is more closely related to distance covered than to the draught force required to pull the implement or cart. Hence, animals doing light work such as pulling a cart can expend more energy in a day than animals doing heavy work such as plowing. Even when oxen are working for six to seven hours a day, their total energy expenditure in a working day is rarely more than two times maintenance requirements. Horses and donkeys can exceed a requirement of two times maintenance in a working day, but this is usually only when they are working steadily for six or more hours per day.

CONSTRAINTS TO PERFORMANCE

Many studies of the husbandry and use of working animals have been undertaken over the last 30 years (e.g., Copeland, 1985; EAAP, 2003; Pearson, Muir, and Farrow, 2008). As well as determining their capabilities, it is important to examine the constraints that can limit the contribution that working animals can make. High ambient temperature and disease (e.g., Jaafar-Furo, Mshelia, and Suleiman, 2008; Pritchard, Burn, Barr, and Whay, 2008) are well-known constraints to performance. However, the constraint most often identified by working animal owners is nutrition. The main problem is how best to meet the nutritional requirements for work with the feed resources available. Location and season determine which feeds are given to work animals.

For most of the year, work animals consume poor-quality forage diets that have a high cell-wall content, low nitrogen content, and poor digestibility. The metabolizable energy (ME) content of these diets is rarely more than 9 MJ ME/kg and crude protein of 90 g/kg dry matter (DM). Research studies have shown that any increase in rate of eating or improvement in digestibility on working days, which results from increased energy demand during working periods, is not sufficient to meet the additional energy requirement for most types of work when animals are fed such diets. In practice, most farmers working with animals expect their animals to lose weight during the work season unless the diet is supplemented with better-quality feed. The start of the cropping season, when animals are required to do the most work, is usually the time when food stocks are at their lowest, particularly in areas that have a long dry or cold season. This further exacerbates the problem of feeding for work.

The need for supplementation is greatest when animals are multipurpose, also being required to maintain weight (if ultimately they are to be sold for meat), or if they are cows used for work and are required to produce a calf.

Various strategies are available to improve feed supply to work animals, dependent upon the financial resources of the owner. The benefits of these techniques are well researched and widely reported (e.g., Pearson, 1995; FAO, 2010), but adoption by draught animal farmers is often poor.

THE FUTURE

Continued mechanization of agricultural practices will occur where it is economically feasible, and work animals will be replaced or used to complement mechanization on those farms that can justify hire or maintenance of two- or four-wheeled tractor power. On steep, inaccessible, or terraced hill-sides, and on mixed farms where farm size and scale of crop production are small, animal power is still a better option than motorized power to supplement manual labor. On small farms of less than 3 ha, animal power can compete economically with gasoline-fueled tractors. Farmers using animal power will have to cope with competition for their land from a growing human population and increasing pressure on natural resources. This is likely to lead to the cultivation of more marginal land and greater use of animals for multiple purposes (e.g., manure, work, and milk, or work and calf production, or meat). Cropping of marginal land will require more attention to soil and water conservation and animal-drawn tillage techniques. Reduction of grazing land may require more farmers to move to a cut-and-carry system of managing their work animals. With the need to use resources more efficiently, it is important to recognize that animal energy can be harnessed to provide several income-generating activities for the smallholder farmer outside of their use in the production of food and cash crops and their role in manure production. More versatile, and therefore more frequent, use of animal power is an ideal way to spread the maintenance costs. A resting draught animal still uses resources, unlike a resting tractor. Hence, broader use of animal power in the areas where it is found should also be encouraged. However, despite the value farmers put on work animals in reducing their drudgery and supporting their food production and trade within communities, as Starkey (2010) points out, animal power continues to have a "poor out-moded image" within governments and many of the organizations and other institutions helping to improve the livelihoods of their farming populations and those people supporting them. This is disappointing in view of the continuing contribution of animal power to food security and farm income on many small farms around the world.

SUMMARY

The use of animals for work and the general contribution that they can make to alleviating drudgery in the livelihoods of the people who use them are discussed in this section. Cattle are the most commonly used animals for work, followed by water buffalo and donkeys, but many other domesticated animals are also worked in suitable environments where the need arises. In some areas, use of working animals goes back many centuries; in other areas, use is more recent commencing within the twentieth century. Outputs, feed requirements, and constraints to performance are also discussed.

REFERENCES

Copeland, J.W., Ed. 1985. *Draught Animal Power for Production*. Australian Centre of International Agricultural Research (ACIAR) Proceedings Series No. 10, Canberra: ACIAR.

European Association of Animal Production (EAAP). 2003. *Working Animals in Agriculture and Transport. A Collection of Some Current Research and Development Observations*. EAAP Technical Series No 6, The Netherlands: Wageningen Academic Publishers.

Food and Agriculture Organisation (FAO). 2010. e-conference—Successes and failures with animal nutrition practices and technologies in developing countries. September 1–30, 2010 (www.fao.org/docrep/014/i2270e/i2270e00.pdf).

Jaafar-Furo, M.R., Mshelia, S.I., and Suleiman, A. 2008. Economic effects of *Fascioliasis* on animal traction technology in Admawa State, Nigeria. *J Appl Sci* 8:1305–1309.

Lawrence, P.R. and Pearson, R.A. 1991. *Feeding Standards for Cattle Used for Work.* Scotland: Centre for Tropical Veterinary Medicine, University of Edinburgh.

Pearson, R.A. 1995. Feeding systems for draught ruminants on high forage diets in some African and Asian countries. In: *Recent Developments in the Nutrition of Herbivores*, Journet, M., Grenet, E., Farce, M.H., Thériez, M., and Demarquilly, C., Eds. Proceedings of the IV International Symposium on the Nutrition of Herbivores, Paris: INRA Editions, pp. 551–567.

Pearson, R.A. 2005. Nutrition and feeding of donkeys. In: *Veterinary Care of Donkeys*, Matthews, N.S. and Taylor, T.S., Eds. Ithaca, NY: International Veterinary Information Service.

Pearson, R.A., Muir, C., and Farrow, M. 2008. Fifth International Colloquium on Working Equines. Proceedings of an International Colloquium held at Addis Ababa University, Addis Ababa, Ethiopia. October 30–November 2, 2006. Devon: The Donkey Sanctuary.

Perez, R., Valenzuela, S., and Merino, V. 1996. Energetic requirements and physiological adaptation of draught horses to ploughing work. *Anim Sci* 63: 343–351.

Pritchard, J.C. 2010. The role of working donkeys, mules and horses in the lives of women, children and other vulnerable groups: A review. In preparation.

Pritchard, J.C., Burn, C.C., Barr, A.R.S., and Whay, H.R. 2008. Validity of indicators of dehydration in working horses: A longitudinal study of changes in skin tent duration, mucous membrane dryness and drinking behaviour. *Equine Vet J* 40: 558–564.

Starkey, P.H. 2010. Livestock for traction: World trends, key issues and policy implications. Paper prepared for Livestock Information, Sector Analysis and Policy Branch (AGAL), Animal Production and Health Division, FAO, Via delle Terme di Caracalla, Rome, Italy.

CROP AND ANIMAL PROCESSING WASTES

Wilson G. Pond and Kevin R. Pond

The human population is expected to increase from the current 6 billion to 8 to 9 billion by 2030. Land available for food production is finite. The dramatic increases in food production resulting from agricultural research and technology and other contributing advances have provided increased, although not adequate, food for a growing world population. A major challenge to society now is to continue to meet the demand for food and other products of agriculture within the constraints of a finite land area and limited natural resources. One factor contributing to a solution is the improved utilization of crop and animal processing wastes. Recycling of wastes from an array of animal and plant sources is used effectively and widely in animal and crop production.

Uses of processing wastes are described as follows:

> … food processing waste generally is either a potential feed ingredient for farm animal or pet food or a potential nutrient source for crops. For example, in cereal processing firms such as breweries, distilleries, and feed mills, by-products are not wasted but marketed as livestock feed ingredients. Similarly, in meat processing firms, poor-quality meat by-products can be converted to better-quality human food-products by means of breakdown and recombination of by-product components. Other by-products such as stomachs, intestines, and fish wastes are converted to pet foods. Finally, poor-quality effluent may be used on cropland as a nutrient source. (CAST, 1995)

In addition to animal feed constituents, inedible animal fats and other animal food processing wastes are used to produce soap, lubricants, cosmetics, candles, floor waxes, paints, varnishes, and other products of value to society.

Crop residues can be utilized in several ways: fuel, animal feed, bulking agents in manure and sewage sludge composting systems to produce organic wastes that are safe, stable, and unobjectionable for land application as fertilizer (CAST, 1995). These and other approaches are being used to reduce crop-processing losses. These advances include the following:

1. Composting of manure, bedding, dead animals, and hatchery wastes for land application.
2. Production of methane and other biogas fuels from the above-composted products by anaerobic fermentation.

3. Improving the digestibility of nutrients in common feedstuffs to reduce levels of carbon (C); nitrogen (N) and phosphorus (P) lost in manure by using new technology (e.g., use of the enzyme phytase to improve utilization of P bound in plant feedstuffs).
4. Developing methods to reduce water volumes used in animal source food production.
5. Continuing pursuit of innovative, safe, and cost-effective ways of utilizing food-processing wastes in food animal production (CAST, 1995) to enhance sustainable agriculture through improved resource utilization. In addition, a worthwhile goal (CAST, 1995) for animal agriculture is to reduce wastes during food processing that currently occur between harvest and delivery to the consumer. Meeting this goal will improve the welfare of food animals on a global basis by enhancing efficiency of utilization and improved nutrition of food animals.

SUMMARY

A major challenge to society in the twenty-first century is the rate of increase in the global population in a finite space on the planet. Large quantities of processing wastes are generated from crop and animal production. These wastes are used to produce soap, cosmetics, candles, paints, methane, ethanol, and many other products that improve the welfare of food animals globally by enhancing efficiency of feed utilization and total food and feed production for a burgeoning human population.

REFERENCES

Council for Agricultural Science and Technology (CAST). 1995. Waste management and utilization in food processing. Ames, IA: Author.

ANIMAL FIBERS, HIDES AND PELTS, AND LEATHER

Wilson G. Pond and Kevin R. Pond

Wide genetic variation exists in mature size and other traits among animals native to different regions and climates in which they are raised. This variation offers an opportunity for breeders to tailor the genetic base of animals to the local environment for improved performance and efficiency. This concept has been adopted for use in temperate and tropical environments. There are now more than 250 registries and associations in the United States and Canada that promote particular species or breeds and that maintain breeding records (Bixby, Christman, Ehrman, and Sponenberg, 1994). Some are concerned with the common breeds of farm animals. Others focus on uncommon breeds of domestic animals and their crosses and on wild species. Worldwide, there is interest in dozens of other species, hybrids, and breeds and their crosses that have potential for commercial or subsistence level of food, hide, and fiber production. The U.S. National Research Council (1991) published a paper on micro-livestock, a term used for species within which some individuals are phenotypically and genetically small compared with the breed average. Such micro-livestock are found in cattle, sheep, goats, pigs, and poultry in which some individuals are less than half the mature sizes of average representatives of the breed. Because of a survey of many animal scientists in 80 countries, it was determined that about 40 breeds and species have sufficient genetic diversity to select for small size to expand micro-livestock populations for use in developing countries. This would allow taking advantage of the ecological interdependence of animal, plant, and human life, the limited amount of the earth's surface that can be safely cultivated, and the innate advantages of small animals to the subsistence family with no refrigeration, and with limited cash, space, and animal feed. Animal well-being would be expected to improve because of a better match of feed supply with animal needs.

Several species of mammals and birds contribute to society through production of wool, hair, feathers, leather, pelts, and other inedible by-products used in the manufacture of clothing, upholstery, carpets, bedding, and other products of the livestock industry. Here we describe briefly examples of the importance of many domesticated mammals and birds in providing leather, fibers, and other by-products of the food animal industry.

MAMMALS

Cattle (beef cattle, dairy cattle, and swine), in addition to their production of meat and milk for food, contribute significantly to the economic value of the animal by yielding hides for leather and hair used in clothing, accounting for approximately 50% of the total by-product value of cattle. Similarly, sheep and goats produce wool and mohair, respectively, widely used in the clothing industry and representing a significant fraction of the total value of the products of the sheep and goat industry, including meat and milk production.

Other mammals used in some cultures for both food and fiber or hides include rabbits, camels, llamas, alpacas, and vicunas (Ullrey and Bernard, 2000). Collectively, camels and llamas are known as camelids, with an even number of toes on each foot and anatomical characteristics that distinguish them from true ruminants. For example, the muscle attachments in the hind legs allow them to rest on their knees when lying down. The Old World camelids include the two-humped Bactrian camel and the one-humped Arabian or dromedary camel. The Bactrian camel is found in the cool desert regions of Central Asia, while the dromedary is found in the hot deserts of North Africa. Both are used for transport, draft, meat, milk, fiber, and hides. The New World camelids include the guanaco, vicuna, and domestic llama and alpaca. The guanaco ranges from the Andean highlands in Ecuador and Peru to the plains of Patagonia. Vicunas live near the snow line of the Andes and have a highly prized fine wool fleece. Alpacas are bred primarily for their wool (Nowak, 1991). Llamas are used mostly as beasts of burden, but their meat may be used for food, fleece for clothing, hair for rope, and hide for leather. The four South American camelids (llamas, alpacas, guanacos, and vicuna) have the same chromosome number (Clutton-Brock, 1987) and will interbreed. Llamas and alpacas have become increasingly numerous in the United States as pets and for production of fibers.

BIRDS

Chickens, Ducks, Geese, and Turkeys

Commercial production of poultry and eggs in the United States began in the early 1800s and gradually evolved into a massive industry in the United States and globally. The poultry industry in the United States involves specialized production units devoted to broilers for meat and layer hens for egg production. Animal welfare concerns are of paramount interest for both industries. Ongoing changes in regulations regarding animal care and welfare of chickens (both broilers and layers) and other poultry continue to receive attention.

Vertically integrated production systems involving thousands of birds have been so successful that today nearly all broilers in the United States are produced under some type of contract arrangement. The system is less frequently used in turkey production; however, if a contract is not used, production is coordinated by some other arrangement between the processor and the growers. Modern chicken meat strains have been developed by cross-breeding layer lines with meat lines.

Turkey growing is similar to growing of broiler chickens, but involves a two-stage system in which day-old turkey poults are started in a brooder house and transferred to a larger growing house at about six weeks of age and marketed weighing 10 to 40 pounds.

Ducks and geese can be raised successfully in confinement on litter floors and do not require swimming water for growth, health, or reproduction. Young ducklings are sometimes started on slatted floors or raised wire. Commercial houses often provide an indoor litter area and an outside run.

Geese are excellent grazers and can be grown on pasture with limited supplemental feeding, although many geese are raised indoors without pasture.

Ostriches

Ostriches are large, flightless birds that are 2 to 2.4 m tall and weigh between 110 and 150 kg. Along with emus and several other large bird species, they are known as ratites. Ostrich feathers were used widely by the fashion industry nearly a century ago, and ostrich leather has been used in boots, shoes, and other leather goods for many years. The commercial ostrich industry began in the mid-nineteenth century in Africa, where the ostrich is indigenous. Ostrich breeding in the United States began in the 1980s. More than one-half of ostrich breeding in the United States is in Texas, California, Arizona, and Oklahoma. Some ostrich meat is imported from South Africa, but most is produced in the United States. A marketing system for ostrich leather is developing in the United States.

Emus

Emus are indigenous to Australia. Emus are 1.5 to 1.8 m tall and weigh between 50 and 65 kg at maturity. Emu production in the United States is relatively new, but is growing steadily. Products include garment leather, plumage, and meat for gourmet restaurants.

SUMMARY

Animals that produce food for people also provide a wide range of non-food products, including wool, mohair, and feathers, as well as hides and pelts used in clothing, shoes, and other leather products. A wide genetic variation within and between breeds and crosses results in opportunities to increase quantity and quality of animal products available for human populations everywhere and also offers new opportunities to enhance the welfare of both humans and animals.

REFERENCES

Bixby, D.E., Christman, C., Ehrman, C.J., and Sponenberg, D.P. 1994. *Taking Stock: The North American Livestock Crisis*. Granville, OH: McDonald & Woodward.

Clutton-Block, J. 1987. *A Natural History of Domestic Animals*. Austin, TX: University of Texas Press.

Nowak, R.M. 1991. *Walker's Mammals of the World*, Volume 2, 5th ed. Baltimore MD: Johns Hopkins University Press.

Ullrey, D.E. and Bernard, J. 2000. Other animals, other uses, other opportunities. In: *Introduction to Animal Science*, Pond, W.G. and Pond, K.R., Eds. New York: John Wiley & Sons, pp. 553–583.

U.S. National Research Council. 1991. *Microlivestock: Little Known Small Animals with a Promising Economic Future*. Washington, D.C.: National Academies Press.

USE OF ANIMALS IN NUTRITIONAL AND PHYSIOLOGICAL RESEARCH

Wilson G. Pond and Kevin R. Pond

The use of farm animals and other animals as surrogates for humans, and animals in agricultural and biomedical research has a long history. Virtually every advance in human and veterinary medicine over the past century has a foundation in animal research. Nutrients, including vitamins, mineral elements, protein, amino acids, fat, and fatty acids known to be required by humans were discovered to a large degree by research in animals, including pigs and other farm animals, along with laboratory animals such as rats, mice, and other small animals and birds. Metabolic processes were defined, and the safety and effectiveness of consumer products, drugs, medical devices, and medical procedures were established.

Continuing research on techniques to repair congenital heart defects, control cancer, cure diabetes, reverse Alzheimer's disease, treat cystic fibrosis and muscular sclerosis, and control HIV and many other diseases requires the use of animals.

Diagnostic tools such as electrocardiography, angiograms, endoscopy, and cataract removal, as well as surgical procedures, organ transplantation (e.g., heart and heart valves), and artificial joint replacement continue to be developed because of animal model research as a vehicle for improved human health and well-being. Major advances have been made in the use of allotransplantation (human-to-human replacement) of kidneys and heart valves. Transplantation of animal organs in human patients (xenotransplantation) is complicated by tissue rejection of the xenograft. The use of pig hearts for xenotransplantation in humans offers promise (Platt, 2005). These well-established approaches for the benefit of humans raise legitimate concerns and questions related to animal welfare. The ethical and social implications of the use of animals as surrogates for humans in biotechnology and biomedical research have been and continue to be addressed by the scientific community. (CAST, 1995; Clutton-Block, 1991; Crawford, 1996; National Research Council, 1996; Pond and Pond, 2000).

Worldwide, it is estimated that 50 to 100 million vertebrate animals are used annually (from zebra fish to nonhuman primates). Invertebrates and vertebrates, including mice, rats, fish, frogs, and animals not yet weaned are not included in the figures. One estimate of mice and rats used in the United States alone in 2001 was 80 million.

SUMMARY

Agricultural and laboratory animals have contributed to major advances in knowledge of human and animal health and progress in knowledge of nutrition and physiology. Most advances in human and veterinary medicine had a foundation in animal research. Metabolic processes were defined and the safety of consumer products was established with animals. The ethical and social implications of the use of animals as surrogates for humans in biotechnology and biomedical research continue to be addressed by scientists and palicymakers. See Chapter 14 for detailed accounts of these advances. Also, see sections titled "Pharmaceutical and Biomedical Products," "Laws, Regulations, and Oversight Mechanisms for Research Studies with Agricultural Animals in the United States," and "The Role of Animal Agriculture in Enrichment of Youth Development Through Organized Hands-On Exposure to High standards of Animal Welfare in Food Animal Production" for additional related information.

REFERENCES

Council of Agricultural Science and Technology. 1995. Waste management and utilization in food production and processing. CAST Task Force Report No. 124.

Clutton-Block, J. 1991. *A Natural History of Domestic Animals*. Austin, TX: University of Texas Press.

Crawford, R.L. 1996. A review of the Animal Welfare Report data: 1973 through 1995. National Agriculture Library, *Animal Welfare Information Center Newsletter*, 7(2): 1–11.

National Research Council. 1996. *Guide for the Care and Use of Laboratory Animals*. Washington D.C.: National Academies Press.

Platt, J.L. 2005. Biotechnology: Xenotransplantation. In: *Encyclopedia of Animal Science*, Pond, W.G. and Bell, A.W., Eds. New York: Marcel Dekker, pp. 152–154.

Pond, K.R. and Pond, W.G. 2000. *Introduction to Animal Science*. New York: John Wiley & Sons.

PHARMACEUTICAL AND BIOMEDICAL PRODUCTS

Christian E. Newcomer

HISTORICAL HIGHLIGHTS OF PROGRESS IN THE USE OF FARM ANIMALS IN BIOSCIENCES

The use of farm animals for scientific advances in the development of pharmaceutical products and in biomedical research has a long historical precedent dating to the antiquities and several important contemporary medical practices had their origins in farm animal studies. Regrettably, animal welfare considerations were not featured in those early studies. Galen, the famous physician (of Greek origin) in Rome during the second century vivisected pigs and goats in an effort to formulate an understanding of the circulatory system, concluding erroneously that there were two separate and unlinked systems. Avenzoar (also known as Ibn Zhur), a Spanish Muslim surgeon and physician of the twelfth century rejected Galen's views and established the general concept of experimental surgery and that the principles of surgery should be proven in animal subjects before being applied to humans (Abdel-Halim, 2005). Among his many other contributions, Avenzoar performed a tracheotomy in a goat to demonstrate the safety of this procedure for use in humans. During the late nineteenth and early twentieth centuries, drawing on the work investigating electrical conductivity of animal tissues, Dutch physiologist Willem Einthoven developed a more sensitive string galvanometer than had previously been used for recording heart muscle conductivity and also successfully imaged and identified the different wave formations of the electrocardiogram (ECG), assigning the letters P, Q, R, S, and T to the various deflections. He later commercialized the first electrocardiograph and described the electrocardiographic features of a number of cardiovascular disorders. Using Einthoven's device, Thomas Lewis, who is credited with introducing cardiology into clinical practice, published a paper detailing his careful clinical and electrocardiographic observations of atrial fibrillation (Lewis, 1912). Lewis had worked with a veterinarian to identify a horse with this condition. Using the string galvanometer's ECG recording, and then following the horse to the slaughterhouse, he could visually confirm the fibrillating atrium. The use of the ECG as a basic medical parameter has now been practiced for decades, and large animal models continue to contribute to the development of new measures for cardiovascular health in humans and animals through the collaborations of physicians, veterinarians, and scientists in various disciplines.

In addition to the role farm animal species have played historically in anatomical and physiological studies of import to the concepts of medicine and surgery, the observations of parallels and associations of contagious diseases in farm animals with humans has stimulated many important medical discoveries. In 1796, William Jenner conclusively documented that material in the crusts of cowpox lesions was capable of inducing protective immunity against smallpox, and introduced the concept of vaccination. Louis Pasteur, along with Robert Koch, is credited with the establishment of the germ theory. They used sheep to demonstrate the role of anthrax bacteria in disease and later to develop a protective vaccine for treatment of anthrax. Pasteur's studies on the elimination of bacterial contamination in fluids, or pasteurization, brought us safe milk products and served as the stimulus for Joseph Lister to develop the principles of aseptic surgery. In the late 1800s in the United States, Theobold Smith, a veterinarian studying cholera in swine, was the first to discover, isolate, and describe organisms in the genus *Salmonella,* a major group of pathogens in humans and animals although not the causative agent of hog cholera.

The speed with which we could identify the retrovirus HIV as the causative agent of AIDS has its origins in studies with farm animals. Retroviruses were detected in solid tumors of chickens in the early twentieth century and have been studied extensively since that time (Medawar, 1997). Scientific efforts to understand the biology of bovine leukemia virus since the 1970s have aided in the identification of HTLV-1 and HTLV-2 retroviruses that cause human cancer. There are many examples of human health improvement resulting from product development for farm animals. For example, ivermectin, an anthelmintic compound, was developed primarily for the elimination of parasites in livestock. However, due to the positive therapeutic effect of ivermectin in equine parasitic (*Onchocerca*) eye infections, the agent was used in human clinical trials for the treatment of river blindness caused by the human parasite *Onchocerca volvulus*. When this program was launched, 1 million people in West Africa alone (and 18 million worldwide) suffered from this parasitic infection; 100,000 of these had serious eye problems (including 35,000 who were blind). Because of this intervention, ocular *Onchocerca* infection has largely been eliminated as a public health problem and as an obstacle to socioeconomic development globally.

CURRENT ADVANCES IN THE USE OF FARM ANIMALS IN THE DEVELOPMENT OF PHARMACEUTICAL AND BIOMEDICAL PRODUCTS

Farm animals continue to play a significant role in pharmaceutical and biomedical product development, both as an extension of the inherent characteristics that made them valuable models initially and now increasingly as a result of the fact that they can be genetically engineered to express novel products of medical and commercial importance (e.g., in the mammary gland to be harvested from milk). Farm animals also have been recognized for several decades to be useful models for spontaneous animal and human disease, many of which have a clear genetic underpinning, and these animal models are invaluable for the elucidation of the basic disease mechanisms (Andrews, Ward, and Altman, 1979). In the era of modern molecular biology and genetic engineering, genetically engineered rodent models have become the favored models for understanding molecular mechanisms and developing therapeutic interventions such as new pharmacological compounds, biopharmaceuticals, small interfering RNAs, and gene therapy. However, once the proof of principle for these compounds is met in small animal models, a resurgence in the use of the larger farm animal models for the demonstration of their clinical efficacy is very likely if relevant animal models are available. A few representative examples of the use of farm animals for the development of pharmaceutical and biomedical products are presented in the following paragraphs.

Birds

Chickens and, to a lesser degree, quail are used for the generation of polyclonal antibodies (the active component in antiserum), which can be simply extracted from the yolk of the immunized bird. The immunization of hens represents an excellent alternative for the generation of polyclonal antibodies and affords a substantial animal welfare benefit because egg collection is noninvasive compared to the usual method of collection of serum for isolation of antibodies that requires repeated blood withdrawal (Hau and Hendriksen, 2005). Moreover, chickens are inexpensive to maintain and produce abundant numbers of eggs. These antibodies can be used as experimental or diagnostic reagents and are showing promise as therapeutic agents in animal and human diseases, particularly for infectious diseases of the gastrointestinal tract. Chickens with ovarian cancer have molecular markers of disease similar enough to those in humans to define a model for predicting the stage of progression of human ovarian cancer (Gonzalez Bosquet et al., 2010). In addition, genetically modified chickens have been developed that fail to propagate avian influenza virus and, therefore, do not perpetuate the cycle of contagion (Lyall et al., 2011). This approach could be used in commercial flocks and thereby eliminate their contribution to the spread of pandemic flu and the emergence of new strains of influenza through interspecies transmission of viral infections.

Mammals

Equine species are used for the production of equine estrogens, which are useful therapeutic agents in the management of some of the conditions and symptoms of the postmenopausal period in women (Stovall, 2010). In addition, the horse has been used historically for the development of antiserum to toxins (e.g., tetanus antitoxin) and to snake and other venoms. Although horse antiserum has been replaced in many instances, especially since its use is highly associated with "serum sickness," which is an immune complex disorder, there are still many types of venom for which it remains the sole therapeutic agent. In many regions of the world, purified horse antiserum is also the primary therapeutic agent for botulism.

Small Ruminants

Sheep and goats are also used in the production of antiserums (antibodies) for use as experimental and diagnostic reagents and, to a lesser degree, as therapeutic antitoxin agents for envenomations (Seger and Krenzelok, 2005). Sheep and goats are also occasionally used as models to train personnel in the techniques of minimally invasive surgery involving the urogenital tract and as models for the study and treatment of urologic conditions. Sheep and goats have been used extensively for the development and testing of artificial joints, bone cements, bone and cartilage replacement products, and therapeutic approaches to osteoarthritis (Martini, Fini, Giavaresi, and Giardino, 2001). Sheep and goats also have been used for the development and testing of various types of cardiac assist devices (Weiss, 2005) and for materials used in vessel surgery and repair. Genetically modified goats have been created to produce valuable novel proteins in their milk, allowing ease of collection and an abundant supply following purification of the desired product. One product reportedly nearing approval by the Food and Drug Administration is produced from goats genetically modified to produce the human form of the protein antithrombin, which prevents blood clotting (http://www.gtc-bio.com/). One in 5000 individuals produces insufficient amounts of antithrombin, and patients prone to clotting following coronary bypass surgery may also benefit from this product to prevent excessive clotting and complications such as stroke. Another genetically modified goat model developed at the University of California-Davis produces lysozyme in its milk; this molecule is important for the destruction of harmful bacteria in the digestive tract, offering some hope of a convenient means for protecting infants in the developing world where diarrheal disease kills 2 million infants annually (Maga et al., 2005). A goat also has been developed that produces the soluble components of spider silk (the material of the spider's web). This material is stronger and more flexible than steel and is a lightweight alternative to carbon fiber (Boyle, 2010). It is important to note that in each of these genetically manipulated goat lines, the animals are behaviorally, clinically, and reproductively normal, which limits the ethical and practical issues related to the expansion and maintenance of their populations (Fahrenkrug et al., 2010).

Cattle

Genetically modified cattle that are otherwise normal in phenotype have been generated using various types of transgenic technology. One genetically modified bovine developed by the USDA secretes the antimicrobial protein lysostaphin in the milk, which confers greater resistance to the development of mastitis in the cow from staphylococcal infection. This achievement marks a significant step toward the development of disease-resistant livestock. Using a different transgenic approach, scientists inserted a human artificial chromosome containing the entire human immunoglobulin loci into the germ line of cows (Robl, 2007). These cattle generate human antibodies in their blood, creating the potential for the generation of a variety of valuable medical therapeutic products. The products have application to the management of antibiotic-resistant infections, immune deficiency, biodefense, and many other immune-mediated conditions simply through immunization of the animal with the agent of interest followed by the collection and purification of the antibodies from the blood of the cattle (http://www.hematech.com/). Bovine calves also have been used extensively since the mid-1960s for the development and testing of artificial hearts, cardiac assist devices, other cardiovascular instruments, and materials to overcome disease conditions of the heart (Delano, Mischler, and Underwood, 2002).

Swine

Swine have been an especially prominent animal model for the investigation of cardiovascular diseases of humans and for the development of apparatus, materials, and approaches used in the medical and surgical management of human cardiovascular diseases. The cardiovascular system of swine has unique anatomical and physiological parallels with that of humans. Swine are omnivores and readily susceptible to dietary-induced atherosclerosis, a major contributing factor to human heart and vascular disease (Swindle, 1998). This has facilitated their extensive use for the development of techniques to treat atherosclerosis and its complications. The skin of pigs also has characteristics very similar to those of humans, making them extremely valuable models for plastic surgery and studies of skin injury and repair and associated therapeutic agents. Swine are proven to be valuable in many other clinical research applications (Laber et al., 2002). Due to their abdominal size and overall comparability of the anatomy of their abdominal organs to those of humans, swine have served as the primary model for surgical training in laparoscopic and endoscopic techniques and the development of new surgical instruments and surgical procedures (Srinivasan, Turs, Conrad, and Scarbrough, 1999; van Velthoven and Hoffmann, 2006). Approximately 1000 articles have been published on the use of swine in this area alone. Pigs also have been genetically modified for various research and future commercial applications. In one of the genetically modified models, the cellular surface marker responsible for the acute rejection of pig organs by humans and other primates has been removed, which offers the prospect that pig organs might one day be available for xenotransplantion into humans (Platt, 2001, 2011a,b). Organs from these pigs have a markedly prolonged survival rate compared to that for normal pig organs transplanted into nonhuman primates (Ekser et al., 2010). Through additional genetic modification to further protect graft survival via modulation of the immune response in the graft recipient (i.e., nonhuman primate or human), these pigs may solve the problem of the critical shortage of human-compatible donor tissues, cells, and organs (http://www.revivicor.com/index.html).

SUMMARY

Farm animals have filled an important niche in our efforts in biological discovery, product and technique development, and product testing historically and into the current era. The use of farm animal species as animal models will likely intensify as cellular and molecular biology advances yield new approaches to disease therapy and leaps in technology provide new products that must be tested in animal models deemed clinically relevant to humans. In addition, the husbandry, management systems, and veterinary care of farm animals are already well established, of high quality, and subject to continuous review and improvement efforts. With due consideration of satisfactory ethical review and outcomes, this facilitates an easy transition from our humane use of farm animals for the natural characteristics we value (i.e., food and fiber) to the pursuit of newly introduced characteristics by transgenic technology that benefit the advancement of medicine and improve patient care.

REFERENCES

Abdel-Halim, R.E. 2005. Contributions of Ibn Zuhr (Avenzoar) to the progress of surgery: A study and translations from his book Al-Taisir. *Saudi Med J* Sept. 26(9): 1333–1339.

Andrews, E.J., Ward, B.C., and Altman, N.H., Eds. 1979. *Spontaneous Animal Models of Human Disease*, Vols. 1 and 2. San Diego, CA: Academic Press.

Boyle, R. 2010. How modified worms and goats can mass-produce nature's toughest fiber, http://www.popsci.com/science/article/2010-10/fabrics-spider-silk-get-closer-reality

Delano, M.L., Mischler, S.A., and Underwood, W.J. 2002. Biology and diseases of ruminants: Sheep, goats and cattle. In: *Laboratory Animal Medicine*, 2nd ed. Fox, J.G., Anderson, L.C., Loew, F.M., and Quimby, F.W., Eds. San Diego, CA: Academic Press, pp. 519–611.

Ekser, B., Echeverri, G.J., Hassett, A.C., Yazer, M.H., Long, C., Meyer, M., Ezzelarab, M., Linm, C.C., Hara, H., van der Windt, D.J., Dons, E.M., Phelps, C., Ayares, D., Cooper, D.K., and Gridelli, B. 2010. Hepatic function after genetically engineered pig liver transplantation in baboons. *Transplantation* 90(5): 483–493.

Fahrenkrug, S.C., Blake, A., Carlson, D.F., Doran, T., Van Eenennaam, A., Faber, D., Galli, C., Gao, Q., Hackett, P.B., Li, N., Maga, E.A., Muir, W.M., Murray, J.D., Shi, D., Stotish, R., Sullivan, E., Taylor, J.F., Walton, M., Wheeler, M., Whitelaw, B., and Glenn, B.P. 2010. Precision genetics for complex objectives in animal agriculture. *J Anim Sci* 88(7): 2530–2539.

Gonzalez Bosquet, J., Peedicayil, A., Maguire, J., Chien, J., Rodriguez, G.C., Whitaker, R., Petitte, J.N., Anderson, K.E., Barnes, H.J., Shridhar, V., and Cliby, W. A. 2010. Comparison of gene expression patterns between avian and human ovarian cancers. *Gynecol Oncol* 120 (2): 256–264.

Hau, J. and Hendriksen, C.F. 2005. Refinement of polyclonal antibody production by combining oral immunization of chickens with harvest of antibodies from the egg yolk. *ILAR J* 46(3): 294–299.

Laber, K.E., Whary, M.T., Bingel, S.A., Goodrich, J.A., Smith, A.C., and Swindle, M.M. 2002. The biology and diseases of swine. In: *Laboratory Animal Medicine,* 2nd ed. Fox, J.G., Anderson, L.C., Loew, F.M., and Quimby, F.W., Eds. San Diego, CA: Academic Press, pp. 615–655.

Lewis, T. 1912, A lecture on the evidences of auricular fibrillation, treated historically. *Br Med J* 1: 57–60, doi:10.1136/bmj.1.2663.57.

Lyall, J., Irvine, R.M., Sherman, A., McKinley, T.J., Núñez, A., Purdie, A., Outtrim, L., Brown, I.H., Rolleston-Smith, G., Sang, H., and Tiley, L. 2011. Suppression of avian influenza transmission in genetically modified chickens. *Science* J331(6014): 223-226.

Maga, E.A., Shoemaker, C.F., Rowe, J.D., Bondurant, R.H., Anderson, G.B., and Murray, J.D. 2006. Production and processing of milk from transgenic goats expressing human lysozyme in the mammary gland. *J Dairy Sci* 89(2): 518–524.

Martini, L., Fini, M., Giavaresi, G., and Giardino, R. 2001. Sheep model in orthopedic research: A literature review. *Comp Med* 51(4): 292–299.

Medawar, P.B. 1997. Historical introduction to the general properties of retroviruses. In: *Retroviruses.* Coffin, J.M., Hughes, S.H., and Varmus, H.E., Eds. Cold Spring Harbor, NY: Cold Spring Harbor Laboratory Press.

Platt, J.L. 2001. Immunology of xenotransplantation. In: *Sampter's Immunologic Diseases.* Philadelphia, PA: Lippincott Williams & Wilkins, pp. 1132–1146.

Platt, J.L. 2011a. Xenotransplantation: Biological barrier. In: *Encyclopedia of Animal Science*, Volume II, 2nd ed. Ullrey, D.E., Baer, C.K., and Pond, W.G., Eds. Boca Raton, FL: CRC Press, pp. 1113–1116.

Platt, J.L. 2011b. Xenotransplantation: Biological barrier. *Encyclopedia of Animal Science*, Volume II, 2nd ed. Ullrey, D.E., Baer, C.K., and Pond, W.G., Eds. Boca Raton, FL: CRC Press, pp. 1117–1120.

Robl, J.M. 2007. Application of cloning technology for production of human polyclonal antibodies in cattle. *Cloning Stem Cells* 9(1): 12–16.

Seger, D., Kahn, S., and Krenzelok, E.P. 2005. Treatment of US crotalidae bites: Comparisons of serum and globulin-based polyvalent and antigen-binding fragment antivenins. *Toxicol Rev* 24(4): 217–227.

Srinivasan, A., Trus, T.L., Conrad, A.J., and Scarbrough, T.J. 1999. Common laparoscopic procedures in swine: A review. *J Invest Surg* 12(1): 5–14.

Stovall, D.W. 2010. Aprela, a single tablet formulation of bazedoxifene and conjugated equine estrogens (Premarin) for the potential treatment of menopausal symptoms. *Curr Opin Investig Drugs* 11(4): 464–471.

Swindle, M.M. 1998. *Surgery, Anesthesia and Experimental Techniques in Swine.* Ames, IA: Iowa State University Press.

van Velthoven, R.F. and Hoffmann, P. 2006. Methods for laparoscopic training using animal models. *Curr Urol Rep* 7 (2): 114-119.

Weiss, W.J. 2005. Pulsatile pediatric ventricular assist devices. *ASAIO J* 51(5): 540-545.

LAWS, REGULATIONS, AND OVERSIGHT MECHANISMS FOR RESEARCH STUDIES WITH AGRICULTURAL ANIMALS IN THE UNITED STATES

Christian E. Newcomer

INTRODUCTION

The legal and regulatory framework for the oversight of research using laboratory animals in the United States is now approaching its 50-year landmark, and the use of agriculturally important mammalian species as animal models pertaining to the exploration of the biology and diseases of humans has fallen under the purview of these regulations for most of that period. The regulatory framework has strengthened over time and has become considerably more focused with the significant and convergent changes that occurred during the mid-1980s. In 1985, working under independent statutory authorities, the Animal Welfare Act Regulations (AWAR) (AWA, 1990) and the Public Health Service Policy on the Humane Care and Use of Laboratory Animals (PHS Policy) (PHS 2002) adopted new progressive provisions emphasizing institutional accountability. The policies and regulations worked together to harmonize the approach and expectations for federal oversight of the care and use of animals used for research in the United States. The convergent interest of these regulations was the manifold considerations of and attention to the promotion of animal welfare and the controls that needed to be in effect to detect and impede potential points of failure in assuring animal welfare within institutions. The key regulatory advancement was the requirement that an organization conducting animal research that fell under regulatory jurisdiction must develop an institutional animal care and use committee (IACUC). The IACUC serves to foster, review, and monitor an institution's program of animal care and use to ensure ongoing regulatory compliance and to provide a thoughtful and deliberative platform for the institution to address emerging needs of animal models and scientists as scientific knowledge advances and new requirements and opportunities become evident. Two excellent professional guidance documents used in conjunction with the regulatory oversight of research in the United States and abroad also re-emphasize the importance of the IACUC in meeting the institution's requirements for the care and use of research animals. These are *The Guide for the Care and Use of Agricultural Animals in Research and Teaching,* 3rd edition (*Ag Guide*) (FASS, 2010) and *The Guide for the Care and Use of Laboratory Animals,* 8th edition (*Guide*) (ILAR, 2011). These two important guidance documents are also used as primary standards for the independent, voluntary, peer-review accreditation program performed by the Association for Assessment and Accreditation of Laboratory Animal Care International. The balance of this section briefly explains the interrelationships and key features of the regulatory and oversight entities, mechanisms, and guidance documents mentioned.

DEFINING THE REGULATORY FRAMEWORK AND GUIDANCE DOCUMENTS

Congress enacted the original legislation in the United States governing research animal care in 1966 under Public Law (P.L.) 89-544 as the Laboratory Animal Welfare Act (LAWA). At that time, the LAWA regulated animal dealers that handled dogs and cats and laboratories that used dogs, cats, rabbits, guinea pigs, hamsters, and nonhuman primates. During the 1970s' amendments under P.L. 91-579, Congress changed the name of the law to the Animal Welfare Act (AWA, 1990) and authorized the Secretary of Agriculture to regulate other warm-blooded animals when used in research, exhibition, or the wholesale pet trade. This was the first time that agricultural animals used in some research applications were included in the regulatory framework. The basis for coverage under the AWA regulations rests with its definition of the term "animal" and there are important exclusions. Specifically, quoting from the section on definitions in the AWAR,

> This term (animal) excludes birds, rats of the genus *Rattus*, and mice of the genus *Mus*, bred for use in research; horses not used for research purposes; and other farm animals, such as, but not limited to, livestock or poultry used or intended for use as food or fiber, or livestock or poultry used or intended for use for improving animal nutrition, breeding, management, or production efficiency, or for improving the quality of food or fiber.

Thus, a vast majority of the research activities currently conducted in agricultural species is not covered today by the AWAR, but with the growth of agriculturally important animal models in a wide variety of facets of biomedical research and product development, the coverage of agricultural animals is increasing. The Research Facility Inspection Guide (APHIS, 2001) provides the criteria and examples used by the Veterinary Medical Officers (VMO) from APHIS's (APHIS, 2006) Animal Care (AC) program to determine whether the farm animals in particular studies at an institution should be included in the inspection process.

An AC VMO inspects institutions registered and licensed as research animal facilities at least annually, and their findings are the basis for evaluating the institution's regulatory compliance. Institutions are expected to have effective IACUCs, personnel training efforts, and programs of veterinary care to ensure ongoing compliance with the AWAR. With regard to compliance with standards, institutions are expected to adhere to Part 3 of the AWAR (Standards), which covers facilities and operating standards, animal health and husbandry standards, and transportation standards. Although the standards are specific and even prescriptive for many of the covered species, the standards in the AWAR for farm animals are written in general terms. In instances where the institution's provisions of oversight are deemed ineffectual, regulatory enforcement is achieved through increased inspections, the opportunity for prompt corrective action in many instances, the issuance of fines for serious or repetitive noncompliance, or the suspension or revocation of licensure.

Institutions that receive funding from the Public Health Service are required to comply with the Public Health Service Policy on the Humane Care and Use of Laboratory Animals (PHS Policy). As authorized by the Health Research Extension Act of 1985, the PHS Policy requires institutions to establish and maintain measures to ensure the appropriate care and use of all vertebrate animals involved in research, research training, and biological testing activities conducted or supported by PHS. Some other federal agencies also expect the programs operating under their jurisdiction to follow PHS Policy standards (e.g., the Veterans Administration Policy requires compliance with the PHS Policy even if PHS funds are not received by the research unit in question). The PHS Policy requires compliance with the *Guide* and the American Veterinary Medical Association Guidelines for Euthanasia. Institutions are required to have an approved Assurance on file with the Office of Laboratory Animal Welfare within the PHS. The Assurance document explains the institution's provisions for compliance with the *Guide*. It is permissible for an institution to delimit the scope of PHS coverage in its Assurance extending compliance with the provisions of the *Guide* only to those studies required by the source of funding, but excluding all other studies. Institutions that choose to take this approach, therefore, could make the claim that many studies conducted in farm animals for the purpose of improving food and fiber production are required to comply with *Guide* standards. On the other hand, if the institution states that all vertebrate animals at the institution are covered by the Assurance, then the PHS will expect the institution to comply with either the *Guide* or the *Ag Guide* when agricultural species are used in research or teaching depending on the source of funding for the activity and other discriminating criteria provided by the institution.

RECENT REVISION OF EXISTING GUIDES

The Guide for the Care and Use of Laboratory Animals has recently been revised, and the release of the *Guide*, 8th edition (ILAR, 2011) has already generated considerable interest and discussion. It is a very comprehensive document that expands the discussion of many issues in animal care and use significantly in comparison to the previous edition published in 1996, and it offers an institution a roadmap to establishing a sound program for the success of biomedical research, testing, and teaching in research animal models. The *Guide* describes the essential components of an institution's overall animal care and use program; considerations and provisions for the animals' environment, housing, and management; multiple facets of a competent program of veterinary care; and the requirements for an adequate physical plant. The *Guide* also addresses the issue of dichotomous treatment of agricultural animals in research depending upon whether their use is aligned with a biomedical inquiry versus an agricultural inquiry. It also notes that the institutions occasionally find that the categorization of research animal studies presents a dilemma. It suggests, therefore, that IACUCs should make the decisions concerning the standards of care for the agricultural animals used in research studies based upon the researcher's goal and the concern for well-being of the animals. The *Guide* also acknowledges that the *Ag Guide* is a useful resource for agricultural animals maintained within typical farm settings.

The *Ag Guide*, 3rd edition, is a scholarly and authoritative professional guidance document published by the Federated Animal Science Societies (FASS) in 2010. Although the document lacks regulatory standing, it carries enormous credibility by virtue of its expert authorship and the careful consideration and extensive review of scientific literature on many topics. As noted previously in this section, there are many circumstances in which agricultural animals could be used in research without any regulatory oversight if neither the funding source for the research nor the category of the research (as non-food and fiber related) dictated. The voluntary adoption of the recommendations of the *Ag Guide* by institutions conducting studies under these circumstances would be an ideal solution for the protection of the quality and integrity of the scientific research, as well as an effective tool in assuaging public concerns about the use of agricultural animals in research. Although it seems fair to speculate that most institutions subscribe to the *Ag Guide* in these situations, the number of outliers is unknown. The *Ag Guide* has many parallels with the *Guide*, especially pertaining to the expectations of an institution's essential policies and provisions for the program of animal care and use. For example, it identifies the need for a properly structured and functioning IACUC with written operating procedures for animal health care, biosecurity, personnel qualifications and training, occupational health, and special considerations. Individual chapters are dedicated to animal health care including husbandry, housing and biosecurity, environmental enrichment, animal handling, and transport, as well as six key animal species areas. There are also several key inconsistencies between the *Ag Guide* and the *Guide* in the areas of space recommendations, sanitation schedules, and environmental conditions, which will require reconciliation by the IACUC through the review of scientific literature and expert opinion or by prevailing regulatory mandates.

Since 1985 when IACUCs were established by U.S. Public Law as noted previously, they have been recognized as a seminal development for the improvement of the welfare of animals used in research. The regulators, the regulatory community, and the professional scientific societies who produce guidance documents have acknowledged the importance of strong internal institutional oversight provisions embodied in the IACUC. In addition, the guidelines or national legislation for animal care and use in research in many other countries mimics this general approach, which further validates its value. There are variations in the committee structure and function of IACUCs across the United States with respect to regulations and the non-regulatory guidelines offered by nongovernmental agencies or professional societies, which are beyond the scope of this discussion. However, the central features are very similar. Committee members should have appropriate training and expertise and represent a variety of perspectives to achieve an appropriate balance in their oversight of the program and the approval of research activities. For example, the *Ag Guide*, which has enhanced membership requirements, specifies that committee members should include an agricultural scientist with teaching or research experience; an animal, dairy, or poultry scientist who has agricultural animal management experience; a veterinarian knowledgeable about agricultural animal medicine; a member whose primary concerns are in an area outside of science; and a person who is not affiliated with the institution and who represents general community interests in the proper care and treatment of animals. The IACUC is required to review and approve, when appropriate, animal use protocols for research and teaching at the institution to ensure that it is justified, scientifically sound, prudent, and conducted under conditions that consider and preserve animal welfare throughout all phases of the activity. In addition to the information in the regulations, the *Guide* and the *Ag Guide* aid IACUCs in conducting a conscientious and competent protocol review process. There are other sources of extensive information on this subject (Silverman, 2007). The IACUC is also empowered to disapprove inappropriate proposals and suspend ongoing activities that prove to compromise animal welfare. In addition to the vital function of protocol review and approval, IACUCs are responsible for evaluating the facilities available for research animal studies and the entire program of animal care and use at the institution. Programmatic review entails knowing and critically assessing the institution's resources pertaining to the following requirements for acceptable animal care and use: Conditions of the physical plant in animal facilities and animal study areas; expertise, training, and staffing levels of personnel supporting or conducting research with animals; occupational health and safety concerns related to animal care and use and experimental conditions; provisions for veterinary care to ensure the health, welfare, experimental reliability, and robustness of animals used in research in accordance with prevailing standards; and assurance that the operations provide the appropriate environment, housing, husbandry, and management of research animals. Through the IACUC's rigorous process of facility and programmatic review, the institution, at a minimum, is afforded the opportunity to plan and take timely, effective, self-corrective actions to correct weaknesses or deficiencies in the institution's resources dedicated to the care and use of animals in research and teaching. Under optimal conditions, the IACUC can play a helpful role in encouraging the institution to be forward thinking in initiatives to meet emerging scientific and educational needs in a contemporary manner.

VOLUNTARY PARTICIPATION OF INSTITUTIONS IN AAALAC

Many institutions choose to participate in a voluntary, confidential, expert peer-review accreditation program developed by the Association for Assessment and Accreditation of Laboratory Animal Care International (AAALAC International). This includes institutions that fall under regulatory mandates in the United States or other regions of the globe, as well as programs that operate in unregulated environments. AAALAC International is a non-profit, nongovernmental organization that has operated its accreditation program for more than 45 years and now accredits more than 830 organizations in 33 nations around the globe. Within the United States, more than 600 organizations are accredited and these include university, pharmaceutical, governmental, commercial, and contract research programs with substantial agricultural components. Among those accredited in the United States are 19 Land Grant Institutions and other universities emphasizing agricultural research and teaching programs. AAALAC International accreditation relies upon three primary standards. These are *The Guide, The Ag Guide,* and the new *European Directive 2010/63/EU on the Protection of Animals Used for Scientific Purposes,* which contains accommodation and care standards from the *European Treaty Series 123.* The peer-review process is comprehensive and entails the thorough review of an institution's facilities, policies, programs, procedures, and personnel qualifications in support of animal care and use programs. Institutions must meet all regulatory requirements that pertain to activities with research animals in their environment as well as relevant portions of the standards identified previously. The experts chosen to conduct the site visit are selected with due regard to the type of institution, the animal models used in research and teaching, the scientific areas emphasized in the institution's research, and the avoidance of any conflicts of interest. Subsequently, the experts on the site visit team must engage a much larger deliberative body, the Council on Accreditation, who determines whether accreditation should be granted. Organizations that attain accreditation must meet or exceed applicable standards and maintain quality programs that ensure animal health, well-being, and welfare as the platform for productive scientific inquiry using animal models for research.

SUMMARY

The regulatory standards and framework governing the use of farm animals in research have improved significantly since the mid-1980s, and many organizations are required to comply with these regulations. In addition to the mandated regulatory standards that are selectively applied, the number of organizations electing to adopt and adhere to the guidelines proposed in the authoritative reference, *The Ag Guide,* and participate in the voluntary, peer-review accreditation program of AAALAC International is increasing. The combination of the mandated and voluntary provisions for the oversight of the use of farm animals in research, teaching, and testing appears to be working well and increases our prospects of ethical and successful outcomes in these endeavors. These measures also help build the public's support and confidence in our use of farm animals in research applications. However, they do not comprise an impervious system of farm animal research oversight sufficient to detect and correct problem areas in every instance.

REFERENCES

APHIS (Animal and Plant Health Inspection Service). 2001. Research Facility Inspection Guide. Available at http://www.aphis.usda.gov/animal_welfare/rig.shtml (Accessed February 10, 2011).

APHIS (Animal and Plant Health Inspection Service). 2006. Animal Care Policy Manual. Available at www.aphis.usda.gov/animal_welfare/policy.shtml (Accessed February 10, 2011).

AWA (Animal Welfare Act). 1990. Animal Welfare Act. PL (Public Law) 89-544. Available at www.nal.usda. gov/awic/legislat/awa.htm (Accessed February 10, 2011).

AWAR (Animal Welfare Act Regulations). Available at http://www.gpo.gov/fdsys/pkg/CFR-2009-title9-vol1/ xml/CFR-2009-title9-vol1-chap1-subchapA.xml (Accessed February 10, 2011).

FASS (Federation of Animal Science Societies). 2010. *The Guide for the Care and Use of Agricultural Animals in Research and Teaching*, 3rd ed. Available in PDF at http://www.fass.org/page.asp?pageID=216&auto try=true&ULnotkn=true (Accessed February 10, 2011).

ILAR. 2011. *The Guide for the Care and Use of Laboratory Animals*, 8th ed. Washington, D.C.: Institute for Laboratory Animal Research, National Academies Press.

PHS (Public Health Service). 2002. *Public Health Service on the Humane Care and Use of Laboratory Animals*. Publication of the U.S. Department of Health and Human Services. National Institutes of Health, Office of Laboratory Animal Welfare. Available at http://grants.nih.gov/grants/olaw/references/phspol.htm (Accessed February 10, 2011).

Silverman, J., Suckow, M.A., and Murthy, S., Eds. 2007. *The IACUC Handbook*, 2nd ed. Boca Raton, FL: CRC Press.

ROLE OF ANIMAL AGRICULTURE IN ENRICHMENT OF YOUTH DEVELOPMENT THROUGH ORGANIZED HANDS-ON EXPOSURE TO HIGH STANDARDS OF ANIMAL WELFARE IN FOOD ANIMAL PRODUCTION

Christopher Boleman, Aaron Alejandro, and Cherie Carrabba

Youth involvement and engagement through organized hands-on experience in animal agriculture has been recognized for many years. A unique interaction takes place between youth and livestock and poultry. This relationship is sometimes challenging to define to the general public, but the life skills, especially an increased sense of responsibility and discipline in youth, gained through this interaction are well documented, and the impact on animal welfare and husbandry practices sustained (Boleman, Cummings, and Briers, 2004).

Before discussing these life skills, we first delve into the history of the relationship between youth and animals. This history can be documented most effectively through the evolution of 4-H and Future Farmers of America (FFA) programs.

4-H CLUBS AND FFA CHAPTERS

The 4-H clubs were beginning to form at the turn of the twentieth century. According to Wessel and Wessel (1982), Cornell University's Liberty Hyde Bailey developed and disseminated educational leaflets on agriculture for youth interested in a career in agriculture as early as 1896. During the early 1900s, several other states developed similar educational pamphlets and distributed them to potential future farmers because of the impact that Bailey's work was having on youth in New York. During this same time, other states began hosting youth corn contests. Illinois, Georgia, Oregon, Missouri, Nebraska, and Indiana were all hosting some type of corn or agricultural exhibit. Not only did these exhibits reveal the highest quality and prize-winning products, but also exhibitors were able to discuss their crops and answer questions concerning farming their crops (Wessel and Wessel, 1982). In 1914, the Smith-Lever Act was passed. The passing of this act provided the financial support for the Cooperative Extension Service to be successful (Wessel and Wessel, 1982). This also allowed the 4-H program to be housed under the Cooperative Extension Service.

According to Reck (1951), World War I was a key contributor to the growth of 4-H. This growth in membership was directly related to the fact that America was at war and needed more food and fiber to sustain itself during that time. In order to ensure that adult farmers were using the best, most effective production practices, County Extension Agents in the field seized the opportunity to work with youth and teach them practices for production of food and fiber. These youth took their knowledge home and helped convince their parents to adopt these new farming practices. As a result, farm production levels increased (Wessel and Wessel, 1982).

During this same period, FFA was being established through the Smith-Hughes Vocational Education Act in 1917. Similar to the origins of 4-H, the idea for what would be known as FFA was initiated with the introduction of agricultural clubs in schools with Virginia being the first to establish such a club. The actual formation of the FFA was in 1928. In terms of membership growth, the trends were the same for FFA as for 4-H. The FFA program experienced tremendous growth during the late 1920s and into the 1930s.

It is also worth pointing out that high school students learned about animal agriculture through agriculture science courses offered in middle and high school. This is separate from 4-H club and FFA chapter experiences. These classes demonstrated academic rigor and relevance related to animal welfare. More than a "club," classroom instruction afforded a focused opportunity of learning and it was then complemented by the "hands-on" aspects of supervised programs for agriculture experience.

Since 1930, both 4-H and FFA have evolved to include even more members and a wide variety of programs and projects. However, the pledges and mottos remain the same. The 4-H motto and pledge are as follows:

In support of the 4-H club motto, to make the best better, I pledge my head to clearer thinking, my heart to greater loyalty, my hands to larger service, and my health to better living, for my club, my community, my country, and my world.

The FFA motto is as follows:

Learning to do, doing to learn, earning to live, and living to serve.

Obviously, these mottos help to reveal the relevance of these organizations in the past, the present, and into the future. In addition, they help to recognize the fact that these youth members who exhibit livestock projects at county, state, and national livestock shows and rodeos are indeed "learning by doing" and "making the best better."

THE GROWTH OF LIVESTOCK PROJECTS

Calf, swine, and dairy clubs increased significantly during World War I. Reck (1951) said that these projects increased because private donors supported these efforts by donating livestock to the youth for their projects. By 1917, states began to have youth shows. According to Wessel and Wessel (1982), the Minnesota State Livestock Breeders Association was the first show to offer youth cash prizes and to help counties hold calf and colt shows. By 1917, two men, T.A. Erickson and W.A. McKerrow, joined this livestock breeders association to establish Minnesota's first junior livestock show (Reck, 1951).

Livestock shows have grown since 1918 and become a symbol of the 4-H and FFA youth organizations. Although it is very challenging to determine the total number of livestock projects exhibited by youth across the nation, a study in Texas in 2000 revealed that Texas 4-H and FFA members accounted for over 70,000 entries for cattle, swine, meat goats, and sheep across the state (Boleman, Howard, Smith, and Couch, 2001).

STUDIES SPECIFIC TO YOUTH LIVESTOCK PROJECTS

According to Boyd, Herring, and Briers (1992), the development of life skills through experiential learning is the cornerstone of the 4-H program and the same can be said for FFA. More specifically, livestock projects are an extremely valuable vehicle for developing life skills.

A study conducted by Ward (1996) asked 4-H alumni to reflect on the impact that exhibiting livestock projects had on their development of life skills. According to respondents, the meaningful life skill impacts were accepting responsibility, relating to others, spirit of inquiry, decision-making, public speaking, maintaining records, and building positive self-esteem.

Rusk, Martin, Talbert, and Balshweid (2002) came to similar conclusions from their study of Indiana 4-H youth that judged livestock. For this study, the most meaningful results noted were that youth learned how to defend a decision, gained knowledge of the livestock industry, and developed oral communication skills, as well as decision-making skills, self-confidence, problem solving, teamwork, self-motivation, self-discipline, and organizational skills.

Finally, Boleman, Cumming, and Briers (2004) ascertained the life skills gained from youth exhibiting beef, swine, sheep, or goat livestock projects. They concluded that the five highest life skills gained were accepting responsibility, setting goals, developing self-discipline, self-motivation, and knowledge of the livestock industry.

THE ROLE OF 4-H AND FFA YOUTH IN ENHANCING WELFARE OF ANIMAL AGRICULTURE AND COMPANION ANIMALS

The learning process about animal care responsibilities begins with the careful example and influence of adult leaders and advisors responsible for training and guiding youth. This influence is fundamental to the continuance of animal-friendly husbandry practices that ensure animal health and well-being. Animal welfare is indeed one of the fundamental educational priorities within youth and animal projects. Over the past 10 years, many state 4-H and FFA programs have implemented quality assurance programs that ensure youth are learning and applying the appropriate quality assurance practices. These include Pork Quality Assurance and Quality Counts (Boleman, Chilek, Coufal, Kieth, and Sterle, 2003).

SUMMARY AND CONCLUSIONS

Youth development is definitely enhanced by hands-on experience gained through interactions with animals. Many people hear testimonials from adults who once raised livestock as youth to learn about their positive experiences and the impact raising these livestock had on their lives. In many instances, the livestock project enhanced the child's relationships with his or her family and friends. The livestock project requires the help and cooperation of family members. Parents, siblings, and grandparents often become involved in the project. It helps the family unit develop common goals and an understanding of the financial side of agriculture. Quite simply, the farm animals they raised helped shape who they are, the character attributes they possess, and the positive life skills they develop and use every day of their lives.

REFERENCES

Boleman, C.T., Chilek, K.C., Coufal, D., Kieth, L., and Sterle, J. 2003. Quality counts: Quality assurance, character education. Texas AgriLife Extension Service. Publication: CHE-1.

Boleman, C.T., Cummings, S.R., and Briers, G.E. 2004. Parents' perceptions of life skills gained by youth participating in the 4-H beef project. *Journal of Extension* 42(5). Retrieved May 30, 2011, http://www.joe.org/joe/2004october/rb6.shtml

Boleman, C.T., Howard, J.W., Smith, K.L., and Couch, M.C. 2001. Trends in market steer, lamb, swine, and meat goat projects based on county participation: A qualitative and quantitative study. *Texas 4-H Research Review* 1: 52–62.

Boyd, B.L., Herring, D.R., and Briers, G.E. 1992. Developing life skills in youth. *Journal of Extension* 30(4). Retrieved October 24, 2002, http://www.joe.org/joe/1992winter/a4.html

Reck, F.M. 1951. *The 4-H Story, a History of 4-H Club Work*. Ames, IA: Iowa State University Press.

Rusk, C.P., Martin, C.A., Talbert, B.A., and Balshweid, M.A. 2002. Attributes of Indiana's 4-H livestock judging program. *Journal of Extension* 40(2). Retrieved June 21, 2002, http://www.joe.org/joe /2002april/rb5.php

Ward, C.K. 1996. Life skill development related to participation in 4-H animal science projects. *Journal of Extension* 34(2). Retrieved April 19, 2000, http://www.joe.org/joe/1996april/rb2.html

Wessel, T. and Wessel, M. 1982. *4-H: An American Idea 1900–1980, A History of 4-H*. Chevy Chase, MD: National 4-H Council.

EDITORS' NOTE TO THIS CHAPTER SECTION

The book editors commend the authors of this chapter section for the historical overview of the important role that 4-H and FFA programs have in youth development and in animal welfare improvements in farm animals.

Not included in this chapter section is a brief account of a major challenge to the ideals and mission of leaders of 4-H and FFA. In the last decade of the twentieth century and extending into the early years of the twenty-first century, episodes of cheating to alter animal appearance or weight have been documented. These cases of animal abuse and unethical behavior among adult and youth exhibitors in show rings have been chronicled in the popular press and consequently the issues and remedies have been addressed by youth leaders, show managers, and judges. A strict code of ethics is required in the show ring. 4-H and FFA are primarily youth development organizations. As emphasized by the authors of this chapter section, the exhibition of projects is only the final stage of a process intended to develop responsibility, goal-setting, and leadership skills. A major role of 4-H and FFA in livestock projects is to advance the concept of improving farm and companion animal welfare as well as personal integrity in future leaders in our society.

3 Contributions of Animals in Human Service

A Two-Way Path

Duane Ullrey

CONTENTS

Temple Grandin has said, "Animals make us human," and used those words in the title of her latest book (Grandin and Johnson, 2009). One might question this opinion when viewing the questionable care and cruelty sometimes visited upon animals by humans, but there is little doubt that animals helped us become—thousands of years ago—warriors, hunters, and farmers. However, have they made us...or will they make us...human?

If to be human is to be humane, let us hope. The Merriam-Webster dictionary defines humane as "marked by compassion, sympathy, or consideration for other human beings or animals." To the extent that animals can help us merit that description, let us, by all means, increase our interactions with them. Most humans who love animals report a personal benefit from those associations. For some, it may be a chance to escape the stress of modern life and revel in the joys of play. For those with special needs, the benefits may be more specific. Guide dogs steer their human masters safely around sidewalk obstacles and across streets. The hearing impaired may be alerted to ringing telephones by trained dogs or cats. For those seeking safety from home invasion, barking dogs can frighten off intruders. These and other benefits help explain why Americans own 93.6 million cats, 77.5 million dogs, 13.3 million horses, ponies, donkeys, and mules, and additional millions of birds, fish, reptiles, and small mammals (2009/2010 National Pet Owners Survey; www.american-petproducts.org).

The human-animal bond has a long history (Walsh, 2009a), and animals have been respected partners in human survival, health, and healing in cultures worldwide since ancient times. Archeological evidence indicates that domesticated wolves, ancestors of the dog, were being used as guardians, guides, and partners in hunting and fishing over 14,000 years ago (Price, 2002). Both dogs and cats were assuming crucial roles in agriculture 5000 years later—dogs in herding, as livestock guardians, and in pulling carts and sleds, and cats in protecting grain stocks from rodents. Some American Indians and indigenous people of Asia and Africa still draw symbolic meaning and teachings from animals. Their historical importance as pets is illustrated by the discovery, in the ruins of Pompeii, of the bones of a dog named Delta—identified by his engraved silver collar—lying next to the bones of a child. Over 63% of U.S. households, and over 75% of those with children, currently have at least one pet (Walsh, 2009b). Some of these pets are highly pampered, receiving presents on holidays, special savory meals, and time off work by their masters to tend them when they are ill.

Pet ownership has been shown to correlate with lower blood pressure, serum triglycerides, and cholesterol levels, and may be more effective in ameliorating the cardiovascular effects of stress than the presence of a spouse or friend (Allen, Blascovich, and Mendes, 2002). Following a heart attack, patients with pets had a significantly higher one-year survival rate than those without, and if the pets were dogs, the patients were 8.6 times more likely to be alive (Friedman and Thomas, 1995). Interactions with companion animals increase blood neurochemical concentrations associated with relaxation and improve function of the human immune system (Charnetsky, Riggers, and Brennan, 2004). A broad range of studies has found that these interactions tend to reduce anxiety, depression, and loneliness among humans in hospitals, eldercare environments, schools, and prisons. Walsh (2009a,b) has summarized these and other positive effects.

Although most dogs are now kept as pets, many have performed—or still perform—duties in addition to those mentioned previously. In the past, some have turned a treadmill connected to a roasting spit or butter churn. Dogs with herding instincts are still used by the stockman in the management of cattle and sheep, and some are used to discourage the presence of geese and seagulls on beaches, park lawns, and airfields. Sled dogs, although now used mostly in sporting competition, still transport supplies and people in arctic regions. Circus and actor dogs provide entertainment by performing for human audiences in person, in movies, or on television. The American Humane Association, founded in 1877, believes that "dogs, books, and kids go together like peanut butter, jelly, and bread" (www.americanhumane.org), and sponsors a children's literacy program that addresses problems of low confidence and poor reading skills by encouraging children to read to their dogs.

Service animals are not legally considered pets, and most undergo extensive training to live and work as partners with humans in specialized roles. Police dogs are trained specifically to assist in law enforcement or military duty, and are often referred to (when using a homophone of canine) as members of a K9 Corp. These dogs fulfill several roles, including officer protection, chasing and detaining suspects, search and rescue of missing persons during natural or man-made disasters, finding cadavers, and detection of drugs and explosives. They may even wear a ballistic vest on dangerous missions and have their own police badge. Popular breeds, with identifiable specialties, include the Argentine Dogo, German Shepherd, Dutch Shepherd, Belgian Malinois, Boxer, Labrador Retriever, Doberman Pinscher, Springer Spaniel, Bloodhound, Beagle, Rottweiler, and Giant Schnauzer. Police dogs were first assigned official responsibilities in Europe in 1859 when the Belgian police force in Ghent began using them to patrol with night-shift personnel. An excerpt from the January 15, 1938, *London Times* quotes Colonel Hoel Llewellyn, Constable of Wiltshire, as follows:

> A good dog with a night duty man is as sound a proposition as you can get. The dog hears what the constable does not, gives him notice of anyone in the vicinity, guards his master's bicycle to the death, and remains mute unless roused. He is easily trained and will go home when told to do so with a message in his collar.

In the United States, the Codes of Federal Regulation for the Americans with Disabilities Act of 1990 (www.ada.gov) defines a service animal as

> any guide dog, signal dog, or other animal individually trained to do work or perform tasks for the benefit of an individual with a disability, including, but not limited to, guiding individuals with impaired vision, alerting individuals with impaired hearing to intruders or sounds, providing minimal protection or rescue work, pulling a wheelchair, or fetching dropped items.

Dogs of many breeds (or crossbreeds) have been used to aid the autistic, the visually or hearing impaired, those requiring mobility assistance, or to alert others of a condition requiring a medical response. However, these services have not been limited to dogs.

Capuchin monkeys have been trained to perform various manual tasks for the seriously handicapped such as retrieving dropped items, microwaving food, opening drink bottles, washing a quadriplegic's face, and turning the pages of a book (www.helpinghandsmonkeys.org). Miniature horses have been trained to guide the blind, pull wheelchairs, and to provide secure walking support for persons with severe Parkinson's disease (www.guidehorse.org). "Comfort animals" may be used as a specific part of therapy designed to improve motivation and the physical, social, emotional, or cognitive function of human patients. This is termed animal-assisted therapy (AAT) and may be provided by a therapist on an individual or group basis. Many animal species have been used in AAT, including dogs, cats, horses, elephants, dolphins, rabbits, birds, lizards, and other small animals. The Dolphin Research Center in Grassy Key, Florida, offers a five-day program for children and adults with special needs, including dockside contact with dolphins and an opportunity to swim with them. Even exotic fish tanks, found frequently in physician waiting rooms, may serve to lessen patient anxiety.

When Liz Hartel, who ordinarily used a wheelchair because of polio, won the silver medal in dressage at the 1952 Olympics, the potential of horses in rehabilitation of human patients began receiving serious attention. In 1969, the North American Riding for the Handicapped Association was founded as a federally registered nonprofit organization. There are now over 3500 certified handicapped riding instructors and 800 member centers around the globe, helping more than 42,000 children and adults face physical, mental, and emotional challenges (www.narha.org).

Therapeutic horse riding has been shown to encourage responsibility and development of new skills, to provide companionship, nonjudgmental acceptance of disabilities, and a variety of physical and neuromuscular benefits (All, Loving, and Crane, 1999; Benda, McGibbon, and Grant, 2003). Kaiser, Smith, Heleski, and Spence (2006b) studied the effects of a therapeutic riding program on psychosocial measurements among children considered at risk of failure or poor performance in school or life because of family circumstances, and among children in special education programs due to emotional impairment or learning disabilities. None of the psychosocial measures for at-risk children was different after completion of the riding program, although three of sixteen measures of motor coordination were significantly improved. Total anger score was significantly reduced by therapeutic riding among special education children, but the greatest psychosocial benefit was seen in boys whose expressions of anger were significantly reduced and whose mothers perceived significant improvements in behavior. It is interesting that the horses used in this riding program exhibited a significant increase in stress-related behaviors when ridden by at-risk children, particularly girls (Kaiser, Heleski, Siegford, and Smith, 2006a). These authors suggested that these children appeared to transfer some of their anger from their family situation to their horses, but because girls tend to repress anger more than boys do, they may have expressed more of that repressed anger or expressed it more intensely.

Finally, if animals make us human, then humans surely have an ethical responsibility for the welfare of those animals. That obligation is particularly clear for the animal companions providing the benefits just described. However, in this writer's view, that obligation extends to the myriad animal and plant species with which we share the earth. Nature's ecosystems nourished our evolution and provided for our needs. If we want that beneficial relationship to continue, we must care for our environment and the creatures that live there, as though our lives depend upon it.

SUMMARY

The historical association of humans with companion animals was discussed, but their contributions to humans in modern times received greatest emphasis. Special attention was given to service animals that assist police and military personnel, the visually and hearing impaired, and those who have severe physical disabilities. Notable are dogs, cats, monkeys, and horses. These and other listed species also play significant roles in decreasing human anxiety and loneliness, and in improving health status, cognitive function, and feelings of self-worth. These benefits for humans warrant reciprocal effort to ensure appropriate care for the animals that provide them. Thus, we have an obligation to understand and meet companion animal needs, just as we expect them to understand and assist us with ours.

REFERENCES

All, A.C., Loving, G.L., and Crane, L.L. 1999. Animals, horseback riding, and implications for rehabilitation therapy. *J Rehabil* 65: 49–57.

Allen, K.M., Blascovich, J., and Mendes, W.B. 2002. Cardiovascular reactivity in the presence of pets, friends, and spouses. *Psychosomatic Med* 64: 727–739.

Benda, W., McGibbon, N.H., and Grant, K.L. 2003. Improvements in muscle symmetry in children with cerebral palsy after equine-assisted therapy (hippotherapy). *J Altern Complement Med* 9: 817–825.

Charnetsky, C.J., Riggers, S., and Brennan, F. 2004. Effect of petting a dog on immune system functioning. *Psychological Reports* 3: 1087–1091.

Friedman, E. and Thomas, S. 1995. Pet ownership, social support, and one-year survival after acute myocardial infarction in the cardiac arrhythmia trial. *Am J Cardiology* 76: 1213–1217.

Grandin, T. and Johnson, C. 2009. *Animals Make Us Human: Creating the Best Life for Animals.* New York: Houghton Mifflin Harcourt.

Kaiser, L., Heleski, C.R., Siegford, J., and Smith, K.A. 2006a. Stress-related behaviors among horses used in a therapeutic riding program. *J Am Vet Med Assoc* 228: 39–45.

Kaiser, L., Smith, K.A., Heleski, C.R., and Spence, L.J. 2006b. Effects of a therapeutic riding program on at-risk and special education children. *J Am Vet Med Assoc* 228: 46–52.

Price, E.O. 2002. *Animal Domestication and Behavior.* New York: CABI publishing.

Walsh, F. 2009a. Human-animal bonds I: The relational significance of companion animals. *Family Process* 48: 462–480.

Walsh, F. 2009b. Human-animal bonds II: The role of pets in family systems and family therapy. *Family Process* 48: 481–500.

Section Two

Treatment of Animals and
Societal Concerns

4 The Opinions and Recommendations of One Particular Study Group
The Pew Commission on Industrial Farm Animal Production

Robert P. Martin

CONTENTS

Every year, between 9 and 10 billion animals are raised and slaughtered in the United States for food. Within the space of only a few decades, the livestock system in the United States has been transformed from one in which most animals were raised in relatively small numbers on small- to mid-size farms, to one in which incredibly large numbers of animals are now produced in concentrated animal feeding operations (CAFOs), or factory farms, that are owned or controlled by large corporations. The impact of these industrial facilities has only recently been realized. Problems associated with CAFOs include air pollution; the contamination of both inland and coastal waters from animal waste; the development of antibiotic-resistant bacteria, stemming from the massive nontherapeutic application of antibiotics to livestock; and the inhumane treatment of many farm animals, which raises ethical considerations for the American public.

The Pew Charitable Trusts (Trusts), a Philadelphia-based public charity and its Pew Environment Group, have a specific interest in how this industrial transformation has affected the environment, public health, and ethics. After nearly a decade of internal planning, the Trusts established the Pew Commission on Industrial Farm Animal Production (PCIFAP) through a $2.6 million grant to Johns Hopkins Bloomberg School of Public Health's Center for a Livable Future to inform and guide the debate over the future of animal production in America.

Recognizing the interrelationship between how animals are raised and the impact on public health, the environment, rural communities, and the actual treatment of the animals, the Commission's purpose was to conduct a comprehensive assessment of the costs, benefits, and issues related to CAFOs in America and to issue a set of thoughtful, consensus-based recommendations on mitigating the negative impacts of factory farms while simultaneously providing quality food products at reasonable prices to American consumers. Its principal product was a final report to the nation released in April 2008 that incorporated 24 basic recommendations, supported by sound research and analysis.

The Commission was comprised of individuals from diverse backgrounds. The chair of the Commission was former Kansas Governor and Archivist of the United States, John Carlin. The vice chair was the Dean of the College of Veterinary Medicine at the University of Tennessee/Knoxville, Dr. Michael Blackwell. Other members of the Commission included representatives of the medical profession, and experts in nutrition, ethics, religion, production agriculture, public health, and the meat industry.[1]

During the first 18 months, the Commission focused primarily on fact-finding and assessment, including conducting site visits to farms and industrial animal production facilities, conducting hearings in various parts of the country, and contracting with scientists and other technical experts to produce up to eight specialized reports that helped inform the commissioners as well as the public. Within the four primary areas of inquiry, the commission determined the critical issues deserving greater scrutiny and analysis, including those that required a specialized report.

The commission determined that separate reports authored by academic experts working as teams would be needed in the areas of antibiotic resistance, animal welfare, environmental impacts of large animal operations, the impact on human health, the impact on animal health, the impact on rural communities, and the economics of industrial swine production to supplement our investigations. In addition, the Commission and staff reviewed hundreds of pages of material submitted by a wide variety of stakeholders, received statements submitted by more than 500 people who attended the two public meetings, and reviewed more than 170 peer-reviewed reports in the areas of the Commission's investigation.

In the final six months, the Commission refined its findings, and discussed and finalized policy recommendations.

COMMISSION FINDINGS

The general finding of the Pew Commission was that the present model of industrial farm animal production is not sustainable and presents an unacceptable level of risk to public health, an unacceptable level of damage to the environment, is harmful to the animals housed in the most restrictive systems, and deters long-term economic activity in the nearby communities.

To solve the problems created by industrial farm animal production, 24 consensus recommendations were developed—12 on public health issues, 4 on environmental problems, 5 in the area of animal welfare, 2 on rural communities, and 1 urging independent research on animal production at universities. As is typical with the consensus process, each primary recommendation was developed with as much detail as Commissioners could agree upon. Therefore, some primary recommendations have several components while others are relatively brief.

Of the 24 recommendations outlined in the Commission's final report, 6 were highlighted in the executive summary of the report and indicated a priority in each of the subject areas studied. The top recommendations in each of the four years studied by the Commission—public health, the environment, animal welfare, and rural communities—will be outlined in each section. The full Pew Commission recommendations will follow the primary recommendation from each area of inquiry.

BACKGROUND ON PUBLIC HEALTH

There can be an impact on human health through a traditional, extensive farming system or the prevalent model today, the intensive system as represented in CAFOs. A significant difference between extensive and intensive animal agriculture production is the large number of the same species of animal in closely confined quarters, creating an atmosphere conducive to the rapid development of pathogens and viruses.

In general, public health concerns associated with industrial farm animal production include heightened risks of pathogens (disease and non-disease-causing) passed from animals to humans; the emergence of microbes resistant to antibiotics and antimicrobials, due in large part to widespread use of antimicrobials for nontherapeutic purposes; foodborne disease; worker health concerns; and dispersed impacts on the adjacent community at large (Pew Commission, 2008, p. 11).

It has been estimated that of the 1400 documented human pathogens, approximately 64% are zoonotic; that is, passed from animals to humans (Pew Commission, 2008, p. 13; Woolhouse and Gowtage-Sequeria, 2005; Woolhouse et al., 2001). In addition to infectious disease and the risk of pathogen transfer, the continuing cycling of viruses and other animal pathogens in large herds or flocks increases opportunities for the generation of novel viruses through mutation or recombinant actions that could result in more efficient human-to-human transmission.

PEW COMMISSION AND PUBLIC HEALTH RECOMMENDATIONS

The Union of Concerned Scientists (UCS) estimates that approximately 70% of antimicrobials used in the United States are used nontherapeutically in industrial farm animal production, including many antibiotics such as penicillin and tetracycline that are used to treat human infections (Mellon et al., 2001; Pew Commission, 2008 p. 13). Estimates of the cost of increased difficult-to-treat infections range from $4 billion to $5 billion per year by the Institute of Medicine in 1998, to an estimated $26 billion in 2010 by the Alliance for the Prudent Use of Antibiotics.

The Commission's primary public health recommendation was to phase out and ban the nontherapeutic use of antimicrobials in food animal production (Pew Commission, 2008, p. 61). In addition, it defined the terms therapeutic, nontherapeutic, and prophylactic use. Therapeutic use was defined as use in the case of diagnosed microbial disease. In other words, antimicrobials were to be used to treat sick animals. Nontherapeutic use was defined as any use in the absence of known or diagnosed microbial disease. Prophylactic use was defined as the use of antimicrobials in healthy animals in advance of an expected exposure to an infectious agent, or after an exposure but before the onset of clinical disease (Pew Commission, 2008, p. 63). The Commission definitions would eliminate the routine, daily, low-level use of antimicrobials for growth promotion or to compensate for the poor animal husbandry conditions that are common in industrial farm animal production systems.

COMPLETE PUBLIC HEALTH RECOMMENDATIONS OF THE PEW COMMISSION

1. RESTRICT THE USE OF ANTIMICROBIALS IN FOOD ANIMAL PRODUCTION TO REDUCE THE RISK OF ANTIMICROBIAL RESISTANCE TO MEDICALLY IMPORTANT ANTIBIOTICS.

- Phase out and ban the use of antimicrobials for nontherapeutic (i.e., growth promoting) use in food animals.
- Immediately ban any new approvals of antimicrobials for nontherapeutic uses in food animals and retroactively investigate antimicrobials previously approved.
- Strengthen recommendations in Food and Drug Administration (FDA) Guidance #152 to be enforceable by the FDA, in particular the investigation of previously approved animal drugs.
- Facilitate reduction in industrial farm animal production (IFAP) use of antibiotics and educate producers on how to raise food animals without using nontherapeutic antibiotics, the U.S. Department of Agriculture's (USDA) extension service should be tasked to create and expand programs that teach producers the husbandry methods and best practices necessary to maintain the high level of efficiency and productivity they enjoy today.

Background

In 1986, Sweden banned the use of antibiotics in food animal production except for therapeutic purposes and Denmark followed suit in 1998. A WHO (2002) report on the ban in Denmark found that

> the termination of antimicrobial growth promoters in Denmark has dramatically reduced the food animal reservoir of enterococci resistant to these growth promoters, and therefore reduced a reservoir of genetic determinants (resistance genes) that encode antimicrobial resistance to several clinically important antimicrobial agents in humans.

The report also determined that the overall health of the animals (mainly swine) was not affected and the cost to producers was not significant. Effective January 1, 2006, the European Union also banned the use of growth-promoting antibiotics (Meatnews.com, 2005).

In 1998, the National Academy of Sciences (NAS) Institute of Medicine (IOM) noted that antibiotic-resistant bacteria increase U.S. health care costs by a minimum of $4 billion to $5 billion annually (IOM, 1998). A year later, the NAS estimated that eliminating the use of antimicrobials as feed additives would cost each American consumer less than $5 to $10 per year, significantly less than the additional health care costs attributable to antimicrobial resistance (NAS, 1999). In a 2007 analysis of the literature, another study found that a hospital stay was $6,000 to $10,000 more expensive for a person infected with a resistant bacterium as opposed to an antibiotic-susceptible infection (Cosgrove et al., 2005). The American Medical Association, American Public Health

Association, National Association of County and City Health Officials, and National Campaign for Sustainable Agriculture are among the more than 300 organizations representing health, consumer, agricultural, environmental, humane, and other interests supporting enactment of legislation to phase out nontherapeutic use of medically important antibiotics in farm animals and calling for an immediate ban on antibiotics vital to human health.

The Preservation of Antibiotics for Medical Treatment Act of 2007 (PAMTA) amends the Federal Food, Drug, and Cosmetic Act to withdraw approvals for feed-additive use of seven specific classes of antibiotics—penicillin, tetracycline, macrolide, lincosamide, treptogramin, aminoglycoside, and sulfonamides—each of which contains antibiotics also used in human medicine (PAMTA, 2007a). PAMTA provides for the automatic and immediate restriction of any other antibiotic used only in animals if the drug becomes important in human medicine, unless FDA determines that such use will not contribute to the development of resistance in microbes that have the potential to affect humans. FDA Guidance #152 defines an antibiotic as potentially important in human medicine if FDA issues an Investigational New Drug determination or receives a New Drug Application for the compound.

Most antibiotics currently used in animal production systems for nontherapeutic purposes were approved before the FDA began considering resistance during the drug approval process. The FDA has not established a schedule for reviewing existing approvals, although Guidance #152 notes the importance of doing so. Specifically, Guidance #152 sets forth the responsibility of the FDA Center for Veterinary Medicine (CVM), which is charged with regulating antimicrobials approved for use in animals:

> Prior to approving an antimicrobial new animal drug application, FDA must determine that the drug is safe and effective for its intended use in the animal. The Agency must also determine that the antimicrobial new animal drug intended for use in food-producing animals is safe with regard to human health. (FDA-CVM, 2003)

The Guidance also says that

> FDA believes that human exposure through the ingestion of antimicrobial-resistant bacteria from animal-derived foods represents the most significant pathway for human exposure to bacteria that has emerged or been selected as a consequence of antimicrobial drug use in animals.

However, it goes on to warn that the

> FDA's guidance documents, including this guidance, do not establish legally enforceable responsibilities. Instead, the guidance describes the Agency's current thinking on the topic and should be viewed only as guidance, unless specific regulatory or statutory requirements are cited. The use of the word 'should' in Agency guidance means that something is suggested or recommended, but not required. (FDA-CVM, 2003)

The Commission believes that the "recommendations" in Guidance #152 should be made legally enforceable and applied retroactively to previously approved antimicrobials. Additional funding for the FDA is required to achieve this recommendation (Pew Commission, 2008, pp. 61–63).

2. CLARIFY ANTIMICROBIAL DEFINITIONS TO PROVIDE CLEAR ESTIMATES OF USE AND FACILITATE CLEAR POLICIES ON ANTIMICROBIAL USE.

- The Commission defines as *nontherapeutic* any use of antimicrobials in food animals in the absence of microbial disease or known (documented) microbial disease exposure; thus, any use of the drug as an additive for growth promotion, feed efficiency, weight gain, routine disease prevention in the absence of documented exposure, or other routine purpose is considered nontherapeutic.

- The Commission defines as *therapeutic* the use of antimicrobials in food animals with diagnosed microbial disease.
- The Commission defines as *prophylactic* the use of antimicrobials in healthy animals in advance of an expected exposure to an infectious agent or after such an exposure but before onset of laboratory confirmed clinical disease as determined by a licensed professional.

Background

In 2000, the WHO, United Nations Food and Agriculture Organization (FAO), and World Organization for Animal Health (OIE, *Fr.* Office International des Épizooties) agreed on definitions of antimicrobial use in animal agriculture based on a consensus (WHO, 2000).

Government agencies in the United States, including USDA and FDA, govern aspects of antimicrobial use in food animals but have varying definitions of such use. Consistent definitions should be adopted for the use of all U.S. oversight groups that estimate types of antimicrobial use and for the development of law and policy. Congress recently revived a bill to address the antimicrobial resistance problem. The Preservation of Antibiotics for Medical Treatment Act of 2007 (PAMTA) defines nontherapeutic use as "any use of the drug as a feed or water additive for an animal in the absence of any clinical sign of disease in the animal for growth promotion, feed efficiency, weight gain, routine disease prevention, or other routine purpose" (PAMTA, 2007a). If the bill becomes law, this will be the legal definition of nontherapeutic use for all executive agencies and, therefore, legally enforceable.

3. IMPROVE MONITORING AND REPORTING OF ANTIMICROBIAL USE IN FOOD ANIMAL PRODUCTION TO ASSESS ACCURATELY THE QUANTITY AND METHODS OF ANTIMICROBIAL USE IN ANIMAL AGRICULTURE.

Require pharmaceutical companies that sell antimicrobials for use in food animals to provide a calendar-year annual report of the quantity sold. Companies currently report antibiotic sales data on an annual basis from the date of the drug's approval, which makes data integration difficult. FDA is responsible for oversight of the use of antimicrobials in food animals and needs consistent data on which to report use.

Require reporting of antimicrobial use in food animal production, including antimicrobials added to food and water, and incorporate the reported data in USDA's National Animal Identification System (NAIS). The FDA-CVM regulates feed additives but does not have the budget or personnel to oversee their disposition after purchase. In addition, CVM and USDA are responsible for monitoring the use of prescribed antimicrobials in livestock production but rely on producers and veterinarians to keep records of the antibiotics used and for what purpose. Institute better integration, monitoring, and oversight by government agencies by developing a comprehensive plan to monitor antimicrobial use in food animals, as called for in a 1999 National Research Council (NRC) report (NAS, 1999). An integrated national database of antimicrobial resistance data and research would greatly improve the organization, amount, and types of data collected and would facilitate necessary policy changes by increasing data cohesion and accuracy. Further, priority should be given to linking data on both antimicrobial use and resistance in the National Antimicrobial Resistance Monitoring System (NARMS). This could be accomplished by full implementation of Priority Action 5 of *A Public Health Action Plan to Combat Antimicrobial Resistance*, which calls for the establishment of a monitoring system and the assessment of ways to collect and protect the confidentiality of usage data (CDC/FDA/NIH, 1999).

Since USDA already provides antimicrobial use data in fruit and vegetable production, it seems logical that usage information can be obtained from either agricultural producers or the pharmaceutical industry without undue burden.

Background

There are no reliable data on antimicrobial use in U.S. food animal production. Rather, various groups have reported estimates of use based on inconsistent standards. For example, in 2001, the

UCS estimated that 24.6 million pounds of antimicrobials were used per year for nontherapeutic purposes (Mellon et al., 2001) in animal agriculture (only cattle, swine, and poultry), whereas the Animal Health Institute (AHI) figure for the same year was only 21.8 million pounds for *all* animals and uses (therapeutic and nontherapeutic) (AHI, 2002). These disparities make it difficult to get a true picture of the state and extent of antimicrobial use and its relationship to antimicrobial resistance in industrial farm animal production (Pew Commission, 2008, p. 64).

4. IMPROVE MONITORING AND SURVEILLANCE OF ANTIMICROBIAL RESISTANCE IN THE FOOD SUPPLY, THE ENVIRONMENT, AND ANIMAL AND HUMAN POPULATIONS IN ORDER TO REFINE KNOWLEDGE OF ANTIMICROBIAL RESISTANCE AND ITS IMPACTS ON HUMAN HEALTH.

- Integrate, expand, and increase the funding for current monitoring programs.
- Establish a permanent interdisciplinary oversight group with protection from political pressure, as recommended in the 1999 NRC report *The Use of Drugs in Food Animals: Risks and Benefits*. The group members should represent agencies involved in food animal drug regulation (e.g., FDA, the CDC, USDA), similar to the Interagency Task Force (CDC/FDA/NIH, 1999). In order to gather useful national data on antimicrobial resistance in the United States, the group should review progress on data collection and reporting, and should coordinate both the organisms tested and the regions where testing is concentrated, in order to better integrate the data. Agency members should coordinate with each other and with the NAIS to produce an annual report that includes integrated data on human and animal antimicrobial use and resistance by region. Finally, the group should receive appropriate funding from Congress to ensure transparency in funding as well as scientific independence.
- Revise existing programs and develop a comprehensive plan to incorporate monitoring of the farm environment (soils and plants) and nearby water supplies with the monitoring of organisms in farm animals.
- Improve testing and tracking of antimicrobial-resistant (AMR) infections in health care settings. Better tracking of antimicrobial-resistant infections will give health professionals and policymakers a clearer picture of the role of AMR organisms in animal and human health and will support decisions about the use of antimicrobials that are more effective.

Background

Monitoring and surveillance of antimicrobial resistance in the United States are covered by the NARMS, a program run by the FDA in collaboration with the Centers for Disease Control (CDC) and USDA. CDC is responsible for monitoring resistance in humans, but other federal agencies also conduct AMR research activities. For instance, USDA's National Animal Health Monitoring System (NAHMS) compiles food animal population statistics, animal health indicators, and antimicrobial resistance data. USDA's Collaboration in Animal Health and Food Safety Epidemiology (CAHFSE) is a joint effort of the department's Animal and Plant Health Inspection Service (APHIS), Agricultural Research Service (ARS), and Food Safety and Inspection Service (FSIS) to monitor bacteria that pose a food safety risk, including AMR bacteria. The United States Geological Survey (USGS) studies the spread of AMR organisms in the environment. To achieve a comprehensive plan for monitoring and responding to antimicrobial resistance in the food supply, the environment, and animal and human populations, these agencies should work together to create an integrated plan with independent oversight, and should upgrade from a passive form of monitoring to an active, comprehensive, uniform, mandatory approach.

The U.S. and state geological surveys (Krapac et al., 2004; USGS, 2006) as well as several independent groups (Batt, Snow et al., 2006; Centner, 2006; Peak, Knapp et al., 2007) have looked closely at the spread of AMR organisms in the environment, specifically in waterways, presumably from runoff or flooding. A recent study by the University of Georgia suggested that even chickens raised without exposure to antibiotics were populated with resistant bacteria. The authors suggested that an incomplete cleaning of the farm environment could have allowed resistant bacteria to persist and reinfect naïve hosts (Idris, Lu et al., 2006; Smith, Drum et al., 2007). In Denmark, it took several years after the withdrawal of antimicrobials for antimicrobial resistance to diminish in farm animal populations. These experiences emphasize the importance of monitoring the environment for antimicrobial contamination and responding with careful and comprehensive planning (Pew Commission, 2008, p. 65).

5. Increase veterinary oversight of all antimicrobial use in food animal production to prevent overuse and misuse of antimicrobials.

- Restrict public access to agricultural sources of antimicrobials.
- Enforce restricted access to prescription drugs. By law, only a veterinarian may order the extra label (i.e., nontherapeutic) use of a prescribed drug in animals, but, in fact, prescription drugs are widely available for purchase online, directly from the distributors or pharmaceutical companies, or in feed supply stores without a prescription. Without stricter requirements on the purchase of antimicrobials, extra label use of these drugs is possible and even probable. For that reason, *no* antibiotics should be available for over-the-counter purchase.
- Enforce veterinary oversight and authorization of all decisions to use antimicrobials in food animal production. The extra label drug use (ELDU) rule under the Animal Medicinal Drug Use Clarification Act (AMDUCA) permits veterinarians to go beyond label directions in using animal drugs and to use legally obtained human drugs in animals. However, the rule does not permit ELDU in animal feed or to enhance production. ELDU is limited to cases in which the health of the animal is threatened or in which suffering or death may result from lack of treatment. Veterinarians should consider ELDU in food-producing animals only when no approved drug is available that has the same active ingredient in the required dosage form and concentration or that is clinically effective for the intended use (1994). North Carolina State University, the University of California-Davis, and the University of Florida run the Food Animal Residue Avoidance & Depletion Program (FARAD) (www.farad.org/), which includes useful information for food animal veterinarians, including vetgram, which lists label information for all food animal drugs. To be effective, AMDUCA and ELDU must be enforced. In addition, as technology allows, the FDA-CVM should compel veterinarians to submit prescription and treatment information on farm animals to a national database to allow better tracking of antibiotic use as well as better oversight by veterinarians. Veterinary education for food animal production should teach prescription laws and reporting requirements.
- Encourage veterinary consultation in these decisions. AMDUCA requires the veterinarian to properly label drugs used in a manner inconsistent with the labeling (i.e., extra label) and to give the livestock owner complete instructions about proper use of the drug. Further, ELDU must take place in the context of a valid, current veterinarian-client-patient relationship—the veterinarian must have sufficient knowledge of the animal to make a preliminary diagnosis that will determine the intended use of the drugs. The producer should be encouraged to work with the veterinarian both to ensure the health of the animal and to conform to antibiotic requirements. For example, the National Pork Board Pork Quality Assurance program encourages consultation with veterinarians to maintain a comprehensive herd health program (NPB, 2005).

Background

Presenters at a 2003 NRC workshop concluded that unlike human use of antibiotics, nontherapeutic uses in animals typically do not require a prescription (certain antimicrobials are sold over the counter and widely used for purposes or administered in ways not described on the label) (Anderson et al., 2003). Before AMDUCA, veterinarians were not legally permitted to use an animal drug in any way except as indicated on the label. After the passage of AMDUCA, veterinarians gained the right to prescribe and dispense drugs for "extra label" use, but FDA limits such use to protect public health (1994). ELDU occurs when the drug's actual or intended use is not in accordance with the approved labeling. For instance, ELDU refers to administration of a drug for a species not listed on the label; for an indication, disease, or other condition not on the label; at a dosage level or frequency not on the label; or by a route of administration not on the label. Over-the-counter sale of antimicrobials opens the door to the nontherapeutic, unregulated use of antibiotics in farm animals (Pew Commission, 2008, p. 66).

6. Implement a disease-monitoring program and a fully integrated and robust national database for food animals to allow 48-hour trace-back through phases of their production.

- Implement a tracking system for animals as individuals or units from birth until consumption, including movement, illnesses, breeding, feeding practices, slaughter condition and location, and point of sale. Use the same numbering system as for USDA's NAIS (see previous text), but expand it to provide more information to appropriate users (NAIS tracks animals based only on their movement).
- Require federal oversight of all aspects of this tracking system, with stringent protections for producers against lawsuits. The tracking arm of the NAIS, which has not yet been implemented, is designed to be administered by private industry in collaboration with state governments. NAIS has garnered support from both, but the program should be expanded significantly and monitored by a separate federal agency to enhance confidentiality for producers. The British Cattle Movement Service (www.bcms.gov.uk) could serve as a model for this system.
- Require registration of premises and animals by 2009 and implement animal tracking by 2010. USDA's aphis has created a voluntary animal ID system in collaboration with the farm animal industry, so implementation of a mandatory federal system should be feasible within a relatively short time.
- Allocate special funding to small farms to facilitate their participation in the national tracking system, which would have a much greater financial impact on them, particularly the costs of the identification method (e.g., ear tag, microchip, retinal scan). Such funding should be made available concurrent with the announcement of mandatory registration.

Background

In May 2005, aphis began implementing an animal tracking system, the NAIS (USDA, aphis 2006), which will track premises and 27 species of animals (including cattle, goats, sheep, swine, poultry, deer, and elk). Data are linked to several databases run by private technology companies, while USDA shops for a technology company with data warehousing expertise to run the full national database in the future. The United Kingdom uses a similar database for its Cattle Tracing System (Doe and Fra, 2001).

NAIS registration is voluntary at the time of this writing, and the Bush administration announced on November 22, 2006, that would not require it of producers. The major industry concerns are about trust and confidentiality, says John Clifford, deputy administrator for aphis veterinary services. However, proposals to make registration mandatory have been floated by USDA; the department has officially stated that, "If the marketplace, along with State and Federal identification programs, does not provide adequate incentives for achieving complete participation, USDA may be required to implement regulations" (USDA, 2006).

The goal of the NAIS is a 48-hour trace-back to identify exposures because the 48-hour period is vital to containing the spread of infection (USDA, 2005). USDA advertises the NAIS as a "valuable tool for other 'non-NAIS' purposes—such as animal management, genetic improvement, and marketing opportunities," and notes that producers could improve the quality of their product and thus increase sales using the tracking. Many industry groups support the NAIS for these reasons, but small producers worry about the costs, oversight of data collection, and maintenance (Western Organization of Resource Councils, May 2006).

The first two phases of the NAIS call for the registration of premises and individual animals using a U.S. Animal Identification Number (USAIN). According to the USDA,

> [t]he US Animal Identification Number (USAIN) will evolve into the sole national numbering system for the official identification of individual animals in the United States. The USAIN follows the International Organization for Standardization (ISO) Standard for Radio Frequency [tracking] of Animals and can thus be encoded in an ISO transponder or printed on a visual tag. (USDA, aphis 2006)

The Wisconsin Livestock ID Consortium developed this USAIN, which has 15 digits, the first three of which are the country code (840 for the United States). The final phase will be the animal tracking phase (Pew Commission, 2008, pp. 67–68).

7. FULLY ENFORCE CURRENT FEDERAL AND STATE ENVIRONMENTAL EXPOSURE REGULATIONS AND LEGISLATION, AND INCREASE MONITORING OF THE POSSIBLE PUBLIC HEALTH EFFECTS OF IFAP ON PEOPLE WHO LIVE AND WORK IN OR NEAR THESE OPERATIONS.

- Because IFAP workers—farmers, caretakers, processing plant workers, veterinarians, federal, state, and private emergency response personnel, and animal diagnostic laboratory personnel—are exposed to and may be infected by zoonotic, novel, or other infectious agents, they should be a priority target population for heightened monitoring, annual influenza vaccines, and training in the use of personal protective equipment. IFAP workers who have the highest risk of exposure to a novel virus or other infectious agent should be priority targets for health information and education, pandemic vaccines, and antiviral drugs.
- IFAP employers and responsible health departments need to coordinate the monitoring and tracking of all IFAP facility employees to document disease outbreaks and prevent the spread of a novel zoonotic disease.
- Occupational health and safety programs, including information about risks to health and information about resources, should be more widely available to IFAP workers.
- Occupational safety and health information must also be disseminated in ways that allow people with little or no education or English proficiency to understand their risks and why precautions must be taken. Because of the well-documented health and safety risks among IFAP workers, the Occupational Health and Safety Administration should develop health and safety standards for IFAP facilities as allowable by law.
- Current legislation and regulations concerning surveillance and health and safety programs should be implemented and should prioritize IFAP workers.

Background

In most jurisdictions, few, if any, restrictions on IFAP facilities address the health of IFAP workers or the public. Localities are therefore often unprepared to properly deal with IFAP impacts on local services and the health of people in the community (Pew Commission, 2008, p. 69).

8. INCREASE RESEARCH ON THE PUBLIC HEALTH EFFECTS OF IFAP ON PEOPLE LIVING AND WORKING ON OR NEAR THESE OPERATIONS, AND INCORPORATE THE FINDINGS INTO A NEW SYSTEM FOR SITING AND REGULATING IFAP.

- Support research to characterize IFAP air emissions and exposures from the handling and distribution of manure on fields—including irritant gases (ammonia and hydrogen sulfide, at a minimum), bioaerosols (endotoxin, at a minimum), and respirable particulates—for epidemiological studies of exposed communities near IFAP facilities. Such research should include characterization of mixed exposures, studies of particulates in rural areas, and standardization and harmonization of exposure assessment methods and instrumentation to the degree possible.
- Support research to identify and validate the most applicable dispersion models for IFAP facilities and their manure emissions. Such modeling research must take into account multiple IFAP facilities and their manure management plans in a given area, meteorological conditions, and chemical transformation of pollutants, and should be evaluated with prediction error determined through comparison of predicted values with actual monitoring data. Such models would be useful to state and federal regulatory agencies to determine the results of best management practices, to assess health impacts on exposed populations, and to model setback distances before the construction of new facilities. There is a further need for models that enable evaluation of concentration/exposure scenarios after an event that triggers asthma episodes or nuisance complaints.
- Support research on the respiratory health and function of populations that live near IFAP facilities, including children and sensitive individuals. Such studies are powerful epidemiological approaches to assess the impact of air pollutants on respiratory health and must include appropriate exposure assessments, exposure modeling, and use of time-activity patterns with personal exposure monitoring to better calibrate modeling of exposures. Exposure assessment data need to be linked with measures of respiratory health outcome and function data, including standardized assessment of respiratory symptoms and lung function, assessment of allergic/immunological markers of response, and measurement of markers of inflammation, including the use of noninvasive approaches such as tear fluid, nasal lavage, and exhaled breath condensate.
- Support systematic and sustained studies of ecosystem health near IFAP facilities, including toxicologic, infectious, and chemical assessments, to better assess the fate and transport of toxicologic, infectious, and chemical agents that may adversely affect human health. Systematic monitoring programs should be instituted to assess private well water quality in high-risk areas, supplemented by biomonitoring programs to assess actual exposure doses from water sources.

Background

While there is an increasing amount of research already taking place on IFAP's impacts on the people who work and live on or near these facilities, there is a need to define more fully the extent to which IFAP poses a threat to those populations. There is clear epidemiological evidence that IFAP facilities are associated with increased asthma risk among those living nearby, but there is a need to develop and understand exposure and health outcome relationships. These topics should be addressed by scientific research (Pew Commission, 2008, p. 69).

9. STRENGTHEN THE RELATIONSHIPS BETWEEN PHYSICIANS, VETERINARIANS, AND PUBLIC HEALTH PROFESSIONALS TO DEAL WITH POSSIBLE IFAP RISKS TO PUBLIC HEALTH.

- To understand the cross-species spread of disease, expand and increase funding for dual veterinary/public health degree programs.

- Fund and implement federal and state training programs to increase the number of practicing food animal veterinarians (2007b).
- Initiate and expand federal coordination between Health and Human Services (HHS), FDA, CDC, and USDA to better anticipate, detect, and deal with zoonotic disease. NARMS is not extensive enough to be effective for outbreak detection; it serves a general monitoring function. Include all the data from the various federal agencies in the IFAP clearinghouse (outlined among the environmental recommendations) for use by a newly created Food Safety Administration (Recommendation 10) and the states.
- Promote international coordination on zoonotic diseases and food safety. As an increasing amount of U.S. food is imported, it is vital to hold this food to the same standards as domestically produced food.
- Provide more training through land-grant universities and schools of public health to producers, community health workers, health professionals, and other appropriate personnel to promote detection of disease as a first line of defense against emerging zoonotic diseases and other IFAP-related occupational health and safety outcomes.

Background

These three groups of health professionals (physicians, veterinarians, and public health professionals) have already begun to collaborate, and such collaboration should be promoted and extended as quickly as possible to protect the public's health as well as that of the food animal population. The American Medical Association and American Veterinary Medical Association's One Health Initiative is a very good beginning, and the Commission recommends the following to further extend this collaboration (Pew Commission, 2008, p.71).

10. CREATE A FOOD SAFETY ADMINISTRATION THAT COMBINES THE FOOD INSPECTION AND SAFETY RESPONSIBILITIES OF THE FEDERAL GOVERNMENT, USDA, FDA, EPA, AND OTHER FEDERAL AGENCIES INTO ONE AGENCY TO IMPROVE THE SAFETY OF THE U.S. FOOD SUPPLY.

Background

The current system to ensure the safety of U.S. food is disjointed and dysfunctional; for example, the FDA regulates meatless frozen pizza whereas the USDA has jurisdiction over frozen pizza with meat. This fractured system has failed to ensure food safety, and a solution requires a thorough national debate about how the most effective and efficient food safety agency would be constructed (Pew Commission, 2008, p. 71).

11. DEVELOP A FLEXIBLE, RISK-BASED SYSTEM FOR FOOD SAFETY FROM FARM-TO-FORK TO IMPROVE THE SAFETY OF ANIMAL PROTEIN PRODUCED BY IFAP FACILITIES.

- Any risk-based, farm-to-fork food safety system must allow for size differences among production systems—a "one-size-fits-all" system will not be appropriate for all operations. The system must be flexible enough for small and local producers to get their products to the marketplace.
- Attack food safety issues at their source, instead of trying to fix a problem after it has occurred, by instituting better sanitary and health practices at the farm level. Ranch operating plans may provide one approach to on-farm food safety; the FDA's 2004 proposed rule for the prevention of *Salmonella enteritidis* in shell eggs is another example (http://www. fda.gov/Food/FoodSafety/Product-SpecificInformation/EggSafety/EggSafetyActionPlan/ucm110169.htm).
- Ensure that diagnostic tools are sensitive and specific and are continuously evaluated to detect newly emerging variants of microbial agents of food origin.

- Make resources available through competitive grants to encourage the development of practical but rigorous monitoring systems and rapid diagnostic tools. Provide resources for the application of newly identified or developed technologies and processes and for the training of inspectors and quality control staff of facilities.
- Introduce greater transparency in feed ingredients. Often producers do not even know what additives they are feeding the animals because the feed arrives premixed from the integrator. One option would be to extend certain provisions of the Food, Drug, and Cosmetic Act to the farm.
- Encourage the food animal production industry (contractors, producers, and integrators) to commit to finding ways to minimize the risk of outbreaks of zoonotic disease and other IFAP-related public health threats to vulnerable communities, such as those where IFAP facilities are the most concentrated and where local citizens are least able to protect their rights (e.g., lower-income or minority areas).
- Include both imported and domestically produced foods of animal origin in the enhanced monitoring systems.

Background

Recent foodborne illness outbreaks and meat recalls have called into question the reliability of our system for ensuring the safety of domestic and imported meat. Fiat facilities can have a variety of effects on public health if precautions are not taken to protect the health of their food animals. Livestock production systems must be assessed for vulnerabilities beyond the naturally occurring disease agents. The U.S. production of food has been a model for the world, but a number of countries have now instituted better practices. The food production system is one of our most vulnerable critical infrastructure systems and requires preparation and protection from possible domestic or foreign bioterrorism. Confidence in the safety of our food supply must be maintained and, in some cases, restored (Pew Commission, 2008, p. 72).

12. IMPROVE THE SAFETY OF OUR FOOD SUPPLY AND REDUCE USE OF ANTIMICROBIALS BY MORE AGGRESSIVELY MITIGATING PRODUCTION DISEASES (DISORDERS ASSOCIATED WITH IFAP MANAGEMENT AND BREEDING).

- More attention should be given to antimicrobial resistance and other diseases on the farm. Too often attempts are made to address the effects of production diseases after they arise (at processing), rather than preventing them from occurring in the first place.
- Research into systems that minimize production diseases should be expanded, implemented, and advocated by the state and the federal governments.

Background

Production diseases are diseases that, although present in nature, become more prevalent because of certain production practices. As production systems increase the number of animals in the same space, preventive health care strategies must be developed in parallel in order to minimize the risks of production-related diseases (Pew Commission, 2008, p. 73).

BACKGROUND ON ENVIRONMENTAL ISSUES

All types of animal agriculture operations present potential environmental problems. However, the structure of intensive animal production operations presents a larger problem. In fact, the storage and disposal of waste presents the single most important and difficult environmental problem facing concentrated animal feeding operations.

The USDA estimates that all livestock and poultry operations in the United States produce an estimated 500 billion tons of manure annually compared to the EPA estimate of 150 million tons of human waste produced annually (EPA, 2007b). The difference is that animal feces and urine are not treated to remove pathogens and contaminants before being applied to cropland, whereas human waste is treated by a wastewater treatment system before being released into the environment.

CAFOs are regulated by the Clean Water Act, where operations of a certain size and those that may possibly discharge require a permit. However, species promotion groups and a general farm organization have challenged the EPA's authority to regulate CAFOs under that act. At the time of this writing, that suit was still pending.

Because of the high volume of concentrated waste, disposal on cropland surrounding these operations can cause problems with phosphorus buildup in the soil and nitrogen runoff in surface waters and into groundwater. Agricultural chemicals and pharmaceuticals used in these operations have also been found in the soil, groundwater, and surface water.

While there is a difference in greenhouse gas emissions between intensive and extensive operations, the EPA estimates that 7.4% of greenhouse gas emissions are from agriculture (EPA, 2007a).

CONCLUSION OF PEW COMMISSION ON THE ENVIRONMENT AND CONCENTRATED ANIMAL FEEDING OPERATIONS

While the number of farms raising animals has declined, the number of animals raised as food remained somewhat constant over the last 50 years, requiring larger and larger numbers of animals to be raised in each operation. Large operations have been able to accomplish this with some gains in production efficiency.

While these large operations have gained some efficiency, the downside of CAFO practices is that they have produced an expanding array of harmful environmental effects, including negative impacts on soil, water, and the air. Those effects impose a cost on society that is not reflected in the retail price of meat; therefore, those costs are called externalized costs.

CAFOs present a major waste management problem with the volume being so large that disposal can be impractical and environmentally damaging.

The Pew Commission believed that to protect against further environmental degradation, management practices must be improved, protective zoning changes must be made, and improved monitoring and enforcement of CAFO regulations are needed (Pew Commission, 2008, p. 29).

COMPLETE ENVIRONMENTAL RECOMMENDATIONS OF THE PEW COMMISSION

1. IMPROVE ENFORCEMENT OF EXISTING FEDERAL, STATE, AND LOCAL IFAP FACILITY REGULATIONS TO IMPROVE THE SITING OF IFAP FACILITIES AND PROTECT THE HEALTH OF THOSE WHO LIVE NEAR AND DOWNSTREAM FROM THEM.

- Enforce all provisions of the Clean Water Act 14 and the Clean Air Act 15 that pertain to IFAP.
- Provide adequate mandatory federal funding to states to enable them to hire more trained inspectors, collect data, monitor farms more closely, educate producers on proper manure handling techniques, write Comprehensive Nutrient Management Plans (CNMPS), and enforce IFAP regulations (e.g., NRCS, EPA Section 106 grants, SBA loans).
- States should enforce federal and state permits quickly, equitably, and robustly. A lack of funding and political will often inhibits the ability of states to adequately enforce existing federal and state IFAP (currently CAFO) regulations. Often, states must rely on general fund appropriations to fund IFAP (CAFO) monitoring and rule enforcement. Dedicated mandatory funding would improve this situation, and additional funding for monitoring and enforcement could be realized if permitting fee funds were dedicated to monitoring and enforcement.
- States should implement robust inspection regimes that are designed to deter IFAP facility operators from ignoring pollution rules. Often, no state-sanctioned official visits an

IFAP facility unless there is a complaint, and then it may be too late to document or fix the problem. Each state should set a minimum inspection schedule (at least once a year), with special attention to repeat violators (Kelly, 2007).

- State environmental protection agencies, rather than state agricultural agencies, should be charged with regulating IFAP waste. This would prevent the conflict of interest that arises when a state agency charged with promoting agriculture is also regulating it (Washington State Department of Ecology, 2006). While environmental protection agencies may not have expertise with food animals, they are generally better equipped than state agriculture agencies to deal with waste disposal because they regulate many other types of waste disposal. Unfortunately, several states are transferring the regulation of IFAP facilities from the department of environment to their department of agriculture.
- The EPA should develop a standardized approach for regulating air pollution from IFAP facilities. IFAP air emissions—including pollutants such as particulate matter, hydrogen sulfide, ammonia, methane, and volatile gases—are unregulated at the federal level.
- Clarify the definition of the types of waste handling systems and number of animals that constitute a regulated IFAP facility (CAFO) in order to bring a greater proportion of the waste from IFAP facilities under regulation. Under currently proposed EPA rules, only 49 to 60% of IFAP waste qualifies for federal regulation (EPA, 2003).
- The federal government should develop criteria for allowable levels of animal density and appropriate waste management methods that are compatible with protecting watershed, air shed, soil, and aquifers by adjusting for relevant hydrologic and geologic factors. States should use these criteria to permit and site IFAP operations.
- Once criteria are established and implemented, EPA should monitor IFAP's effects on entire watersheds, not just on a per farm basis, since IFAP can have a cumulative effect on the health of a watershed.
- Grant permits only to new IFAP facilities that comply with local, state, and federal regulations.
- Require existing IFAP facilities to comply and shut down those that cannot or do not.
- The federal and state governments should increase the number of IFAP operations (currently restricted to EPA-defined CAFOs) to be regulated under federal and state law (NMPS, effluent restrictions, national pollutant discharge elimination system [NPES] permits) and provide robust financial and technical support to smaller producers included in the expanded IFAP (CAFO) definition to help them comply with these regulations. Under the current definition of a CAFO, only 5% of animal feeding operations (AFOs) are CAFOs, yet they raise 40% of U.S. livestock. Only approximately 30% (4000) of the 5% have federal permits (Copeland, 2006). If the current final rule (1000 animal units, or au) were lowered to the original rule proposed in 2000, which would regulate CAFOs between 300 and 999 au or a 500-animal threshold (EPA, 2003), 64 to 72% more waste would be covered under the federal permitting process.
- Require operations that do not obtain a permit to prove they are not discharging waste into the environment. Test wells for groundwater monitoring, and require surface water monitoring for those who wish to opt out of obtaining a permit. This would expand the number of AFOs subject to regulation. Currently, many operations that meet IFAP facility (CAFO) size thresholds do not obtain permits or fall outside state and federal regulation because they claim they do not discharge. Claiming no discharge exempts IFAP facilities from federal regulation, although they are often still subject to state laws, which vary greatly from state to state (as noted in the National Conference of State Legislatures study [NCSL, 2008]).

Background

Too few IFAP operations are monitored, regulated, or even inspected on a regular basis. It is imperative that all levels of government thoroughly enforce existing IFAP laws for all IFAP facilities.

Funding should be increased to enable federal and state authorities to enforce IFAP regulations in order to reduce the number of large operations negatively affecting the soil, air, and water (Pew Commission, 2008, p. 77).

2. DEVELOP AND IMPLEMENT A NEW SYSTEM TO DEAL WITH FARM WASTE (THAT WILL REPLACE THE INFLEXIBLE AND BROKEN SYSTEM THAT EXISTS TODAY) TO PROTECT AMERICANS FROM THE ADVERSE ENVIRONMENTAL AND HUMAN HEALTH HAZARDS OF IMPROPERLY HANDLED IFAP WASTE.

- Congress and the federal government should work together to formulate laws and regulations outlining baseline waste handling standards for IFAP facilities. These standards would address the minimum level of mandatory IFAP facility regulation as well as which regulations states must enforce to prevent IFAP facilities from polluting the land, air, and water. States could choose to implement regulations that are more stringent if they considered them necessary. Our diminishing land capacity for producing food animals, combined with dwindling freshwater supplies, escalating energy costs, nutrient overloading of soil, and increased antibiotic resistance, will result in a crisis unless new laws and regulations go into effect in a timely fashion. This process must begin immediately and be fully implemented within 10 years.
- Address site-specific permits for the operation of all IFAP facilities and include the monitoring of air, water, and soil total maximum daily loads (TMDLs),16 site specific NMPS,17 comprehensive nutrient management plans (CMPs),18 inspections, data collection, and self reporting to the clearinghouse (see fifth item under Recommendation 3 in this section).
- Require the use of environmentally sound treatment technologies for waste management (without specifying a particular technology that might not be appropriate for all conditions).
- Mandate shared responsibility and liability for the disposal of IFAP waste between integrators and producers proportional to their control over the operation (instead of this burden being solely the responsibility of the producer [Arteaga, 2001]).
- Include baseline federal zoning guidelines that set out a framework for states. Require a pre-permit/construction environmental impact study. Such a requirement would not prevent states and counties from enacting their own, more comprehensive, zoning laws if necessary (see Recommendation 1 under Competition and Community Impacts).
- Establish mechanisms for community involvement to provide neighbors of IFAP facilities opportunities to review and comment on proposed facilities, and allow them to take action in cases where federal or state regulations have been violated in the absence of enforcement of those laws by the appropriate authority. Individuals who have had their private property contaminated through no fault of their own must have access to the courts to obtain redress.
- Ensure that all types of IFAP waste (e.g., dry litter, wet waste) are covered by regulations (EPA, 2003).
- Establish standards that protect people, animals, and the environment from the effects of IFAP waste on and off the operation's property (Arteaga, 2001; EPA, 2003; Schiffman, Studwell et al., 2005; Sigurdarson and Kline, 2006; Stolz, Perera et al., 2007).
- Phase out the use of lagoon and spray systems in areas that cannot sustain their use (e.g., fragile watersheds, floodplains, certain geologic formations, areas prone to disruptive weather patterns).
- Require new and expanding IFAP facilities in vulnerable areas to use primary, secondary, and tertiary treatment of animal waste (similar to the treatment associated with human waste) until lagoon and spray systems can be replaced by safe and effective alternative technologies.
- Require minimal water use in alternative systems to protect the nation's dwindling freshwater resources, balanced with the system's effect on air and soil quality. Liquid manure

handling systems should be used only if another system is not feasible or would have greater environmental impact than a liquid system. The sustainability of alternative systems in relation to water resources and carbon use should be a major focus during their development.

- Prohibit the installation of new liquid manure handling systems and phase out their use on existing operations as technology allows.
- Require states to implement a robust inspection regime that combines adequate funding for annual inspections with additional risk-based inspections where necessary. It is important that all IFAP facilities be inspected on a regular basis to ensure compliance with state and federal waste management regulations. Additionally, some IFAP facilities may need special attention because of the type of manure handling system in use, the facility's age, its size, or its location. These high-risk operations should be inspected more often than lower-risk operations.

Background

Most animal production facilities in the United States and increasingly in the world have become highly specialized manufacturing endeavors and should be viewed as such. The regulatory system for oversight of IFAP facilities is flawed and inadequate to deal with the level and concentration of waste produced by current food animal production systems, which were not well understood or even foreseen when the laws were written. A new system of laws and regulations that applies specifically to modern IFAP methods is needed.

IFAP facilities have become more concentrated in certain geographic areas. New regulations must address the zoning and siting of IFAP facilities, particularly with regard to the topography, demographics, and climate of the suggested region. They must also take into account an individual's right to property free from pollution caused by neighboring IFAP facilities. IFAP facility owners and integrators do not have a right to pollute their neighbors' land. Property owners or tenants must have the right to take legal action or petition the government to do so on their behalf if their property is polluted by a neighboring IFAP facility.

Waste from IFAP facilities contains both desirable and undesirable by-products. Desirable by-products include nutrients that, when applied in appropriate amounts, can enhance production of food crops and biomass to produce energy. Undesirable components include excess pathogenic bacteria, antibiotic-resistant bacteria, viruses, industrial chemicals, heavy metals, and other potentially problematic organic and inorganic compounds. New IFAP laws and regulations must mandate development of sustainable waste handling and treatment systems that can use the beneficial components and render the less desirable components benign. These new laws should not mandate specific systems for producers; rather they should set discharge standards that can be met using a variety of systems that accommodate the local climate and geography.

Congress should work with the EPA, USDA, and FDA to establish a clear and consistent definition of which IFAP facilities should be regulated and to develop regulations (Pew Commission, 2008, pp. 77–79).

3. INCREASE AND IMPROVE MONITORING AND RESEARCH OF FARM WASTE TO HASTEN THE DEVELOPMENT OF NEW AND INNOVATIVE SYSTEMS TO DEAL WITH IFAP WASTE AND TO BETTER OUR UNDERSTANDING OF WHAT IS HAPPENING WITH IFAP TODAY.

- All IFAP facilities should have, at a minimum, a nutrient management plan (NMP) for the disposal of manure. An NMP describes appropriate methods for the handling and disposal of manure and for its application to fields. The plan should also include records of the method and timing of manure disposal.
  - State and federal governments should provide funds through state regulatory agencies and the National Resources Conservation Service (NRCS) to help producers write and implement NMPs.

- The EPA should set federal minimum standards for the extent of NMPs and specify what monitoring data should be kept.
- Allow the Environmental Quality Incentives Program (EQIP) to (1) fund the writing of NMPs to expedite their implementation and (2) provide business plans for alternative systems to equalize access to government funds for non-IFAP and IFAP (CAFO)-style production.
- The federal, state, and local governments should begin collecting data on air emissions, ground and surface water emissions, soil emissions, and health outcomes (e.g., cardiovascular disease, heart disease, injuries, and allergies) for people who live near IFAP facilities and for IFAP workers. These data should be tabulated and combined with existing data in a national IFAP data clearinghouse that will enable the EPA and other agencies to keep track of air, water, and land emissions from IFAP facilities and evaluate the public health implications of these emissions. The EPA and other state and federal agencies should use these comprehensive data both to support independent research and to better regulate IFAP facilities. Currently, FDA, EPA, and other federal agencies each keep extensive records for different industries as a way to track changes and regulate each industry. The clearinghouse would consolidate data from around the country, thereby giving producers the chance to improve their operation by providing access to information about better technologies and improved waste systems. It would also allow researchers, regulators, and policymakers to evaluate changing environmental and public health impacts of agriculture and adjust regulations accordingly. The EPA, FDA, and USDA should take the following actions:
  - Add data collected on farm waste handling systems to the clearinghouse for use in assessing and evaluating the sustainability of animal production models and farm waste handling systems by region.
  - Link data to their collection location to facilitate regional comparisons, given different environmental and geological conditions.
  - Implement data protection procedures to ensure that only authorized agencies and personnel can access personal information (e.g., information that could be used by identity thieves) for official purposes.
  - Include comprehensive USDA Agriculture Census data in the national clearinghouse to provide a context for the data and thus improve their utility.
  - Include data on individual violations of state and federal IFAP facility (CAFO) regulations in the public portion of the national clearinghouse. Currently, it is difficult to determine compliance with IFAP (CAFO) laws because states may or may not keep good records of violations and may make them extremely difficult for the public to access (NASDA, 2001).
- Expand our understanding of how to deal with concentrated IFAP waste, and the health and environmental effects of this waste through more diversely funded and well-coordinated research, as well as to move the United States toward more sustainable systems for dealing with farm waste. National standards for alternative waste systems are needed to guide development of improvements to existing waste handling systems as well as the development of alternative/new waste handling systems.
  - Require states to report basic data (general location, number of animals, NMP, etc.) on all IFAP facilities in the public portion of the national clearinghouse.
  - Federal and state governments should fund research into alternative systems to replace existing, insufficient waste handling systems, similar to the recent research done at North Carolina State University. They should also increase funding for research on the effects of IFAP waste on public health, the environment, and animal welfare.
  - Establish a national clearinghouse for data on alternative systems. The clearinghouse would be the repository of regionally and topographically significant data on

economic performance, environmental performance (air, water, and soil), and overall sustainability for potentially useful alternative waste handling systems.

- Improve and standardize research methods for data collection and analysis for the clearinghouse. Standardized methods would allow states and the federal government to compare regionally relevant data in the clearinghouse and facilitate evaluation of new waste handling systems.
- Increase funding for research to effectively assess and improve the economic performance, energy balance, risk assessment, and environmental sustainability of alternative waste handling systems.
- Increase funding for research focused on comprehensive systems to deal with waste, rather than those focused on one process to deal with one aspect of waste (such as using a digester to reduce volume, which does little to reduce the levels of certain toxic components). Dealing with only one component of waste may have the unintended consequence of causing greater harm to the environment.
- Expand the type and number of entities researching farm waste handling by expanding the public funding of research at both land grant and non-land-grant institutions, and other research entities. In addition, transparency of funding source in agricultural research should be standard.

Background

A robust monitoring system should be instituted to improve knowledge about IFAP facilities' current waste management practices as the basis for development of cleaner and safer methods of food animal production (Pew Commission, 2008, pp. 79–81).

4. Increase funding for research into improving waste handling systems and standardize measurements to allow better comparisons between systems.

- Develop a central repository for information on how to best facilitate rapid adoption of new air and water pollution reduction technologies that currently exist or are under development across the country. Research to develop effective means of assistance to pay for them (EQIP should be part of this) should be a component of this repository. (Examples of technologies include biofilters, buffer strips, dehydration, injection, digesters, and reduced feed wastage.)
- Increase funding for the creation and expansion of programs for implementing improved husbandry and technology practices on currently existing facilities including funding conversions to alternative farming practices. (Examples of such programs include, but are not limited to EQIP, cooperative extension, NRCs, cost share, loans, grants, and accelerated capital depreciation.) Sign-up and application information for these types of programs should be included in the clearinghouse so that producers only have to go to one place to get information and sign up for a program. A dollar amount cap should be placed on the cost-share program to prevent large-scale operators from using the program to externalize their costs. These funds should not be used for the physical construction of new facilities.
- Target increased assistance and information to small producers who are least able to afford implementation of new practices and deal with increased regulation, but still have the potential to pollute. Air emission technologies, such as biofilters, that are used in other parts of the world should be considered for use in IFAPs in the United States.

Background

Data from research into alternative systems should be linked to the IFAP information clearinghouse to facilitate and expedite access and use. Greater financial and technical assistance must be provided to those who wish to implement alternative systems (Pew Commission, 2008, p. 81).

BACKGROUND ON ANIMAL WELFARE

In 1964, Ruth Harrison published the landmark book, *Animal Machines,* in the United Kingdom detailing the conditions in many large-scale industrial farms. A year later, the Brambell Commission Report outlined criteria for the scientific investigation of farm animal welfare. This blue ribbon panel was composed of veterinarians, animal scientists, and biologists, and defined welfare in tems of both physical and mental well being. It was a radical statement for the time and still contributes to the debate on farm animal welfare today. In 1997, the Farm Animal Welfare Council in the United Kingdom adopted the "five freedoms" outlined in the Brambell Commission Report.

Today, it is universally accepted that we consider that good animal welfare implies both fitness and a sense of well-being. Any animal kept by humans must at least be protected from unnecessary suffering. An animal's welfare, whether on farm, in transit, at market, or at a place of slaughter, should be considered in terms of the five freedoms. These freedoms define ideal states rather than standards for acceptable welfare. They form a logical and comprehensive framework for analysis of welfare within any system together with the steps and compromises necessary to safeguard and improve welfare within the proper constraints of an effective livestock industry.

1. Freedom from hunger and thirst—by ready access to fresh water and a diet to maintain full health and vigor.
2. Freedom from discomfort—by providing an appropriate environment including shelter and a comfortable resting area.
3. Freedom from pain, injury or disease—by prevention or rapid diagnosis and treatment.
4. Freedom to express normal behavior—by providing sufficient space, proper facilities, and company of the animals' own kind.
5. Freedom from fear and distress—by ensuring conditions and treatment that avoid mental suffering.[2]

Sound animal husbandry systems that are designed to accommodate the five freedoms do so at minimal cost to the consumer. In recent years, the European Union has further refined the five freedoms to clarify and add detail to the original criteria.

European Union Criteria for Animal Well-Being

Welfare Criteria	Welfare Principles	Meaning
Good feeding	Absence of prolonged hunger	Animals should not suffer from prolonged hunger
	Absence of prolonged thirst	Animals should not suffer from prolonged thirst
Good housing	Comfort around resting	Animals should be comfortable, especially within their lying areas
	Thermal comfort	Animals should be in good thermal environment
	Ease of movement	Animals should be able to move around freely
Good health	Absence of injuries	Animals should not be physically injured
	Absence of disease	Animals should be free of disease
	Absence of pain induced by management procedures	Animals should not suffer from pain induced by inappropriate management
Appropriate behavior	Expression of social behaviors	Animals should be allowed to express natural, non-harmful, social behaviors.
	Expression of other behaviors	Animals should have the possibility of expressing other intuitively desirable natural behaviors, such as exploration and play

European Union Criteria for Animal Well-Being

Welfare Criteria	Welfare Principles	Meaning
	Good human–animal relationship	Good human–animal relationships are beneficial to the welfare of animals
	Absence of general fear	Animals should not experience negative emotions such as fear, distress, frustration, or apathy

Source: European Union Animal Welfare Quality Program: http://www.welfarequality.net/everyone/36059/5/0/22).

As the members of the Pew Commission considered animal welfare issues during its inquiry, it applied the principles of the five freedoms and the enhanced criteria that built on the five freedoms to determine the quality of animal husbandry in industrial operations. In addition, Pew Commissioners reviewed scientific analyses of the industrial production system and how it affects animal welfare. The Commission secured a technical report, titled *Animal Well Being*, by four leading animal welfare researchers[3] specifically for Commission deliberations. However, perhaps the most significant information that informed the Commission and influenced its final recommendations on animal welfare issues was the first-hand, visual evidence gained by visiting industrial animal production facilities.

It is helpful to look at the conditions in the industrial production system of two species that led the Commission to recommend the phase out and ban of gestation crates, restrictive farrowing crates, battery cages, and restrictive veal crates.

SWINE

The Commission had two opportunities to view industrial swine production—one in Iowa and one in North Carolina. In Iowa, Commissioners visited the Iowa State University (ISU) swine teaching farm. At the ISU farm, the Commission was able to see the industrial model utilizing gestation crates, the hoop barn system, and an open lot system. All three systems were on a small scale as part of the overall teaching farm. In North Carolina, the Commission visited a large CAFO that utilized both the gestation crate system and the group pen system.

Gestation crates and liquid waste management systems are the two components most common in industrial swine production operations. According to the USDA, a majority of breeding sows in the United States are housed in gestation crates. In fact, 81% of breeding sows are housed in these restrictive crates (USDA, 2007).

Gestation crates typically have metal railings to enclose the sow and concrete, slatted floors to allow the animal's waste to drop through the floor into a collecting trough. The troughs are then flushed periodically to wash the feces and urine into an open settling pond, sometimes referred to as a lagoon. The crates are only slightly larger than the animals are, measuring 0.6 to 0.7 m (2.0 to 2.3 ft) by 2.0 to 2.1 m (6.6 to 6.9 ft)[4] and, as a result, the sow cannot turn around and can lie down only with difficulty. The sows are kept in these crates from insemination until approximately one week before birth of a litter when they are moved to somewhat less restrictive farrowing crates.

In comparison, pen systems allow sows to be housed in groups of 10 to 12 with room to move and socialize. The pens vary in sizes depending on the number of sows housed in them. In addition, floor space allowance, group management, feeding systems, and bedding are among the variable factors in pen systems.

Hoop barns, similar to the example viewed by the Commissioners at the ISU teaching farm, are being used more frequently to house pen systems. A hoop barn consists of 4-ft-high sidewalls fitted with steel tubular arches covered with an opaque UV-resistant polypropylene tarp. Most of the floor area inside the hoop is bedded with cornstalks or other crop residues. The remaining floor is a concrete slab where feeders and waterers are located. Pigs typically are housed in groups ranging from 75 to 250 head, with each building holding one group of pigs. Occasionally, the building is divided lengthwise to accommodate two groups. Sows also can be housed in hoop barns.[5]

Visiting and reviewing the North Carolina facility was instructive in viewing the gestation crates and pen system in a large-scale production. The facility gestation crate system and pen system were industry standard as outlined previously.

The contrast between the two systems within one operation was dramatic. The sows in the three gestation crate buildings were vocalizing constantly and moving forward and backward in their limited space creating a chaotic atmosphere in the buildings. Many were chewing on the bars in the front of their stalls and made aggressive moves toward Commissioners as they walked in the narrow aisle between the rows of stalls. The animals' contact with people was limited due to the automated watering and feeding systems and contact with other swine was almost non-existent due to the restrictions of the crates.

By contrast, the sows in the three buildings configured to house sows in a pen system were more docile and not as vocal, with the exception of brief, periodic episodes to establish the social order. Sows were more inquisitive about the Commissioners and exhibited none of the aggressive, agitated behavior seen in the crate buildings.

LAYERS

Commissioners visited a concentrated egg laying facility in eastern Colorado, the largest in the state, to view both caged and cage-free egg production. The company produces caged eggs, cage-free eggs, and organic eggs that are marketed under several brands in the western United States. Of the 12 large barns housing laying hens, 4 housed the cage-free system.

The outside configuration of the cage-free and caged buildings was identical, constructed of cinderblock, measuring approximately 300 ft long and 100 ft wide. Owners began converting the facility to include cage-free production in 2002 based on increased consumer demand for cage-free eggs and the higher market value for cage-free eggs. The cage-free production model used by this facility is based on a model used extensively in Germany.

The interiors of the two types of production differed dramatically. In the caged buildings, wire cages housing the laying hens were stacked approximately 40 ft high with cages resting on top of one another. Five to six birds were housed in each cage roughly the size of a standard office file drawer. Industry standards allow for approximately 67 in.^{2} for each bird, roughly the size of an 8 × 10 in. piece of paper. The floors of the cages were wire, raising welfare concerns for hens that lived their entire lives in the small, confined cages.

Feed and water were supplied to one side of the wire cages and the hens would stick their heads out through the wire to eat and drink. Litter and eggs dropped out the back of the cage and were taken to the end of the line of cages via a conveyor belt. Feed and water were delivered by an automated system. Birds could only move around with difficulty and could not spread their wings. There was no ability to dust bathe or to exhibit other normal behaviors.

The cage-free buildings were enclosed with no access to the outside. However, there were no battery cages, but instead two levels. The lower level was a dirt floor where birds could move freely about, dust bathe, and socialize. The upper level was accessible by ramps, and contained perching areas and enhanced housing for the hens to lay their eggs. Feed and water were available on both levels. Freedom of movement and expression of natural behaviors was much greater in the cage-free system.

PEW COMMISSION CONCLUSION ON RESTRICTIVE CONFINEMENT SYSTEMS

After reviewing the literature, visiting production facilities, and listening to producers themselves, the Commission determined that the most intensive confinement systems, such as restrictive veal crates, hog gestation pens, restrictive farrowing crates, and battery cages for poultry, all prevent the animal from a normal range of movement, and constitute inhumane treatment.

COMPLETE ANIMAL WELFARE RECOMMENDATIONS
OF THE PEW COMMISSION

The welfare of the animals we consume for food should be an integral component in the decisions we make about how they are raised. The industrial model of animal production has reduced talking about sentient beings to production units as opposed to living creatures. While it delivers a uniform meat and egg product, that comes at a cost to the animals, and eventually to those people raising the animals in an industrial system.

The full Pew Commission recommendations dealing with animal welfare concerns follow.

1. THE ANIMAL AGRICULTURE INDUSTRY SHOULD IMPLEMENT FEDERAL PERFORMANCE-
BASED STANDARDS TO IMPROVE ANIMAL HEALTH AND WELL-BEING.

The federal government should develop performance-based (not resource-based) animal welfare standards. Animal welfare has improved in recent years based on industry research and consumer demand; the latter has led, for example, to the creation of the United Egg Producers' certification program and the McDonald's animal welfare council. However, in order to fulfill our ethical responsibility to treat farm animals humanely, federally monitored standards that ensure at least the following minimum standards for animal treatment should be enacted:

- Good feeding: Animals should not suffer prolonged hunger or thirst.
- Good housing: Animals should be comfortable especially in their lying areas, should not suffer thermal extremes, and should have enough space to move around freely.
- Good health: Animals should not be physically injured and should be free of preventable disease related to production; in the event that surgical procedures are performed on animals for the purposes of health or management, modalities should be used to minimize pain.
- Appropriate behavior: Animals should be allowed to perform normal non-harmful social behaviors and to express species-specific natural behaviors as much as reasonably possible; animals should be handled well in all situations (handlers should promote good human–animal relationships); negative emotions such as fear, distress, extreme frustration, or boredom should be avoided.

Implement a government oversight system similar in structure to that used for laboratory animal welfare.

Each CAFO facility would be certified by an industry-funded, government-chartered, not-for-profit entity accredited by the federal government to monitor the CAFO. Federal entities would audit CAFO facilities for compliance. Consumers could look for the third-party certification as proof that the production process meets federal farm animal welfare standards.

Change the system for monitoring and regulating animal welfare, recommend improvements in animal welfare as a science, and encourage consumers to continue to push animal welfare policy. Improved animal husbandry practices and an ethically based view of animal welfare will solve or ameliorate many CAFO animal welfare problems.

Federal standards for farm animal welfare should be developed immediately based on a fair, ethical, and evidence-based understanding of normal animal behavior (Pew Commission, 2008, p. 83).

2. IMPLEMENT BETTER ANIMAL HUSBANDRY PRACTICES TO IMPROVE
PUBLIC HEALTH AND ANIMAL WELL-BEING.

Change breeding practices to include attributes and genetics besides productivity, growth, and carcass condition (Appleby and Lawrence, 1987); for example, hogs might be bred for docile behavior, fowl for bone strength and organ capacity, and sows, dairy and beef cattle for "good" mothering. In recent decades, farm animals have been selectively bred for specific physical traits (e.g., fast growth, increased lean muscle mass, increased milk production) that have led to greater incidence of and susceptibility to transmissible diseases, new genetic diseases, a larger number and scope of mental or behavioral abnormalities, and lameness.

Improve and expand the teaching of animal husbandry practices at land-grant universities.

Federal and state governments should fund (through tax incentives and directed education funding, including for technical colleges) the training of farm workers and food industry personnel in sustainable, ethical animal husbandry.

Diversify the type of farm animal production systems taught at land-grant schools beyond the status quo CAFO system. Increase funding for the teaching of good husbandry and alternative production techniques through local extension offices. Work to reduce and eliminate "production diseases," defined as diseases caused by production management or nutritional practices; liver abscesses in feedlot cattle are an example of a production disease (Pew Commission, 2008, p. 85).

3. PHASE OUT THE MOST INTENSIVE AND INHUMANE PRODUCTION PRACTICES WITHIN A DECADE TO REDUCE IFAP RISKS TO PUBLIC HEALTH AND IMPROVE ANIMAL WELL-BEING.

- Gestation crates where sows are kept for their entire 124-day gestation period. The crates do not allow the animals to turn around or express natural behaviors, and they restrict the sow's ability to lie down comfortably. Alternatives such as open feeding stalls and pens can be used to manage sows.
- Restrictive farrowing crates, in which sows are not able to turn around or exhibit natural behavior. As an alternative, farrowing systems (e.g., the Freedom Farrowing System, Natural Farrowing Systems) provide protection to the piglets while allowing more freedom of movement for the sow.
- Any cages that house multiple egg-laying chickens (commonly referred to as "battery cages") without allowing the hens to exhibit normal behavior (e.g., pecking, scratching, roosting).
- The tethering or individual housing of calves for the production of white veal. This practice is already rare in the United States, so its phase-out can be done quickly.
- Forced feeding of fowl to produce foie gras.
- Tail docking of dairy cattle.
- Forced molting by feed removal for laying hens to extend the laying period (for the most part, this has been phased out by UEP standards implemented in 2002) (Pew Commission, 2008, p. 85).

4. IMPROVE ANIMAL WELFARE PRACTICES AND CONDITIONS THAT POSE A THREAT TO PUBLIC HEALTH AND ANIMAL WELL-BEING.

- Flooring and housing conditions in feedlots and dairies: Cattle kept on concrete, left in excessive amounts of feces, and not provided shade or misting in hot climates.
- Flooring and other housing conditions at swine facilities: Hogs that spend their entire lifetime on concrete are prone to higher rates of leg injury (Andersen and Boe, 1999; Brennan and Aherne, 1987).
- The method of disposal of unwanted male chicks and of adult fowl in catastrophic situations that require the destruction of large numbers of birds.
- Hand-catching methods for fowl that result in the animals' broken limbs, bruising, and stress.
- Body-altering procedures that cause pain to the animals, either during or afterward.
- Air quality in IFAP buildings: Gas buildup can cause respiratory harm to animal health and to IFAP workers through exposure to gas buildup, toxic dust, and other irritants.

- Ammonia burns on the feet and hocks of fowl due to contact with litter.
- Some weaning practices for piglets, beef cattle, and veal calves: The shortening of the weaning period or abrupt weaning to move the animals to market faster can stress the animals and make them more vulnerable to disease.

The federal government should act on the following recommendations to improve animal welfare:

- Strengthen and enforce laws dealing with the transport of livestock by truck. Transport laws should also address the over-packing of livestock during transportation, long-distance transport of farm animals without adequate care, and transport of very young animals.
- The federal government must include fowl under the Humane Methods of Slaughter Act (Pew Commission, 2008, p. 86).

5. Improve animal welfare research in support of cost-effective and reliable ways to raise food animals while providing humane animal care.

There is a significant amount of animal welfare research being done, but the funding often comes from special interest groups. Some of this research is published and distributed to the agriculture industry, but without acknowledgment of the funding sources. Such lack of disclosure taints mainstream animal welfare research. To improve the transparency of animal research, there needs to be disclosure of funding sources for peer-reviewed published research. Much of today's agriculture and livestock research, for example, comes from land-grant colleges with animal science and agriculture departments that are heavily endowed by special interests or industry. However, a lot of very good research on humane methods of stunning and slaughter has been funded by the industry.

More diversity in the funding sources for animal welfare research is also needed. Most animal welfare research takes place at land-grant institutions, but other institutions should not be barred from engaging in animal welfare research due to lack of research funds. The federal government is in the best position to provide unbiased animal welfare research; therefore, federal funding for animal welfare research should be revived and increased.

Focus research on animal-based outcomes relating to natural behavior and stress, and away from physical factors (e.g., growth, weight gain) that do not accurately characterize an animal's welfare status except in the grossest sense.

Include ethics as a key component of research into the humaneness of a particular practice. Scientific outcomes are critical, but whether a practice is ethical must be taken into account (Pew Commission, 2008, p. 86).

BACKGROUND ON THE IMPACT ON RURAL COMMUNITIES

The nature of agriculture is changing and with that change, the social fabric of rural America is changing as well. The family-owned farm producing a diverse mix of crops and food animals is largely gone as an economic entity, replaced by large farm factories for animals and monoculture cropping for grains and fiber.

Since the 1930s, research consistently has shown the social and economic well-being of rural communities benefits from larger numbers of farmers rather than fewer farms producing increased volumes. The more farmers there are producing our food and fiber, the more support there is for Main Street business, religious institutions, schools, and social organizations.

Researchers in Michigan documented the magnitude of this difference by tracking local purchases of supplies for swine production. Abeles-Allison and Connor (1990) found that local expenditures per hog were $67 for the small, locally owned farms and $46 for the larger, industrialized farms. The $21 difference is largely due to the larger farms' purchases of bulk food from outside the community.

COMPLETE RURAL COMMUNITY RECOMMENDATIONS

1. STATES, COUNTIES, AND LOCAL GOVERNMENTS SHOULD IMPLEMENT ZONING AND SITING GUIDANCE GOVERNING NEW IFAP OPERATIONS THAT FAIRLY AND EFFECTIVELY EVALUATE THE SUITABILITY OF A SITE FOR THESE TYPES OF FACILITIES.

Regulatory agencies should consider the following factors for inclusion in their IFAP plans, and should adopt such guidelines regardless of whether an IFAP facility currently exists in their jurisdiction. (Please note that each of the following components should take climate, soil type, prevailing winds, topography, air emissions, operation size, noise levels, traffic, designated lands, and other criteria deemed relevant into account.)

- Setback distances: IFAP facilities pose environmental and public health risks to the areas in which they are sited. Determining an exact distance from the production facility at which risks begin and end is very difficult, but it is important to consider. Distances from schools, residences, surface and groundwater sources, religious institutions, parks, and areas designated to protect wildlife should all be factored into the proposed location of a food animal production facility. Waterways are particularly crucial as any waste that seeps into water sources may travel great distances. Proximity, size, available environmental monitoring data, and state regulations for setbacks or other industries also must be taken into account. Setback distances should be significant enough to alleviate public health and environmental concerns. Local officials should make determination of appropriate distances because state regulators cannot take into account every particular factor—they typically set a minimum base standard, which localities should follow, and make more stringent where necessary.
- Method of production: Every type of livestock and poultry production has positive and negative aspects. Zoning officials should consider the economic, environmental, and health effects of, for example, cage-free versus caged facilities, hoop barn versus crate facilities, operations with outdoor or pasture access versus permanent indoor confinement, or any other systems.
- Concentration: Each locality should take into account the number of IFAP facilities already in existence, particularly per watershed. A surge in the number of IFAP facilities in North Carolina led to devastating environmental effects, including serious environmental justice issues. Growth there and in other places has been so rapid that potential concerns were not fully recognized until they had already created problems. Too many IFAP facilities in one area can destroy land and waterways and devastate entire communities. No facility should be sited that cannot coexist with the land, water, environment, or community in a sustainable manner.
- Waste disposal: One of the most important issues concerning IFAP facilities is the method of waste handling. If manure is properly applied to land or injected using an approved manure management plan, there should be enough land available to avoid runoff into surface or groundwater or seepage into groundwater. Many states have already become aware of the potentially hazardous nature of lagoons and, therefore, have made the decision to prohibit them for new facilities. The aforementioned criteria are very important in ensuring waste can be handled properly. Consideration should be given to the fact that animal waste can be as dangerous, if not more so, than untreated human waste and some industrial wastes. Further, localities should operate under the premise that every IFAP facility has the potential for runoff and, therefore, should prepare accordingly. Plans to prevent and deal with this situation are part of the NMP, discussed later.
- Agency capabilities: Local officials should fully fund the costs associated with the review of zoning applications.

- Public input: Because IFAP facilities affect the entire community, advance public input should factor into the decision of whether to site a facility. This should not be only in cases where there is controversy. Public input is important to a community's well-being as it allows all citizens, regardless of economic or social status, to participate in the decision-making process. Neighbors and other citizens should also have access to redress when IFAP facilities fail to comply with standards.
- Local control: Again, localities will have to deal with IFAP impacts and should therefore be the authority on facilities sited within community boundaries. Local officials and citizens tend to have the best knowledge about potential impacts, positive or negative, whereas state officials are more likely to make decisions based on generalizations. Further, local officials are more directly accountable for decisions than state officials are.
- Inspections: The relationship between inspections and zoning is twofold. First, zoning officials should conduct an on-site inspection before siting an operation to evaluate adequately the criteria mentioned previously. Second, operators should be aware that inspections would take place as determined by the state in order to ensure all operations follow established regulations as well as their NMPs.
- Proof of financial responsibility: All operations should be bonded for performance and remediation.
- Permit fees: Fees are suggested in order to help the state or locality fund inspections, enforcement, and the day-to-day function of the local agency. Such fees can range from approximately \$100 up to any amount the agency deems appropriate, and should reflect a sliding scale based on the size of the operation. Two specific components the Commission believes should be mandatory in zoning permits are:
  - Environmental impact statement: The IFAP facility owner and the animal grower must establish the potential impact of the facility on the land, water, and general environment. The statement should include best practice information for maintaining soil, water, and air quality, as well as descriptions of chemical management (e.g., use of fertilizers), manure management, carcass management, storm water response, and an emergency response plan, at a minimum.
  - NMP: All IFAP facilities must comply with USDA-NRCS Standard 590, which requires an NMP. NMPs outline appropriate methods for handling and disposing of manure, including land application issues. Producers should be able to indicate clearly in their NMP that the facility will implement all possible best practices to minimize the potential for runoff, and that they will minimize runoff during catastrophic events (e.g., floods).

Background

Regulations governing the siting and zoning of IFAP facilities vary tremendously across the country. In fact, many states, counties, and local governments have little or no regulations on the books for dealing with new IFAP facilities. Questions often arise on how to establish zoning and siting regulations, how to enforce them, and how to reconcile the needs of the producers and integrators with the lifestyle and health of their neighbors and environmental maintenance of the land. Without well-developed and thought-out regulations, governments are often unable to regulate the siting of IFAP facilities in a way that protects the rights of both the community and the producers. Compliance with all criteria of a zoning permit ensures protection of communities, producers, and the environment (Pew Commission, 2008, pp. 89–91).

2. IMPLEMENT POLICIES TO ALLOW FOR A COMPETITIVE MARKETPLACE IN ANIMAL AGRICULTURE TO REDUCE THE ENVIRONMENTAL AND PUBLIC HEALTH IMPACTS OF THE IFAP.

The Commission recommends the vigorous enforcement of current federal antitrust laws to restore competition in the farm animal market. If enforcing existing antitrust laws is not effective in restoring competition, further legislative remedies should be considered, such as more transparency in price reporting and limiting the ability of integrators to control the supply of animals for slaughter.

Background

The current food animal production system is highly concentrated and exhibits conditions that suggest monopsony, in which there are very few buyers for a large number of suppliers. Under monopsonistic conditions, fewer goods are sold, prices are higher in output markets and lower for sellers of inputs, and wealth is transferred from the party without market power to the party with market power. For example, the top four pork-producing companies in the United States control 60% of the pork market, and the top four beef packers control over 80% of the beef market. Farmers have little choice but to contract with those few producers if they are to sell the food animals they grow.

Vigorous market competition is of vital importance to consumers: They benefit most from an open, competitive, and fair market where the values of democracy, freedom, transparency, and efficiency are in balance. Rural communities and consumers suffer from a loss of competitive markets as wealth is transferred from the party without market power to the party with market power. These situations require robust remedy.

The consolidation in the food animal industry, as well as the continued growth of completely integrated operations (where the processor owns the farm, the animals, and the processing plant), has led to a situation where independent producers, whether contracting or selling on the open market, are beholden to big corporations. Growers often take out large loans to pay for land and equipment in anticipation of a contract from a big corporate integrator. Because the contracts are often presented in "take-it-or-leave-it" terms, the producer may end up with a large loan and no way to pay it off if the integrator revokes the contract (Pew Commission, 2008, p. 93).

SUMMARY

This chapter has introduced the goals and impacts of the Pew Commission on Industrial Farm Production and has provided a background for Chapter 5, which explores the varied viewpoints and approaches to animal welfare issues expressed from differing perspectives. In addition to animal welfare issues, the PEW Report also identified public health, environmental, and rural community concerns.

REFERENCES

Abeles-Allison M, Connor L (1990). An Analysis of Local Benefits and Costs of Michigan hog operations experiencing environmental conflicts. Department of Agricultural Economics, Michigan State University, East Lansing.

AHI (2002). Survey Shows Decline in Antibiotics Use in Animals. Animal Health Institute.

Andersen IL, Boe KE (1999). Straw bedding or concrete floor for loose-housed pregnant sows: Consequences for aggression, production and physical health. *Acta Agriculturae Scandinavica Section a-Animal Science* 49: 190–195.

Anderson AD, McClellan J, Rossiter S, Angulo FJ (2003). Public health consequences of use of antimicrobial agents in agriculture. In: *The Resistance Phenomenon in Microbes and Infectious Disease Vectors: Implications for Human Health and Strategies for Containment Workshop*. Knobler SL, Lemon SM, Najafi M, Burroughs T, (eds). The National Academies Press: Washington, DC, pp. 231–243.

Appleby MC, Lawrence AB (1987). Food restriction as a cause of stereotypic behavior in tethered gilts. *Animal Production* 45: 103–110.

Arteaga ST (2001). National pollutant discharge elimination system compliance challenges. In: *American Society of Agricultural Engineers Annual Meeting*: St. Joseph, MI.

Batt AL, Snow DD, Aga DS (2006). Occurrence of sulfonamide antimicrobials in private water wells in Washington County, Idaho, USA. *Chemosphere* 64: 1963–71.

Brennan JJ, Aherne FX (1987). Effect of floor type on the severity of foot lesions and osteochondrosis in swine. *Canadian Journal of Animal Science* 67: 517–523.

Centner TJ (2006). Governmental oversight of discharges from concentrated animal feeding operations. *Environmental Management* 37: 745–52.

Copeland C (2006). Animal Waste and Water Quality: EPA Regulation of Concentrated Animal Feeding Operations (CAFOs). Service CR, (ed). United States Congress: Washington, DC, pp. CRS-1-CRS-23.

Cosgrove SE, Qi Y, Kaye KS, Harbarth S, Karchmer AW, Carmeli Y (2005). The impact of methicillin resistance in *Staphylococcus aureus* bacteremia on patient outcomes: mortality, length of stay, and hospital charges. *Infection Control & Hospital Epidemiology* 26: 166–74.

CDC/FDA/NIH (1999). A Public Health Action Plan to Combat Antimicrobial Resistance. Interagency Task Force on Antimicrobial Resistance.

DOE, FRA (2001). British Cattle Movement Service.

EPA (2003). NPDES Permit Regulation and Effluent Limitations Guidelines for Concentrated Animal Feeding Operations.

EPA (2007a). Inventory of U.S. Greenhouse Gas Emissions and Sinks: 1990–2005. National Center for Environmental Publications, pp. 393.

EPA (2007b). U.S. EPA 2008 Compliance and Enforcement: Clean Water Act. pp. 1–3

FDA-CVM (2003). Guidance for Industry #152.

Idris U, Lu J, Maier M, Sanchez S, Hofacre CL, Harmon BG, Maurer JJ, Lee MD (2006). Dissemination of fluoroquinolone-resistant Campylobacter spp. within an integrated commercial poultry production system. *Applied and Environmental Microbiology* 72: 3441–7.

IOM (1998). *Antimicrobial drug resistance: Issues and Options.* National Academy Press: Washington, DC.

Kelly N (March 20, 2007). Senate panel cool to feed-farm curbs. In: *The Journal Gazette*: Ft. Wayne.

Krapac IG, Koike, S., Meyer, M.T., Snow, D.D., Chou, S.F.J., Mackie, R.I., Roy, W.R., and Chee-Sanford, J.C. (2004). Long term monitoring of the occurrence of antibiotics residues and antibiotic resistance genes in groundwater near swing confinement facilities. In: *4nd International Conference on Pharmaceuticals and Endocrine Disrupting Chemicals in Water.* National Ground Water Association: Minneapolis, Minnesota, pp. 158–174.

Meatnews.com (2005). Europe Bans Antibiotic Growth Promoters.

Mellon MG, Benbrook C, Benbrook KL, Union of Concerned Scientists. (2001). *Hogging it: Estimates of antimicrobial abuse in livestock.* Union of Concerned Scientists: Cambridge, MA.

Pew Charitable Trusts. (2008). *Putting Meat on the Table: Industrial Farm Animal Production in America. Final Report*, Pew Commission on Industrial Farm Animal Production, April 2008. Washington, DC: Author.

NAS (1999). *The Use of Drugs in Food Animals: Benefits and Risks.* National Academies Press: Washington, DC.

NASDA (2001). Comments on Pollutant Discharge Regulation / Guidelines & Standards for Animal Feeding Operations. Graves L, (ed). National Association of State Departments of Agriculture.

NCSL (2008). Concentrated Animal Feeding Operations: A Survey of State Policies. In: *A Report to the Pew Commission on Industrial Farm Animal Production.* PCIFAP, (ed). National Conference of State Legislatures: Washington, DC.

NPB (2005). Take Care: A Producer's Guide to Using Antibiotics Responsibly. Des Moines, Iowa.

Peak N, Knapp CW, Yang RK, Hanfelt MM, Smith MS, Aga DS, Graham DW (2007). Abundance of six tetracycline resistance genes in wastewater lagoons at cattle feedlots with different antibiotic use strategies. *Environmental Microbiology* 9: 143–51.

(2007a). Preservation of Antibiotics for Medical Treatment Act. (2007b) should be changed. Both citations reference the same legislation.

Schiffman SS, Studwell CE, Landerman LR, Berman K, Sundy JS (2005). Symptomatic effects of exposure to diluted air sampled from a swine confinement atmosphere on healthy human subjects. *Environmental Health Perspectives* 113: 567–76

Sigurdarson ST, Kline JN (2006). School proximity to concentrated animal feeding operations and prevalence of asthma in students. *Chest* 129: 1486–91.

Smith JL, Drum DJ, Dai Y, Kim JM, Sanchez S, Maurer JJ, Hofacre CL, Lee MD (2007). Impact of antimicrobial usage on antimicrobial resistance in commensal *Escherichia coli* strains colonizing broiler chickens. *Applied and Environmental Microbiology* 73: 1404–14.

Stolz JF, Perera E, Kilonzo B, Kail B, Crable B, Fisher E, Ranganathan M, Wormer L, Basu P (2007). Biotransformation of 3-nitro-4-hydroxybenzene arsonic acid (roxarsone) and release of inorganic arsenic by Clostridium species. *Environmental Science Technology* 41: 818–23.

USDA (2005). National Animal Identification System: Draft Strategic Plan. USDA, (ed).

USDA, APHIS (2006). National Animal Identification System (NAIS).

USGS (2006). Pharmaceuticals and Other Emerging Contaminants in the Environment-Transport, Fate, and Effects Ecology.

WSDo (2006). Agency-Wide Notices, Orders & Penalties 1985–2005. Report WSDoEE, (ed), Drop in inspections when authority given to Department of Agriculture in 2003, as compared with previous years when Department of Ecology had dairy inspection authority.

Western Organization of Resource Councils (May 2006). National Animal Identification System: The Unanswered Questions.

Woolhouse ME, Gowtage-Sequeria S (2005). Host range and emerging and reemerging pathogens. *Emerging Infectious Diseases* 11: 1842–7.

Woolhouse ME, Taylor LH, Haydon DT (2001). Population biology of multihost pathogens. Science 292: 1109–12.

WHO (2000). Report on Infectious Diseases, Geneva, Switzerland.

WHO (2002). Impacts of antimicrobial growth promoter termination in Denmark. In: *International Invitational Symposium: Beyond Antimicrobial Growth Promoters in Food Animal Production*. Panel WIR, (ed). WHO Department of Communicable Diseases, Prevention and Eradication: Foulum, Denmark.

ENDNOTES

1. Commission members were: former Kansas Governor John Carlin, chair; Michael Blackwell, DVM, MPH former Dean of the College of Veterinary Medicine at the University of Tennessee/Knoxville, Assistant Surgeon General, USPHS (Ret.), vice chair; Brother David Andrews, CSC, JD; Fedele Bauccio, MBA, Founder and CEO of Bon Appetite Management Company; Tom Dempster, South Dakota State Senator; Dan Glickman, JD, former United States Department of Agriculture Secretary; Alan M. Goldberg, PhD, Johns Hopkins Bloomberg School of Public Health; John Hatch, DrPH, University of North Carolina; Dan Jackson, Montana Cattle Rancher; Frederick M. Kirschenmann, PhD, Distinguished Fellow, Leopold Center for Sustainable Agriculture, Iowa State University; James Merchant, MD, DrPH, Dean, University of Iowa School of Public Health; Marion Nestle, PhD, MPH, Department of Nutrition, Food Studies, and Public Health, New York University; Bill Niman, founder of Niman Ranches, Inc.; Bernard Rollin, PhD, Distinguished Professor of Philosophy, Colorado State University; and Mary Wilson, MD, Associate Professor, Harvard School of Public Health, Associate Clinical Professor, Harvard Medical School.

2. Source: fawc, 2007 at http://www.fawc.org.uk/freedoms

3. Joy A. Mensch, Professor, Department of Animal Science, University of California, Davis; Harvey James, Associate Professor, Department of Agricultural Economics, University of Missouri; Edmond A. Pajor, Associate Professor, Department of Animal Sciences, Purdue University; and Paul D. Thompson, W.K. Kellogg Professor of Agriculture, Food, and Community Ethics, Michigan State University.

4. Commission of the European Communities. 2001. COM(2001) 20 final 2001/0021 (CNS) Communication from the Commission to the Council and the European Parliament on the welfare of intensively kept pigs in particular taking into account the welfare of sows reared in varying degrees of confinement and in groups. Proposal for a Council Directive amending Directive 91/630/EEC laying down minimum standards for the protection of pigs.

5. http://www.abe.iastate.edu/hoop_structures/

5 Defining Agricultural Animal Welfare

Varying Viewpoints and Approaches

Bernard E. Rollin, Donald M. Broom, David Fraser,
Gail C. Golab, Charles Arnot, and Paul Shapiro

CONTENTS

FIRST VIEWPOINT: AN ETHICIST'S AND PHILOSOPHER'S PERSPECTIVE

Bernard E. Rollin

As anyone can tell from Robert Martin's chapter on the Pew Commission findings, or simply by reading the news, animal agriculture faces myriad socio-ethical problems, including concerns about the effects of manure created by confined animal feeding operations on the environment, concerns about human health effects of such operations, concerns about zoonotic diseases, and concerns about modern agriculture's effect on small agriculture and rural communities. These issues are quite clear; the problem becomes finding reasonable solutions that meet societal concerns on the one hand, and are practicable on the other.

In the case of animal welfare, however, the conceptual basis of the problem is not well understood by industry. What seems to completely elude both the agriculture industry and the veterinarians who leap to its defense is that "animal welfare" contains an irreducibly ethical or moral component. Universally, however, the animal agricultural industry and the veterinarians who serve it see the issue of animal welfare as a strictly empirical notion, with no ethical component, open to being resolved by, to use the recurrent industry phrase, "sound science." I shall argue that as long as industry fails to understand the irreducibly ethical component of "animal welfare," it can make no progress toward satisfying social-ethical concerns.

Those of us serving on the Pew Commission, better known as the National Commission on Industrial Farm Animal Production, encountered this erroneous response regularly during our dealings with industry representatives. This commission studied intensive animal agriculture in the United States (Pew Trusts, 2008). For example, one representative of the Pork Producers, testifying before the Commission, answered that while people in her industry were quite "nervous" about the Commission, their anxiety would be allayed were we to base all of our conclusions and recommendations on "sound science." Hoping to rectify the error in that comment, as well as educate the numerous industry representatives present, I responded to her as follows: "Madame, if we on the Commission were asking the question of *how* to raise swine in confinement, science could certainly answer that question for us. But that is *not* the question the Commission, or society, is asking. What we are asking is, *ought* we to raise swine in confinement? And to this question, science is not relevant." Judging by her "huh," I assume I did not make my point.

Questions of animal welfare are at least partly "ought" questions, questions of ethical obligation. The concept of animal welfare is an ethical concept to which science brings relevant data. When we ask about an animal's welfare, or about a person's welfare, we are asking about *what* we owe the animal, and to *what extent*. A document called the CAST report, first published by U.S. agricultural scientists in the early 1980s, discussed animal welfare, and eloquently illustrated the limitation of the "sound science" approach when it affirmed that the necessary and sufficient conditions for attributing positive welfare to an animal were represented by the animals' productivity. A productive animal enjoyed positive welfare; a nonproductive animal enjoyed poor welfare (CAST, 1981).

This notion was fraught with many difficulties. First, productivity is an economic notion predicated on a whole operation; welfare is predicated on individual animals. An operation, such as caged laying hens, may be quite profitable if the cages are severely over-crowded yet the individual hens do not enjoy good welfare. Second, as we shall see, equating productivity and welfare is, to some significant extent, legitimate under husbandry conditions, where the producer does well if and only if the animals do well, and square pegs, as it were, are fitted into square holes with as little friction as possible. Under industrial conditions, however, animals do not naturally fit in the niche or environment in which they are kept, and are subjected to "technological sanders" that allow for producers to force square pegs into round holes—antibiotics, feed additives, hormones, air handling systems—so the animals do not die and produce more and more kilograms of meat or milk. Without these technologies, the animals could not be productive. We will return to the contrast between husbandry and industrial approaches to animal agriculture.

The key point to recall here is that even if the CAST report definition of animal welfare did not suffer from the difficulties we outlined, it is still an ethical concept. It essentially says, "What we owe animals and to what extent is simply what it takes to get them to create profit." This in turn would imply that the animals are well off if they have only food, water, and shelter, something the industry has sometimes asserted, but clearly does not satisfy societal concerns. Even in the early 1980s and before, however, there were animal advocates and others who would take a very different ethical stance on what we owe farm animals. Indeed, the famous five freedoms articulated in Britain by the Farm Animal Welfare Council during the 1970s (even before the CAST report) represents quite a different ethical view of what we owe animals, when it affirms that:

> The welfare of an animal includes its physical and mental state and we consider that good animal welfare implies both fitness and a sense of well-being. Any animal kept by man must, at least, be protected from unnecessary suffering.
>
> We believe that an animal's welfare, whether on farm, in transit, at market or at a place of slaughter, should be considered in terms of "five freedoms" (see www.fawc.org.uk and Chapter 4, Martin, this book).

1. **Freedom from Hunger and Thirst**—by ready access to fresh water and a diet to maintain full health and vigor.
2. **Freedom from Discomfort**—by providing an appropriate environment including shelter and a comfortable resting area.
3. **Freedom from Pain, Injury, or Disease**—by prevention or rapid diagnosis and treatment.
4. **Freedom to Express Normal Behavior**—by providing sufficient space, proper facilities, and company of the animal's own kind.
5. **Freedom from Fear and Distress**—by ensuring conditions and treatment that avoid mental suffering.

Clearly, the two definitions contain very different notions of our moral obligation to animals (and there is an indefinite number of other definitions). Which is correct, of course, cannot be decided by gathering facts or doing experiments—indeed, as we shall see, which ethical framework one adopts will in fact determine the shape of one's science studying animal welfare!

Science tells us about the physical world. It uncovers empirical facts and creates theoretical models to explain those facts. As the great twentieth-century philosopher Ludwig Wittgenstein once remarked, "If one takes an inventory of all the facts in the universe, one does not find it a fact that killing is wrong. Nor does one find any ethical facts." As Hume famously pointed out, statements of fact are *is* statements; statements of ethics are *ought* statements. A huge conceptual gulf yawns between *is* and *ought*. Since, as we just explained, the concept of animal welfare is at root ethics-laden, it is impossible for issues of welfare to be resolved without appeal to ethics, that is, by referring them to *science*. Ironically, Smithfield Farms understood this when they sampled public opinion and agreed to phase out gestation crates, while the American Veterinary Medical Association tellingly said that there is no *scientific* way of validating a preference for one system of sow housing over another.

Let us return to the notion raised earlier, that far from sound science determining animal welfare, the ethical component of animal welfare will determine the nature of your science answering questions about various aspects of animal welfare.

To clarify, suppose you hold the view that an animal is well off when it is productive, as per the CAST report. The role of your welfare science in this case will be to study what feed, bedding, temperature, etc. are most efficient at producing the most meat, milk, or eggs for the least money— much what animal and veterinary science does today. On the other hand, if you take the FAWC view of welfare, your efficiency will be constrained by the need to acknowledge the animal's natural behavior and mental states, and to assure that there is minimal pain, fear, distress, or discomfort— not factors in the CAST view of welfare unless they have a negative impact on economic productivity. Thus, actually, sound science does not determine your concept of welfare; rather, your concept of animal welfare determines what counts as relevant sound science! Indeed, in one version of animal welfare, that of Ian Duncan, Marian Dawkins, and the author, which views animal welfare as based on subjective experiences and feelings of animals, mainstream science helps very little if at all, being agnostic about the knowability of animal thoughts and feelings.

The failure to recognize the inescapable ethical component in the concept of animal welfare leads inexorably to those holding different ethical views talking past each other. Thus, producers ignore questions of animal pain, fear, distress, confinement, truncated mobility, bad air quality, social isolation, and impoverished environment unless any of these factors negatively affect the "bottom line." Animal advocates, on the other hand, give such factors primacy, and are very unimpressed with how efficient or productive the system may be.

A major question obviously arises here. If the notion of animal welfare is inseparable from ethical components, and people's ethical stance on obligations to farm animals differs markedly across a highly diverse spectrum, whose ethic is to predominate and define, in law or regulation, what counts as "animal welfare"? This is of great concern to the agriculture industry, worrying as it does about "vegetarian activists hell-bent on abolishing meat." In fact, of course, such concern is misplaced, for the chance of such an extremely radical thing happening is vanishingly small. Largely, however, the ethic adopted in society reflects a *societal consensus*, what most people either believe to be right and wrong or are willing to accept upon reflection.

All of us have our own personal ethics, which rule a goodly portion of our lives. Such fundamental questions as what we read, what we eat, to whom we give charity, what political and religious beliefs we hold, and myriad others are answered by our personal ethics. These derive from many sources—parents, religious institutions, friends, reading books, movies, and television. One is certainly entitled to believe ethically as do some PETA members, that "meat is murder," that one should be a vegan, that it is immoral to use products derived from animal research, and so on.

Clearly, a society, particularly a free society, contains a bewildering array of such personal ethics, with the potential for significant clashes between them. If my personal ethic is based on fundamentalist religious beliefs and yours is based on celebrating the pleasures of the flesh, we are destined to clash, perhaps violently. For this reason, social life cannot function simply by relying on an individual's personal ethics, except perhaps in singularly monolithic cultures where all members share overwhelmingly the same values. One can find examples of something resembling this in small towns in rural farming areas, where there is no need to lock one's doors, remove one's keys from the car, or fear for one's personal safety. However, such places are few, and are probably decreasing in number. In larger communities, the extreme case being New York City or London, one finds a welter of diverse cultures and corresponding personal ethics crammed into a small geographical locus. For this reason alone, as well as to control those whose personal ethic may entail taking advantage of others, a *social consensus ethic* is required, one that transcends personal ethics. This social consensus ethic is invariably articulated in law, with manifest sanctions for its violation. As societies evolve, different issues emerge, leading to changes in the social ethic.

My claim then is that beginning roughly in the late 1960s, the treatment of animals has moved from being a paradigmatic example of personal ethics to ever-increasingly falling within the purview of societal ethics and law. How and why has this occurred, and to what extent?

If one looks to the history of animal use in society back to the beginning of domestication some 11,000 years ago, one finds very little social ethics dictating animal treatment. The one exception to this generalization is the prohibition against deliberate, purposeless cruelty, that is, needless infliction of pain and suffering or outrageous neglect, such as failing to provide food or water. This mandate is well illustrated in the Old Testament, where many injunctions illustrate its presence. For example, one is told that when collecting eggs from a bird's nest, one should leave some eggs so as not to distress the animal. Kosher and halal slaughter accomplished by a trained person using a very sharp knife was clearly intended as a viable alternative to the much more traumatic bludgeoning. (That is not of course to suggest that such slaughter remains welfare-friendly in high throughput industrialized slaughterhouses.) The rule of Kashrut prohibiting the eating of milk and meat—"thou shalt not seethe a kid in his mother's milk" (Exodus 34:26)—seems to be aimed at avoiding loss of sensitivity to animal suffering.

In the middle ages, St. Thomas Aquinas provided a more anthropocentric reason for prohibiting cruelty, based in the prescient psychological insight that those who would abuse animals would inexorably progress to abusing humans. Aquinas does not see animals as direct objects of moral concern, but nonetheless, strongly prohibits their abuse.

In the late eighteenth century in Britain, and in subsequent years elsewhere, the prohibition against deliberate, sadistic, deviant, willful, malicious cruelty, that is, inflicting pain and suffering on animals to no reasonable purpose, or outrageous neglect such as not providing food or water, were encoded in the anti-cruelty laws of all civilized societies. While adopted in part out of a moral notion of limiting animal suffering, an equally important reason was the Thomistic one—to ferret out individuals who might graduate to harming humans; case law in the United States and elsewhere make this manifest.

In one revealing case in the nineteenth century, a man was charged with cruelty after throwing pigeons into the air and shooting them to demonstrate his skill. After killing the birds, he ate them. The court ruled that the pigeons were not "needlessly or unnecessarily killed" because the killing was done "in the indulgence of a healthful recreating during an exercise tending to promote strength, bodily agility and courage" (*The State v. Bogardus*, 4 MO. App. 215, 219, Mo. Ct. App. 1877). In discussing a similar nineteenth-century case of a tame pigeon shoot in Colorado, the court affirmed that "every act that causes pain and suffering to animals is not prohibited. Where the end or object in view is reasonable and adequate, the act resulting in pain is…necessary and justifiable, as…where the act is done to protect life or property, or to minister to the necessities of man" (*Waters v. the People*, Supreme Court of Colorado 23 Colo. 33, 46, p. 112, 1896 Colo.). To the credit of the Colorado court, it did not find that such tame pigeon shoots met the test of "worthy motive" or "reasonable object." Even today, however, there are jurisdictions where tame pigeon shoots and "canned hunts" do not violate the anti-cruelty laws.

It is certainly true that cruelty to animals is closely linked to psychopathic behavior—animal cruelty, along with fire starting and bed-wetting, are signs of future psychopaths. The majority of children who shoot up their schools have early histories of animal abuse, as do 80% of the violent offenders in Leavenworth Prison and most serial killers. Animal abusers often abuse wives and children (Ascione et al., 2007). Most battered women's shelters must make provisions for keeping the family pet, as the abuser will hurt the animal to hurt the woman. Several studies have shown a relationship between childhood animal cruelty and violence toward people (Miller, 2001).

However, these anti-cruelty laws conceptually provide little protection for animals. Animal cruelty accounts for only a tiny fraction of the suffering that animals undergo at human hands. For example, the United States produces around 9 billion broiler chickens a year, and many have bruises and fractures or other skeleto-muscular injuries that occur during catching. Before restaurant companies started doing animal welfare audits, careless rough handling of chickens resulted in 5% of the birds suffering broken wings, which is a shocking 450,000,000 birds with an injury as severe as a broken arm. If even 1% of chickens are so injured (a ridiculously low number), then we have 90,000,000 suffering animals there alone—there is nothing like 90,000,000 incidents of cruelty, and those chickens are legally unprotected. In the United States, they are not even subject to humane slaughter law! In Europe and Canada, humane slaughter laws include poultry.

In short, over the last 40 years society has come to realize the need for an ethic that expresses its concern for all animal suffering, not just the relatively small amount resulting from deliberate cruelty.

The obvious question that presents itself is this: What has occurred during the last half century that led to social disaffection with the venerable ethic of anti-cruelty and to strengthening of the anti-cruelty laws, which now make cruelty a felony in almost 40 states?

In a study commissioned by USDA to answer this question, the author distinguished a variety of social and conceptual reasons (Rollin, 1995):

1. Changing demographics and consequent changes in the paradigm for animals.

   Whereas at the turn of the century, more than half the population was engaged in producing food for the rest, today only some 1.5% of the U.S. public is engaged in production agriculture (AMC, 2003). One hundred years ago, if one were to ask a person in the street, urban or rural, to state the words that come into their mind when one says "animal," the answer would doubtless have been "horse," "cow," "food," "work," etc. Today, however, for the majority of the population, the answer is "dog," "cat," "pet." Repeated studies show that between 90 and 100% of the pet-owning population view their animals as "members of the family" (*The Acorn*, 2002), and virtually no one views them as an income source. Divorce lawyers note that custody of the dog can be as thorny an issue as custody of the children!

2. We have lived through a long period of ethical soul-searching.

   For almost 50 years, society has turned its "ethical searchlight" on humans traditionally ignored or even oppressed by the consensus ethic—blacks, women, the handicapped, other minorities. The same ethical imperative has focused attention on our treatment of the nonhuman world—the environment and animals. Many leaders of the activist animal movement in fact have roots in earlier movements—civil rights, feminism, homosexual rights, children's rights, labor, etc.

3. The media has discovered that "animals sell papers."

   One cannot channel-surf across normal television service without being bombarded with animal stories, real and fictional. (A *New York Times* reporter recently told me that more time on cable TV in New York City is devoted to animals than to any other subject.) Recall, for example, the extensive media coverage a decade ago of some whales trapped in an ice floe and freed by a Russian icebreaker. This was hardly an overflowing of Russian compassion. Rather, someone in the Kremlin was bright enough to realize that liberating the whales was an extremely cheap way to score points with U.S. public opinion.

4. Strong and visible arguments by philosophers, scientists, and celebrities have been advanced in favor of raising the status of animals (Singer, 1975; Rollin, 1981; Regan, 1983; Sapontzis, 1987).

5. Changes in the nature of animal use demanded new moral categories.

In my view, while all of the reasons just discussed are relevant, the most important reasons are the dramatic and precipitous changes in animal use that occurred after World War II. These changes were, first, the huge conceptual changes in the nature of agriculture, which was discussed in Chapter 1, and second, the rise of vast amounts of animal research and testing.

Chapter 1 discussed the circumstances leading agriculture from husbandry to industry. Clearly, those who developed modern agriculture were not motivated by cruelty. Rather, they aimed at providing cheap and plentiful food in the face of social, economic, and cultural changes. They did not see the threat industrial agriculture posed to animal welfare because they assumed that what was true in husbandry agriculture carried over to industrial agriculture; namely, that if animals were productive, their welfare was assured. While this is true in husbandry, where all aspects of animal needs must be met to assure productivity, it is not true of industrial agriculture, where technological fixes such as antibiotics and vaccines allow the animals' nature to be violated despite their remaining productivity. Society eventually became aware of the new kinds of suffering—not cruelty—engendered by modern agriculture on at least four fronts: production diseases, lack of attention to individual animals, physical and psychological deprivation in confinement, and lack of "animal smart" employees.

These sources of suffering are not captured by the vocabulary of cruelty. In addition, people began to realize that biomedical and other scientific research, toxicological safety testing, uses of animals in teaching, pharmaceutical product extraction from animals, and so on all produce far more suffering than does overt cruelty. This suffering comes from creating disease, burns, trauma, fractures, and the like in animals in order to study them; producing pain, fear, learned helplessness, aggression, and other states for research; poisoning animals to study toxicity; and performing surgery on animals to develop new operative procedures. In addition, the housing of research animals engenders suffering. Indeed, it has been argued by Dr. Tom Wolfle and I that the discomfort and suffering experienced by animals used in research by virtue of being housed under conditions that are convenient for us, but inimical to their biological natures—for example, keeping rodents, which are nocturnal, burrowing creatures, in polycarbonate crates under artificial, full-time light—far exceed the suffering produced by invasive research protocols.

Now it is clear that farmers and researchers are not intentionally cruel—they are motivated by plausible and decent intentions: To cure disease, advance knowledge, ensure product safety, provide cheap and plentiful food. Nonetheless, they may inflict great amounts of suffering on the animals they use. Furthermore, the traditional ethic of anti-cruelty and the laws expressing it had no vocabulary for labeling such suffering because researchers were not maliciously intending to hurt the animals. Indeed, this is eloquently marked by the fact that the cruelty laws exempt animal use in science and standard agricultural practices from their purview. Therefore, a new set of concepts beyond cruelty and kindness was needed to discuss the issues associated with burgeoning research animal use and industrial agriculture.

Given that the old anti-cruelty ethic did not apply to animal research or confinement agriculture, society needed new ethical concepts to express its concern about these new uses. However, ethical concepts do not arise *ex nihilo*.

Plato taught us a very valuable lesson about effecting ethical change. If one wishes to change another person's—or society's—ethical beliefs, it is much better to *remind* than it is to *teach*. In other words, if you and I disagree ethically on some matter, it is far better for me to show you that what I am trying to convince you of is already implicit—albeit unnoticed—in what you already believe. Similarly, we cannot force others to believe as we do; we can, however, show them that their own assumptions, if thought through, lead to a conclusion different from what they currently entertain. These points are well exemplified in twentieth century U.S. history. Prohibition was an attempt to forcefully impose a new ethic about drinking on the majority by the minority. As such, it was doomed to fail, and in fact, people drank *more* during Prohibition.

So society was faced with the need for new moral categories and laws that reflect those categories in order to deal with animal use in science and agriculture and to limit the animal suffering with which it is increasingly concerned. At the same time, recall that western society has gone through almost 50 years of extending its moral categories for *humans* to people who were morally ignored or invisible—women, minorities, the handicapped, children, citizens of Third-World countries. As noted earlier, new and viable ethics do not emerge *ex nihilo.* Therefore, a plausible and obvious move is for society to continue in its tendency and *attempt to extend the moral machinery it has developed for dealing with people, appropriately modified, to animals.* This is precisely what has occurred. Society has taken elements of the moral categories it uses for assessing the treatment of people and is in the process of modifying these concepts to make them appropriate for dealing with new issues in the treatment of animals, especially their use in science and confinement agriculture.

What aspect of our ethic for people is being so extended? One that is, in fact, quite applicable to animal use, is the fundamental problem of weighing the interests of the individual against those of the general welfare. Different societies have provided different answers to this problem. Totalitarian societies opt to devote little concern to the individual, favoring instead the state or whatever their version of the general welfare may be. At the other extreme, anarchical groups such as communes give primacy to the individual and very little concern to the group—hence they tend to enjoy only transient existence. In our society, however, a balance is struck. Although most of our decisions are made to the benefit of the general welfare, fences are built around individuals to protect their fundamental interests from being sacrificed to the majority. Thus, we protect individuals from being silenced even if the majority disapproves of what they say; we protect individuals from having their property seized without recompense even if such seizure benefits the general welfare; we protect individuals from torture even if they have planted a bomb in an elementary school and refuse to divulge its location. We protect those interests of the individual that we consider essential to being human, to *human nature,* from being submerged, even by the common good. Those moral/legal fences that so protect the individual human are called *rights* and are based on plausible assumptions regarding what is essential to being human.

It is this notion to which society in general is looking in order to generate the new moral notions necessary to talk about the treatment of animals in today's world, where cruelty is not the major problem but where such laudable, general human welfare goals as efficiency, productivity, knowledge, medical progress, and product safety are responsible for the vast majority of animal suffering. People in society are seeking to "build fences" around animals to protect the animals and their interests and biological and psychological natures from being totally submerged for the sake of the general welfare, and are trying to accomplish this goal by going to the legislature.

It is necessary to stress here certain things that this ethic, in its mainstream version, is *not* and does not attempt to be. As a mainstream movement, it does not try to give human rights to animals. Since animals do not have the same natures and interests flowing from these natures as humans do, human rights do not fit animals. Animals do not have basic natures that demand speech, religion, or property; thus, according them these rights would be absurd. On the other hand, animals have natures of their own and interests that flow from these natures, and the thwarting of these interests matters to animals as much as the thwarting of speech matters to humans. For mainstream society, the agenda is not making animals have the same rights as people. Rather, it is preserving the common sense insight that "fish gotta swim and birds gotta fly," and suffer if they don't.

This new ethic is *conservative*, not radical, harking back to the animal use that necessitated and thus entailed respect for the animals' natures. It is based on the insight that what we do to animals *matters* to them, just as what we do to humans matters to them, and that consequently we should respect that mattering in our treatment and use of animals as we do in our treatment and use of humans. Moreover, since respect for animal nature is no longer automatic as it was in traditional husbandry agriculture, society is demanding that it be encoded in law. Significantly, in 2004, no fewer than 2100 bills pertaining to animal welfare were proposed in U.S. state legislatures. More than 90 law schools now teach animal law. The same point is evidenced by the referenda at state level abolishing sow stalls, battery cages, and veal crates.

About animal agriculture, the pastoral images of animals grazing on pasture and moving freely are iconic. As the 23rd Psalm indicates, people who consume animals wish to see the animals live decent lives, not lives of pain, distress, and frustration. It is for this reason in part that industrial agriculture conceals the reality of its practices from a naïve public—witness Perdue's advertisements about raising "happy chickens," or the California "happy cow" ads. As ordinary people discover the truth, they are shocked. When I served on the Pew Commission and other commissioners had their first view of sow stalls, many were in tears and all were outraged.

Just as our use of people is constrained by respect for the basic elements of human nature, people wish to see a similar notion applied to animals. Animals, too, have natures, what I call *telos* following Aristotle—the "pigness of the pig," the "cowness of a cow." Pigs are "designed" to move about on soft loam, not to be in gestation crates. If this no longer occurs naturally, as it did in husbandry, people wish to see it legislated. This is the mainstream sense of "animal rights," an attempt to restore fairness and husbandry to the use of animals in agriculture.

As property, strictly speaking, animals cannot have legal rights. However, a functional equivalent to rights can be achieved by limiting property rights. When others and I drafted the U.S. federal laws for laboratory animals, we did not deny that research animals were the property of researchers. We merely placed limits on the use of their property. I may own my car, but that does not mean I can drive it on the sidewalk or at any speed I choose. Similarly, our law states that if one hurts an animal in research, one must control pain and distress. Thus, research animals can be said to have the *right* to have their pain controlled.

In the case of farm animals, people wish to see their basic needs and nature, *teloi*, respected in the systems in which they are raised. Since this no longer occurs naturally as it did in husbandry, it must be imposed by legislation or regulation. A Gallup poll conducted in 2003 shows that 75% of the public wants legislated guarantees of farm animal welfare. This is what I call "animal rights as a mainstream phenomenon." Legal codification of rules of animal care respecting animal *telos* is thus the form animal welfare takes where husbandry has been abandoned.

Thus, in today's world, the ethical component of animal welfare prescribes that the way we raise and use animals must embody respect and provision for their psychological needs and natures. It is, therefore, essential that industrial agriculture phase out those systems that cause animal suffering by violating animals' natures and replace them with systems respecting their natures.

REFERENCES

The Acorn. 2002. Survey says pets are members of the family. January 31.

Ascione, F.R., Weber, C.V., Thompson, T.M., Heath, J., Maruyama, M., and Hayashi, K. 2007. Battered pets and domestic violence: Animal abuse reported by women experiencing intimate violence and by nonabused women, *Violence Against Women,* 13: 354–373.

CAST (Council for Agricultural Science and Technology). 1981. *Scientific Aspects of the Welfare of Food Animals,* Report #91.

Miller, C. 2001. Childhood animal cruelty and interpersonal violence, *Clinical Physiological Review,* 21: 735–749.

Pew Charitable Trusts. 2008. Report of the Pew Commission on Industrial Farm Animal Production, PCIFAP. org.

Regan, T. 1983. *The Case for Animal Rights.* Berkeley, CA: University of California Press.

Rollin, B. 1981. *Animal Rights and Human Morality.* Buffalo, NY: Prometheus Books.

Rollin, B. 1995. *Farm Animal Welfare: Social, Bioethical and Research Issues.* Ames, IA: Iowa State University Press.

Sapontzis, S. 1987. *Morals, Reason and Animals.* Philadelphia: Temple University Press.

Singer, P. 1975. *Animal Liberation.* New York: New York Review Press.

SECOND VIEWPOINT: FROM A SUSTAINABILITY
AND PRODUCT QUALITY PERSPECTIVE

Donald M. Broom

INTRODUCTION

The scientific study of animal welfare has developed rapidly in recent years. The concept is defined here and its relationship with other concepts, such as health, stress, and needs, is discussed.

The welfare of animals is a matter of substantial public concern and is an aspect of our decisions about whether animal usage systems are sustainable. A system that results in poor welfare is unsustainable because it is unacceptable to many people. The various criteria for sustainability are briefly discussed. The quality of animal products is now judged in relation to the ethics of production, including impact on the welfare of the animals, as well as on price, taste, and consequences for consumers.

Animal welfare is a term that describes a potentially measurable quality of a living animal at a particular time and hence is a scientific concept. It requires strict definition if it is to be used effectively and consistently. A clearly defined concept of welfare is needed for use in precise scientific measurements, in legal documents, and in public statements or discussion. Welfare refers to a characteristic of the individual animal rather than something given to the animal by people (Duncan, 1981). Broom (1986) defined the welfare of an individual as its state as regards its attempts to cope with its environment. It has been emphasized (Duncan, 1981; Broom, 1988, 1991a,b; Broom and Johnson, 2000; Fraser, 2008) that welfare can be measured scientifically, independently of any moral considerations. Once the welfare has been objectively assessed, ethical decisions can be taken about what is to be done about it. This definition of welfare refers to a characteristic of the individual at the time, that is, how well it is faring (Broom and Fraser, 2007; Broom, 2008). This state of the individual will vary on a scale from very good to very poor. Welfare will be poor if there is difficulty in coping or failure to cope so that the individual is harmed. One or more coping strategies may be used to attempt to cope with a particular challenge so a wide range of measures of welfare may be needed to assess welfare.

Feelings, such as pain, fear, and pleasure, are often a part of a coping strategy and they are a key part of welfare (Duncan and Petherick, 1991; Broom, 1991b, 1998). They are adaptive aspects of an individual's biology that must have evolved to help in survival just as aspects of anatomy, physiology, and behavior have evolved. Fear and pain can play an important role in the fastest acting urgent coping responses, such as avoidance of predator attack or risk of immediate injury. Positive and negative feelings, as well as other brain processes that involve no affect, are among the causal factors determining what decisions are taken in longer time-scale coping procedures, where various risks to the fitness of the individual are involved. Aspects of suffering also contribute significantly to how the individual tries to cope in attempts to deal with very long-term problems that may harm the individual. In the organization of behavior to achieve important objectives, pleasurable feelings and the expectation that these will occur have a substantial influence.

Coping with pathology is necessary if welfare is to be good so health is an important part of the broader concept of welfare, not something separate (Dawkins, 1980; Webster, 1994; Broom, 2006; Broom and Fraser, 2007). However, health is not all of welfare, as those with a medical or veterinary background have sometimes assumed. Health is the state of the individual as regards its attempts to cope with pathology. This refers to body systems, including those in the brain, which combat pathogens, tissue damage, or physiological disorder.

When considering how to assess the welfare of animals, it is necessary to start with knowledge of the biology of the animal and of all of its needs. It is important to be aware that needs have a biological basis, but this does not mean that degree of naturalness is a part of the definition of welfare (Fraser, 2008). Some events that occur in nature, such as starvation or predation, result in very poor welfare. The needs of individuals will vary according to genotype and will be affected by conditions during development. It is more useful to consider the needs of animals of a given species, using scientific information about them, than to use the vaguer concept of freedoms.

The word "stress" should be used for the part of poor welfare that involves failure to cope, as the common public use of the word refers to a deleterious effect on an individual (Broom and Johnson, 2000). Reference to stress as just a stimulation that could be beneficial, or as an event that elicits adrenal cortex activity, is of no scientific or practical value. One indicator of adversity is whether there is an effect on biological fitness. Stress can be defined as an environmental effect on an individual that over-taxes its control systems and reduces its fitness or seems likely to do so. Using this definition, the relationship between stress and welfare is very clear. First, while welfare refers to a range in the state of the animal from very good to very poor, whenever there is stress welfare is poor. Second, stress refers only to situations in which there is failure to cope, but poor welfare refers to the state of the animal, both when there is failure to cope and when the individual is having difficulty in coping.

In the early 1990s and later, Broom's definition was referred to by some as a functional definition and was contrasted with the feelings-related definition of Duncan (see also Broom, 2008). Duncan argued that welfare is wholly about feelings (e.g., Duncan and Petherick, 1991). A more common position was that of Dawkins (1990), who stated that the feelings of the individual are the central issue in welfare but other aspects such as the health of that individual are also important. As explained earlier, feelings are biological mechanisms that form part, but not all, of the set of coping systems. The term welfare means essentially the same as well-being but, in most of the world, welfare is used as the scientific term.

Sustainability

A central question, when decisions are made about whether a system for exploiting resources should be used, is whether the system is sustainable (Aland and Madec, 2009). The fact that something is profitable and there is a demand for the product is not sufficient reason for the continuation of production. A system or procedure is sustainable if it is acceptable now and if its effects will be acceptable in future, in particular in relation to resource availability, consequences of functioning, and morality of action (Broom, 2001, 2010). A system might not be sustainable for several possible reasons. For animal usage systems, examples of such reasons are: (1) because it involves so much depletion of a resource that it will become unavailable to the system, (2) because a product of the system accumulates to a degree that prevents the functioning of the system, or (3) because members of the public find an action involved in it unacceptable. Where there is depletion of a resource or accumulation of a product, the level at which this is unacceptable, and hence the point at which the system is unsustainable, is usually considerably lower than that at which the production system itself fails. Other reasons for unacceptability are exemplified in the following. A system could be unsustainable because of harms to the perpetrator, other people, the environment, or other animals (Table 5.1).

No system or procedure is sustainable if a substantial proportion of the local or world public finds aspects of it now, or of its consequences in the future, morally unacceptable. Each of the examples in Table 5.2 is unsustainable. Adverse effects on people or animals can be reported in the media around the world and there are now consequences of unacceptable practices in manufacturing, animal production, or other human activities because of increased efficiency of communication.

Media reports of activities or events that the public find unacceptable may result in consumers in many countries refusing to buy animal and other products from the companies or countries involved (Table 5.3; Broom 2002).

Consumers drive legislation and retail company codes of practice for animal production (Bennett, 1994; Bennett, Anderson, and Blaney, 2002). Legislation on animal welfare has developed in the European Union and in many countries because of pressure from voters (Broom 2002, 2009). In general, the standards of retail companies have a substantially greater effect on the welfare of farm

TABLE 5.1
Reasons for Lack of Sustainability of a System

1. Resource depletion   to a level that is unacceptable

           to a level that prevents system function

2. Product accumulation   to a level that people detect and find unacceptable

           to a level that affects other systems in an unacceptable way

           to a level that affects the system itself, perhaps blocking its function

3. Other effect   to a level that is unacceptable

 The consequences of acts or of system functioning (in 1, 2, and 3) could be unacceptable because of immediate or later:

  [a] Harm to the perpetrator: resource loss or poor welfare

  [b] Harm to other humans: resource loss

  [c] Harm to other humans: poor welfare

  [d] Harm to other animals: poor welfare

  [e] Harm to the environment including that of other animals.

Modified after Broom, 2010.

TABLE 5.2
Unsustainability — Categories of Unacceptable Harms and Examples That Led to Headlines in Newspapers

1. Harm to perpetrator: Resource loss or poor welfare

 [a] System for energy production uses more energy than it produces.

 [b] Machinery for process made of poor quality materials so injury to working person likely.

 [c] Toxic insecticide spread on fields — spreaders poisoned by insecticide in China.

2. Harm to other humans: Resource loss

 [a] Factory/agricultural system outflow into lake or river — fishing industry lost because of the pollution by manure of a river in Thailand.

 [b] Heavy metals from industry — reduces farm production.

 [c] Radiation from energy production system — reduces farm production.

3. Harm to other humans: Poor welfare

 [a] Dioxin released from factory — people become sick, some die.

 [b] Cheap cattle protein fed to other cattle — bovine spongiform encephalopathy in cattle and people catching new-variant Creutzfeldt-Jacob disease by eating beef in the U.K. Also, consumer health risk from slaughtered sick cattle in United States.

 [c] Work that is too demanding — some workers become injured, depressed, or psychotic.

4. Harm to other, nonhuman, animals: Poor welfare

 [a] Traditional entertainment for people, for example, bull-fight, dog-fight, cock-fight, bear-bait, throw goat off church tower.

 [b] Use leg-hold trap for pests or fur-bearing animals.

 [c] Veal production from calves kept in small crates and fed only milk.

 [d] Sheep on an Australian ship dying in large numbers en route to Saudi Arabia.

 [e] Slaughterhouse cruelty in the United States.

 [f] Chickens killed by inhumane methods during avian influenza control in Indonesia.

5. Harm to environment including that of other animals

 [a] Use of CFCs in refrigerators — ozone layer damage.

 [b] Use of chlorinated hydrocarbon insecticides — birds, which are insectivores, or top predators killed or unable to reproduce.

 [c] Produce too much carbon dioxide and other greenhouse gases — global warming.

Modified after Broom, 2010.

TABLE 5.3
Examples of Actions that Led to Consumers Refusing to Buy Products

Action Reported by Media	Consequences
Dolphins being killed in nets set for tuna.	The sales of tuna dropped sharply. This was a long-term effect and resulted in a permanent change in fishing practices.
In France, poor welfare of calves kept in small crates for veal production.	In U.K., a drop in the sales of all French products, including unrelated products such as wine. For most consumers, this was temporary but for some it continued until the introduction of European Union legislation banning the production of veal using crate-housing and low iron and low fiber diets.
The death of thousands of sheep on an Australian ship going to Saudi Arabia.	In several countries, a temporary drop in sales of Australian products.
Very low payments to poor coffee farmers in Third World countries supplying a coffee shop chain reported in many countries.	Temporary and permanent loss of customers at coffee shop chain.
Rainforest destruction for beef production for restaurant company.	A drop in sales of company in many countries. Some permanent loss of customers.
Cruelty to poultry in slaughterhouse shown in one television program and cruelty to cattle in another.	Temporary reduction in poultry sales. Reduction in beef consumption, duration not known. A few people respond to information about poor welfare in animals by becoming vegetarian but a much larger number make some changes to their food purchasing practices.

animals than legislation. The codes of practice of food companies have international impact. For example, many pig producers in Brazil have to comply with the animal welfare standards of United Kingdom supermarkets in order to sell to them, and egg producers in Thailand have to rear their birds according to the standards of the increasing numbers of U.S. food chain companies who have animal welfare standards.

WHAT IS FOOD PRODUCT QUALITY?

The idea of quality for the goods that people buy has changed in the last 10 to 20 years. Quality formerly referred to immediately observable aspects, that is, for an animal food product, its visual qualities and taste. These aspects of quality are still important, and expectations about taste are tending to become more refined, but other factors are now becoming incorporated into what constitutes good quality. Consumption has consequences and a higher proportion of these are now considered. If a food causes people to become sick, the quality is considered poor. If the food tends to make you fat, for some people the quality is considered poor. If food has added nutrients, some consider the quality better. In addition, a major recent change is that the ethics of the production method are taken into account. Factors considered by purchasers include: (1) the welfare of the animals used in production, (2) any impact on the environment, including conservation of wildlife, (3) ensuring a fair payment for producers, especially in poor countries, (4) the preservation of rural communities so that the people there do not go to live in towns, and (5) the carbon footprint of each product as factors leading to global warming are now high on the agenda of many discriminating consumers.

If food is not safe, in that it contains damaging levels of toxins or pathogens, most consumers will never buy it no matter how cheap it is. Individual food production companies are expected to be responsible for this aspect of food quality, but the public expects their government to ensure that adequate standards and adequate checking systems exist. National governments have fallen and companies have gone bankrupt because of known failure on this issue.

In parallel with the FDA in the United States, in the European Union the European Food Safety Authority (EFSA) has been set up. A difference from the FDA is that (1) many aspects of sustainability are part of the work of EFSA and (2) the major part of its work is done by independent scientists, appointed solely on scientific expertise and not as representatives of countries or interest groups. In producing scientific reports, a significant part of their work is the assessment of risks and benefits. The subject area covered by EFSA is wide, reflecting the public concern. One panel deals with animal disease and animal welfare. The reports that it produces has led to changes in EU legislation and scientifically based standards in Europe and elsewhere in the world. A scientific committee producing reports on animal welfare is of value in any major country. Measures to check that there is compliance with legislation exist in the member states of the EU and in other countries, such as the United States with regard to food content.

In order that the ethics of the production method can be properly taken into account, products must be traceable. If foods can be traced, it is less likely that toxins, other poor quality materials, or pathogens will be in them. If animals can be traced, the sources of animal disease outbreaks are more likely to be found and places where injuries or other causes of poor welfare occurred are more likely to be found (Broom, 2007). Legislation and industry initiatives ensuring traceability are important.

Aspects of Sustainability and Product Quality

Consumers will refrain from purchasing animal products if they judge that the production procedures are unsustainable and thus not of good quality. The quality may be judged poor based on negative effects of the production or the product on human health, human diet, the acceptability of genetic modification, animal welfare, environmental effects such as pollution, conservation and carbon footprint, the efficient use of world food resources, fair trade, that is, considering poor producers, and preserving rural communities. Each of these factors, now an aspect of both product quality and the sustainability of the production method, is considered here.

Human Disease Resulting from a Food Product, Sustainability, and Product Quality

Some examples of human health issues that affect views of product quality are *Salmonella* in eggs and meat, *Campylobacter* in chicken carcasses, and avian influenza (H5N1 or H1N1) and bovine spongiform encephalopathy (BSE) in beef products. In the late 1980s and early 1990s, the British government failed to initiate measures that would prevent the large-scale mortality of people from new-variant Creutzfeldt Jacob Disease (CJD) if they ate meat products from animals with BSE. Luckily, for the British public, the number dying is likely to be a few thousand rather than hundreds of thousands. Eventually, with scientific expertise from EU committees, an appropriate policy was developed. The one good consequence of this has been the development of the risk assessment approach in disease management and in animal welfare in Europe. However, the subsequent unwillingness of other governments, faced with an unknown amount of BSE, to damage their beef production industries is disturbing. Recent actions in the United States make it clear that cattle showing abnormal locomotion and other behavior on arrival at the slaughterhouse must still be considered a BSE risk.

Human Diet, Sustainability, and Product Quality

In recent years, there have been large effects on animal production because of concern about human diet. In particular, saturated fats increase risks of heart disease and farm livestock are a major source. Because of the benefits of consuming fish oils, fish production is increasing rapidly. The production of fish that consume vegetable matter, rather than predators like salmonids, which have to be fed mainly fish products, is likely to increase the most because much of the fish product fed to the salmonids could have been consumed by humans and less resource wastage occurs if the fish are herbivorous. The value of farmed fish production is already larger than that of open water fish production, and the weight of farmed fish will be greater than that of fish from open water within a few years.

Genetic Modification, Sustainability, and Product Quality

In some countries, genetically modified plants are not accepted because of ethical concerns, the issue being whether living things should be modified in the laboratory as opposed to genetic changes that occur naturally. There is also concern because protein changes can cause allergies. Genetic modifications in animals can benefit the animals (e.g., confer disease resistance), help to treat human disease (e.g., a blood clotting factor in the milk of a sheep), develop new products for other purposes, or increase efficiency of animal production. Some people accept all of these but others accept some or none as sufficient justification for genetic modification. A major reason for this is that, in some cases, animal welfare may be poorer because of the modification. The conclusion of many people is that any production of genetically modified animals should occur only if it has been demonstrated by scientific studies of animal welfare that the welfare of the animals is not poorer than that of unmodified animals.

Animal Welfare, Sustainability, and Product Quality

Poor welfare of animals that are used in the production system is a major reason why the public regards some animal production systems as unacceptable. Hence, these systems become unsustainable unless there is some modification to them. Animal welfare is becoming more important to members of the public as a reason for demanding change from farmers, food retail companies, and governments. Members of the European Parliament receive more letters about animal welfare than about any other subject (Broom, 1999). However, most people think about animal welfare issues infrequently, unless their attention is drawn to it by media coverage. When the information is drawn to public attention, there is a point at which the welfare of the animals becomes so poor that the majority consider the system to be unacceptable. Hence, animal welfare and public attitudes toward it must be considered wherever the sustainability of an animal production system is evaluated. In order to produce laws or codes of practice, scientific evidence is needed.

Conservation, Carbon Footprint, Sustainability, and Product Quality

A major harm that results from agriculture is that it normally reduces biodiversity as compared with the original natural vegetation. Where wild or semi-wild areas are cleared for animal production, substantial harm can be done to populations of animals and plants. Hence, some animal production is not considered acceptable and products are not bought because these harms have been done. One solution to this problem, for animals that currently consume pasture plants, is to keep the animals in areas where they can browse on bushes and trees as well as grazing (Murgueitio et al., 2009, 2010).

A second solution is the creation of significant areas of nature reserve, as demanded by the public in most countries. Preservation of wildlife can sometimes result in greater income through eco-tourism than would have been possible by farming. The purchase of land to conserve natural resources can often stimulate local economies and lead to a sense of regional pride that would not have existed if low-level animal production had continued. A further example of a possible adverse impact of animal production on conservation is the inappropriate use of antimicrobials and other medicines. The numbers of several species of vultures in India have declined by 96.8 to 99.9% in 15 years (Prakash et al., 2007). This is a consequence of poisoning by the pain killer Diclofenac and the Indian government has recently banned its use (Pain et al., 2008).

Mismanagement of resources and production of effluents that can result in contamination of water supplies, loss of plant nutrients, greenhouse gas production, and increased human disease are also a cause of unsustainability. The animal producer should pay any costs of environmental pollution and, wherever possible, animal waste should be efficiently recycled.

Efficient Use of World Food Resources, Sustainability, and Product Quality

Many people consider that the inefficient use of world food resources is unsustainable. However, animal production activities can be changed to exploit existing resources. Some animals used for food production can eat food that humans cannot eat (see Chapter 13). Hence, keeping grazers and browsers will often be more advantageous than raising pigs or poultry, since the latter do compete with humans for food. There will be energy loss if we eat animals that consume food that we could have eaten. There is also an effect on greenhouse gas production because carbon dioxide and other greenhouse gases are emitted in the course of production of animals such as poultry and pigs, for example because of the combustion of materials in the course of food production and the transport of food and animals. The advantage of using grazers or browsers can be weighed against any adverse consequences for greenhouse gas emissions of methane production by ruminants.

Fair Trade, Preserving Rural Communities, Sustainability, and Product Quality

Many traditions and ways of life for people are associated with animal agriculture. Many human communities exist as they do because of particular animal production systems. If that production is changed so that the number of farms is greatly reduced in the original areas, or the whole production system is moved away from those areas, there are social and environmental consequences. The destruction of rural communities is thus another factor that is taken into account by those considering whether animal production systems are sustainable (see Chapter 6). A central aim of the EU's Common Agricultural Policy was to preserve rural communities and to reduce the number of people who leave country areas and move to large cities, thus increasing their size. That policy has been successful in minimizing such movement and some U.S. government agricultural policies that prevented the prices of certain agricultural goods from falling to a low level have had this effect. In many other countries, in contrast, cities have become much bigger and rural communities have declined or disappeared. Similar destruction of rural communities has occurred where the number of people employed on farms has been drastically reduced because machinery, often with high consumption of energy, has replaced the people. When all of the real costs of agriculture are evaluated properly, major changes will ensue. Areas for change include the welfare of agricultural animals, energy usage, conservation of natural environments, the welfare of human consumers and agricultural workers, and the preservation of rural communities. Sustainable agriculture is the only way forward.

REFERENCES

Aland, A. and Madec, F., Eds. 2009. *Sustainable Animal Production*. Wageningen: Wageningen Academic Publishers.
Bennett, R.M., Ed. 1994. *Valuing Farm Animal Welfare*. Reading, PA: University of Reading.
Bennett, R.M., Anderson, J., and Blaney, R.J.P. 2002. Moral intensity and willingness to pay concerning farm animal welfare issues and the implications for agricultural policy. *Journal of Agricultural and Environmental Ethics* 15: 187–202.
Broom, D.M. 1986 Indicators of poor welfare. *British Veterinary Journal* 142: 524–526.
Broom, D.M. 1988. The scientific assessment of animal welfare. *Applied Animal Behavior Science* 20: 5–19.
Broom, D.M. 1991a. Animal welfare: Concepts and measurement. *Journal of Animal Science* 69: 4167–4175.
Broom, D.M. 1991b. Assessing welfare and suffering. *Behavioral Processes* 25:117–123.
Broom, D.M. 1998. Welfare, stress and the evolution of feelings. *Advances in the Study of Behavior* 27: 371–403.
Broom, D.M. 1999. Welfare and how it is affected by regulation. In: *Regulation of Animal Production in Europe*, M. Kunisch and H. Ekkel, Eds. Darmstadt: K.T.B.L., pp. 51–57.
Broom, D.M. 2001. The use of the concept Animal Welfare in European conventions, regulations and directives. *Food Chain 2001*. Uppsala: SLU Services. pp. 148–151.
Broom, D.M. 2002. Does present legislation help animal welfare? *Landbauforschung Völkenrode* 227: 63–69.
Broom, D.M. 2003. *The Evolution of Morality and Religion*. Cambridge: Cambridge University Press.
Broom, D.M. 2006. The evolution of morality. *Applied Animal Behavior Science* 100: 20–28.
Broom, D.M. 2007. Traceability of food and animals in relation to animal welfare. In: *Proceedings of the Second International Conference on Traceability of Agricultural Products*. EMBRAPA: Brasilia.
Broom, D.M. 2008. Welfare assessment and relevant ethical decisions: key concepts. *Annual Review of Biomedical Science* 10: T79–T90.

Broom, D.M. 2009. Animal welfare and legislation. In: *Welfare of Production Animals: Assessment and Management of Risks*, F. Smulders and B.O. Algers, Eds. Wageningen: Wageningen Pers., pp. 341–354.

Broom, D.M. 2010. Animal Welfare: An aspect of care, sustainability, and food quality required by the public. *Journal of Veterinary Medical Education*, 37, 83–88.

Broom, D.M. and Fraser, A.F. 2007. *Domestic Animal Behavior and Welfare,* 4th ed. Wallingford: CABI.

Broom, D.M. and Johnson, K.G. 2000. *Stress and Animal Welfare*. Dordrecht: Kluwer/Springer (first published Chapman and Hall 1993).

Dawkins, M.S. 1980. *Animal Suffering: The Science of Animal Welfare*. London: Chapman and Hall.

Dawkins, M.S. 1990. From an animal's point of view: Motivation, fitness and animal welfare. *Behavioral and Brain Sciences*, 13, 1–61.

Duncan, I.J.H. 1981. Animal rights — animal welfare, a scientist's assessment. *Poultry Science* 60: 489–499.

Duncan, I.J.H. and Petherick, J.C. 1991. The implications of cognitive processes for animal welfare. *Journal of Animal Science* 69: 5017–5022.

Fraser, D. 2008. *Understanding Animal Welfare: The Science in Its Cultural Context*. Oxford: Wiley Blackwell.

Murgeitio, E., Calle, Z., Uribe, F., Calle, A., and Solorio, B. 2010. Native trees and shrubs for the productive rehabilitation of tropical cattle ranching lands. *Forest Ecology and Management* doi:10.1016/j.foreco.2010.09.027.

Murgueitio, E., Cuartas C., and Naranjo J., Eds. 2009. *Ganadería del Futuro: Investigación para el Desarrollo*. Segunda edición. Cali, Colombia: Fundación CIPAV.

Pain, D.J., Bowden, C.G.R., Cunningham, A.A., Cuthbert, R., Das, D., Gilbert, M., Jakati, R.D., Jhala, Y., Khan, A.A., Naidoo, V., Oaks, J.L., Parry-Jones, J., Prakash, V., Rahmani, A., Ranade, S.P., Baral, H.S., Sanacha, K.R., Saravanan, S., Shah, N., Swan, G., Swarup, D., Taggart, M.A., Watson, R.T., Virani, M.Z., Wolter, K., and Green, R. 2008. The race to prevent the extinction of South Asian vultures. *Bird Conservation International,* 18: S30–S48.

Prakash, V., Green, R.E., Pain, D.J., Ranade, S.P., Saravanan, S., Prakash, N., Venkitachalam, R., Cuthbert, R., Rahmani, A.R., and Cunningham, A.A. 2007. Recent changes in populations of resident *Gyps* vultures in India. *Journal of the Bombay Natural History Society* 104: 129–135.

Webster, J. 1994. *Animal Welfare: A Cool Eye towards Eden*. Oxford: Blackwell.

THIRD VIEWPOINT: UNDERSTANDING ANIMAL WELFARE FROM A RESEARCH SCIENTIST'S PERSPECTIVE*

David Fraser

INTRODUCTION

The treatment of animals has been a topic of ethical concern since classical times (Sorabji, 1993) and showed a major resurgence during the 1700s and 1800s in Europe and the English-speaking countries (Harwood, 1928; Radford, 2001). In the 1900s, during the span of the two World Wars and the Great Depression, concern about the welfare of animals seemed to take a back seat. However, as human prosperity and security returned in the 1950s, concern about animals began to regain its former prominence. Both the United States and Canada passed their first humane slaughter legislation in 1958 and 1960, respectively, and some jurisdictions added humane animal transport requirements soon after.

As long as the focus was on slaughter and transport, the nature of the concern seemed clear enough: To protect animals from avoidable pain, distress, and injury after they left the safe confines of the farm. Beginning in the 1960s, however, attention also turned to the relatively new "confinement" systems for raising farm animals, and here the nature of the concerns was less easy to define. As debate over these systems developed, it became apparent that different people were raising somewhat different issues, all under the umbrella term of "animal welfare."

---

* This section brings together material from several of my previous publications, especially my book *Understanding Animal Welfare: The Science in its Cultural Context* (Fraser, 2008), which gives a much more detailed treatment of the issues. I am grateful to Wiley-Blackwell (Oxford) and the Universities Federation for Animal Welfare for allowing me to re-work some of that material here.

THREE VIEWS OF ANIMAL WELFARE

The first major criticism of confinement production systems came in the book *Animal Machines*, by the English animal advocate Ruth Harrison. She described cages for laying hens and crates for veal calves, and she claimed that these highly restrictive systems caused animals to lead miserable and unhealthy lives. She asked:

> How far have we the right to take our domination of the animal world? Have we the right to rob them of all pleasure in life simply to make more money more quickly out of their carcasses? (Harrison, 1964)

Later, in *Animal Liberation*, Australian philosopher Peter Singer based his criticism of animal production on the principle that actions should be judged right or wrong based on the pain or pleasure that they cause, and he claimed:

> There can be no moral justification for regarding the pain (or pleasure) that animals feel as less important than the same amount of pain (or pleasure) felt by humans. (Singer, 1990)

In these and many other criticisms of modern animal production, concerns centered around words like "pleasure," "pain," "suffering," and "happiness." There is no simple English word to capture this class of concepts. They are sometimes called "feelings,"but the word seems too insubstantial for states like pain and suffering. They are sometimes called "emotions," but emotions do not include states like hunger and thirst. Perhaps the most accurate (if rather technical) term is "affective states," a term that refers to emotions and other feelings that are experienced as either pleasant or unpleasant rather than hedonically neutral.

In discussing confinement systems, however, some people put the main emphasis elsewhere. A British committee that was formed to evaluate the issues raised by Ruth Harrison concluded:

> In principle we disapprove of a degree of confinement of an animal which necessarily frustrates most of the major activities which make up its natural behaviour. (Brambell, 1965)

Astrid Lindgren, the famous author of the Pippi Longstocking stories and a driving force behind animal welfare reform in Sweden, proposed:

> Let [farm animals] see the sun just once, get away from the murderous roar of the fans. Let them get to breathe fresh air for once, instead of manure gas. (Anonymous, 1989)

American philosopher Bernard Rollin (1993) insisted that we need:

> ... a much increased concept of welfare. Not only will welfare mean control of pain and suffering, it will also entail nurturing and fulfilment of the animals' natures.

In these quotations, although affective states were often involved implicitly or explicitly, the central concern was for a degree of "naturalness" in the lives of animals: That animals should be able to perform their natural behavior, that there should be natural elements in their environment, and that we should respect the "nature" of the animals themselves.

All of the previous quotations reflected the views of social critics and philosophers, but when farmers and veterinarians engaged in the debate, they brought a different focus. For example, one veterinarian defended the early confinement systems this way:

> My experience has been that ... by-and-large the standard of welfare among animals kept in the so called 'intensive' systems is higher. On balance I feel that the animal is better cared for; it is certainly much freer from disease and attack by its mates; it receives much better attention from the attendants, is sure of shelter and bedding and a reasonable amount of good food and water. (Taylor, 1972)

On the other hand, as the veterinary educator David Sainsbury (1986) put it:

Good health is the birthright of every animal that we rear, whether intensively or otherwise.

Here the primary emphasis is on the traditional concerns of veterinarians and animal producers that animals should have freedom from disease and injury, plus food, water, shelter, and other necessities of life—concerns that we might sum up as the basic health and functioning of the animals.

In these various quotations, then, we see a variety of concerns that can be grouped roughly under three broad headings: (1) the affective states of animals, (2) the ability of animals to lead reasonably "natural" lives, and (3) basic health and functioning.

These are not, of course, completely separate or mutually exclusive. Allowing a pig to wallow in mud on a hot day improves its welfare because it can use its natural cooling behavior (a natural living criterion), because it will feel more comfortable (an affective state criterion), and because its bodily processes will be less disrupted by heat stress (a basic health criterion).

Nonetheless, the different concerns are sufficiently independent that the pursuit of any one does not necessarily improve animal welfare as judged by the others. An intensive pig producer may feel that the most important elements of animal welfare are basic health and functioning as reflected by neonatal survival, longevity of sows, rapid growth, and low incidence of disease. For such a person, a well-run, high-health confinement unit might seem to provide the best welfare for pigs. An organic pig producer, in contrast, may feel that for pigs to have a good life, it is most important that they are free to live in fresh air and sunlight with ample space to roam and socialize. For such a person, a free-range system is far better for animal welfare than any confinement unit is, even if parasites are not as well controlled and rates of growth are lower. An animal protectionist might attach particular importance to affective states and not be too concerned whether pigs are indoors or outdoors, so long as fear, pain, and hunger are minimized. Thus, different beliefs about what is important for animals to have a good life can lead to very different conclusions.

These disagreements are not, of course, disagreements about facts. The intensive producer and the organic producer may agree on factual matters such as the rate of mortality in a herd or the concentration of ammonia in the air. Their disagreement is about values—about what they consider most important for animals to have good lives.

The situation can perhaps be captured by a simple Venn diagram (Figure 5.1), which serves to summarize three points: (1) most of the concerns that people express about animal welfare can be grouped roughly under three main headings; (2) these involve considerable but incomplete overlap;

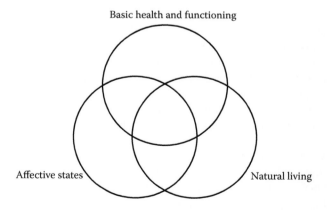

FIGURE 5.1   Three conceptions of animal welfare. (Adapted from Appleby, M.C. 1999. *What Should We Do about Animal Welfare?* Oxford: Blackwell Science; and Lund, V. 2006. Natural living — a precondition for animal welfare in organic farming. *Livestock Science* 100: 71–83.)

and (3) the pursuit of animal welfare as defined by any one criterion does not guarantee a high level of welfare as judged by the others.

ANIMAL WELFARE AND SCIENCE

When these differences began to emerge in the debate about confinement production systems, many people looked to scientific research as the way to decide among the different, value-based interpretations of animal welfare and thus turn the assessment of animal welfare into an objective, value-free scientific process. What actually happened, however, proved to be much more interesting.

Some scientists focused on the basic health and functioning of animals as a basis for assessing and improving animal welfare. In one classic example, Ragnar Tauson and co-workers improved the welfare of laying hens by studying the basic health of birds in cages of different types, and then developing cage designs that would prevent the various health problems they observed. The scientists found that the birds developed foot lesions if the floor was too steeply sloped, and neck lesions if the feed trough was too deep and installed too high for comfortable access. There was often feather damage that could be reduced by using solid side partitions and overgrown claws that could be prevented by installing abrasive strips on the cage floor. Thus, just by focusing on injuries it was possible to make large improvements in animal welfare and, coincidentally, in the productivity of the flock. These results formed the basis of the early animal welfare standards for cage design in Sweden and later in the European Union (Tauson, 1998).

Other scientists tried to improve animal welfare by creating living conditions that were more natural for animals. For example, in an effort to design better housing for pigs, Alex Stolba and David Wood-Gush began by observing pigs that had been released in a hilly, wooded area. They found that the pigs showed certain characteristic types of behavior: They rooted in the soil; they exercised their neck muscles by levering against fallen logs; they built nests in secluded areas before giving birth; and they used dunging areas well removed from their resting areas. The scientists then designed a complex commercial pen that allowed the animals to behave in these ways. It included an area with peat moss for rooting, logs for levering, a separate dunging area, and secluded areas where a sow could be enclosed to build a nest and farrow (Stolba and David Wood-Gush, 1984). The authors claimed that the animals' welfare was significantly improved by the complex pen; however, because some aspects of basic health (especially neonatal survival) were not as good in this system as in well run confinement systems, some people disagreed with that conclusion.

In less radical approaches, scientists have incorporated simple elements of natural behavior into existing rearing systems. On many commercial dairy farms, calves are separated from their mothers on the first day after birth, and are then fed milk by bucket, usually twice per day. This, of course, is highly unnatural. Under natural conditions, calves would stay close to the cow for the first two weeks, and would consume many small meals per day by sucking rather than drinking. Although normally it is not feasible to leave calves with the cow on a diary farm, feeding systems can still be made to correspond more closely to the animals' natural behavior. First, if the calves suck from an artificial teat rather than drinking from a pail, the sucking action seems to stimulate certain digestive processes more effectively (de Passillé, Christopherson, and Rushen, 1993). Second, if the teat system allows the calves to feed with a more natural frequency and meal size, then they can gain substantially more weight than calves fed twice daily by bucket (Appleby, Weary, and Chua, 2001).

In other cases, scientists have used animal welfare research to reduce unpleasant affective states in animals. Many dairy calves are subjected to "hot-iron disbudding." This involves the use of a ring-shaped iron heated to 600°C and pressed against the head of the calf to burn through the nerves and blood vessels that would allow the horn-bud to develop. In some countries, this procedure is commonly done without any form of pain management. A research group in New Zealand used levels of cortisol (a stress-related hormone) in the blood as an indicator of the pain caused by disbudding. They found that disbudding is followed immediately by a large increase in cortisol, but that the reaction is blocked if calves are treated with a local anesthetic to freeze the area first. However, the treated calves showed a later rise in cortisol level, several hours after the disbudding, probably because the injury remained inflamed and painful when the anesthetic had worn off. This later rise in cortisol could be eliminated by giving the calves an analgesic. Thus, the research showed that management of the pain of disbudding requires both a local anesthetic and a longer-acting analgesic (Stafford and Mellor, 2005).

All of the approaches described previously—some designed to improve basic health, others incorporating natural behavior, and others focused on affective states—have been useful for identifying and solving animal welfare problems. However, rather than the science providing an objective means of arbitrating among the different views of animal welfare, the different views of animal welfare were actually adopted by the scientists as the rationale for their scientific work. In fact, the different views of animal welfare enriched the science by providing a wide and complementary range of ways in which animal management could be improved, often with benefits to animal producers as well as to the animals.

Clarifying and Applying The Views

If the science has not arbitrated among the different views of animal welfare, it has nonetheless done a great deal to clarify the different views and put them into practice.

For one thing, science has helped clarify how "naturalness" relates to animal welfare. Clearly, modern methods of keeping animals raise concerns because they are so unnatural, but how should we determine what is natural for these animals? For example, because sows living outdoors typically wean their young at three to four months of age, critics often assume that "natural" weaning means delaying weaning until this late age, and that sows and litters should be left together throughout this time. In fact, research shows that starting about 10 days after farrowing, many sows choose to spend less and less time with their young and thus force the offspring to start using a solid diet. Hence, although removing the piglets from the sow at two to three weeks is not natural, leaving them confined together in a pen for many weeks is not natural either. On this basis, "get-away" farrowing systems have been designed that allow sows to initiate the weaning process and better prepare the young for transition to solid food (Pajor et al., 1999).

One problem in invoking natural behavior to improve animal welfare is that natural behavior falls, very roughly, into two types: Behavior that animals generally want to do, such as eating and playing, and behavior that animals generally do not want to do, such as shivering in the cold and fleeing from predators. When we encounter a type of natural behavior, how do we know in which category it belongs?

One way is simply to ask the animals. Hens, for example, can be trained to perform "instrumental" tasks, such as pecking a key or pushing against a weighted door, for rewards such as food or the opportunity to perform such natural behavior as dust bathing or roosting on a perch. By determining the amount of work a bird will do to obtain a given reward, we can better understand the nature and strength of their motivation (Duncan, 1992; Dawkins, 1998). Using such methods, it has been shown that hens are motivated to obtain a modest space allowance (somewhat more than is provided in standard commercial cages), a perch where they can roost at night, a nest box where they can retreat to lay eggs, and litter for dust bathing and feather care. Based on such research, the European Union will soon require that caged hens have some form of "furnished" environment with 750 cm^2 of floor space per bird, plus a perch, a nest-box, and litter (Appleby, 2003).

Animal welfare science has also provided many ways of using research to understand better the affective states of animals. As one example, Francis Colpaert and co-workers have done many studies in which rats had the opportunity to self-administer analgesics. In one case, they gave arthritic and non-arthritic rats a choice of drinking from two water bottles, one of which contained sweetened water and the other a dilute but unpalatable solution of the opiate analgesic fentanyl. Healthy rats consumed very little of the fentanyl, but arthritic rats consumed substantial amounts, and the time course of self-administration corresponded with changes in the severity of the arthritis. Based on this and other lines of evidence, Colpaert et al. (2001) concluded that self-administration of fentanyl provides an objective indicator of chronic pain in rats.

Finally, science has helped to clarify the relationship between health, productivity, and animal welfare. It is uncontroversial to say that preventing disease and injury is fundamental to animal welfare, but some people have made much bolder claims. Some have proposed, for example, that "suffering of any kind is reflected by a corresponding fall in productivity" (Brambell, 1965, pp. 10–11), and that "the goal of maximum profitability pursued by animal producers (and others) leads automatically to improved welfare" (CAST, 1981, p. 1). Scientific analysis has shown the need for caution over such claims. For example, modern hens have been bred so strongly for egg production that they will mobilize calcium from their bones to create eggshells. This can lead to significant weakness in the leg bones and a high frequency of broken bones when the birds are removed from their cages for slaughter (Knowles, and Wilkins, 1998). Genetic selection of beef cattle for very large muscles has produced certain breeds whose carcasses have high commercial value, but these breeds are prone to difficult calving and poorer calf survival, and some animals react to heat stress with an excessive build-up of lactic acid in the muscles, sometimes to the point of paralysis (Gregory, 1998). Many dairy cows are bred and fed for very high levels of milk production, but this is associated with a high incidence of certain diseases and short life span (Sandøe et al., 1999). Hence, arguments linking productivity and animal welfare need to be treated with caution, especially if genetics, diet, or hormones have been manipulated in ways that enhance one aspect of functioning to the detriment of others.

Arguments linking animal welfare and profitability are especially suspect. Profit requires a certain level of productivity, but profit can also be increased by limiting input costs. Reducing space allowance, staff time, bedding, veterinary care, and other amenities can help to reduce costs; and even if these cutbacks reduce productivity to some extent, the net result may still be greater profit. A striking example was provided by Adams and Craig (1985), who analyzed how space allowance for hens in cages is associated with different levels of productivity and profit. Their analysis showed that if egg prices are high and feed costs are low, profit could often be increased by adding extra birds to a facility so that crowding is severe, even though the death rate is increased and the birds' individual rate of egg production declines.

As we see in these examples, research and thoughtful scientific analysis can do a great deal to improve our understanding of animal welfare. Specifically, research can show what elements of natural behavior are important to the animals themselves; research can put the affective states of animals on a scientific footing so that we do not just project human emotional reactions onto other species; and scientific thinking can clarify the complex relationship between animal welfare, health, and productivity.

Concluding Remarks

The idea of applying science to a value-based concept may sound strange to some scientists. Surely (they might argue) when scientists confront a new term—whether it be metabolic rate, feed efficiency, or animal welfare—they should first agree on how to define the term, and then they can measure it in a purely objective and value-free way.

In fact, many of the concepts studied by scientists incorporate values in a fundamental way. "Food safety," "environmental integrity," "agricultural sustainability," "mental health," "animal welfare"—each of these topics contains a word (safety, integrity, etc.) that invokes notions of better or worse. To say that safety or integrity has increased implies not simply a change, but a change for the better. We might call these "evaluative concepts" (Fraser, 1999). We can certainly use scientific methods in the assessment of evaluative concepts, but the empirical work is underlain by value-based presuppositions about what constitutes a better or worse situation.

Animal welfare is also an "everyday" concept. Unlike concepts such as atomic weight and metabolic rate, which arose in science and took their meaning from science, many evaluative concepts arose in everyday language and acquired a meaning (or meanings) in everyday life before scientists began paying attention to them. When society calls on science to help resolve questions about animal welfare, food safety, or other topics that are the subject of everyday concern and policy-making, the scientists need to understand and respect the everyday meanings of the concepts that they study. If they do not—if, for example, they try to give the term a new, technical meaning that does not correspond to its everyday meaning—then their conclusions may be irrelevant or (worse yet) misleading to the very issues that the scientists were trying to address.

SUMMARY

Science can make major contributions to understanding and improving animal welfare, and to finding constructive solutions to animal welfare debates; but in defining animal welfare and in selecting corresponding research methods, scientists need to be attentive to the everyday meaning of the term and to the underlying value-based presuppositions.

REFERENCES

Adams, A.W. and Craig, J.V. 1985. Effect of crowding and cage shape on productivity and profitability of caged layers: A survey. *Poultry Science* 64: 238–242.

Anonymous. 1989. How Astrid Lindgren Achieved Enactment of the 1988 Law Protecting Farm Animals in Sweden. Washington: Animal Welfare Institute.

Appleby, M.C. 1999. *What Should We Do about Animal Welfare?* Oxford: Blackwell Science.

Appleby, M.C. 2003. The EU ban on battery cages: History and prospects. In: *The State of the Animals II*, D.J. Salem and A.N. Rowan, Eds. Washington: Humane Society of the United States, pp. 159–174.

Appleby, M.C., Weary, D.M., and Chua, B. 2001. Performance and feeding behaviour of calves on ad libitum milk from artificial teats. *Applied Animal Behaviour Science* 74: 191–201.

Brambell, F.W.R. (chairman) 1965. *Report of the Technical Committee to Enquire into the Welfare of Animals Kept under Intensive Livestock Husbandry Systems*. London: Her Majesty's Stationery Office.

CAST. 1981. *Scientific Aspects of the Welfare of Food Animals*. Report 91. Ames, IA: Council for Agricultural Science and Technology (CAST).

Colpaert, F.C., Tarayre, J.P., Alliaga, M., Slot, L.A.B., Attal, N., and Koek, W. 2001. Opiate self-administration as a measure of chronic nociceptive pain in arthritic rats. *Pain* 91: 33–45.

Dawkins, M.S. 1998. Evolution and animal welfare. *Quarterly Review of Biology* 73: 305–328.

de Passillé, A.M.B., Christopherson, R., and Rushen, J. 1993. Nonnutritive sucking by the calf and postprandial secretion of insulin, CCK, and gastrin. *Physiology & Behaviour* 54: 1069–1073.

Duncan, I.J.H. 1992. Measuring preferences and the strength of preferences. *Poultry Science* 71: 658–663.

Fraser, D. 1999. Animal ethics and animal welfare science: Bridging the two cultures. *Applied Animal Behaviour Science* 65: 171–189.

Fraser, D. 2008. *Understanding Animal Welfare: The Science in its Cultural Context*. Oxford: Wiley-Blackwell.

Gregory, N.G. 1998. *Animal Welfare and Meat Science*. Wallingford: CABI Publishing.

Harrison, R. 1964. *Animal Machines*. London: Vincent Stuart Ltd.

Harwood, D. 1928. *Love for Animals and How it Developed in Great Britain*. Republished in 2002 as *Dix Harwood's Love for Animals and How it Developed in Great Britain (1928)*. R. Preece and D. Fraser, Eds. Lewiston: Edwin Mellen Press.

Knowles, T.G. and Wilkins, L.J. 1998. The problem of broken bones during the handling of laying hens — a review. *Poultry Science* 77: 1798–1802.

Lund, V. 2006. Natural living — a precondition for animal welfare in organic farming. *Livestock Science* 100: 71–83.

Pajor, E.A., Weary, D.M., Fraser, D., and Kramer, D.L. 1999. Alternative housing for sows and litters: 1. Effects of sow-controlled housing on responses to weaning. *Applied Animal Behaviour Science* 65: 105–121.

Radford, M. 2001. *Animal Welfare Law in Britain: Regulation and Responsibility*. Oxford: Oxford University Press.

Rollin, B.E. 1993. Animal welfare, science, and value. *Journal of Agricultural and Environmental Ethics* 6 (Suppl. 2): 44–50.

Sainsbury, D. 1986. *Farm Animal Welfare. Cattle, Pigs and Poultry*. London: Collins.

Sandøe, P., Nielsen, B.L., Christensen, L.G., and Sørensen, P. 1999. Staying good while playing god — the ethics of breeding farm animals. *Animal Welfare* 8: 313–328.

Singer, P. 1990. *Animal Liberation,* 2nd ed. New York: Avon Books.

Sorabji, R. 1993. *Animal Minds and Human Morals: The Origins of the Western Debate*. Ithaca, NY: Cornell University Press.

Stolba, A. and Wood-Gush, D.G.M. 1984. The identification of behavioural key features and their incorporation into a housing design for pigs. *Annales de Recherches Vétérinaires* 15: 287–298.

Stafford, K.J. and Mellor, D.J. 2005. Dehorning and disbudding distress and its alleviation in calves. *Veterinary Journal* 169: 337–349.

Tauson, R. 1998. Health and production in improved cage designs. *Poultry Science* 77: 1820–1827.

Taylor, G.B. 1972. One man's philosophy of welfare. *Veterinary Record* 91: 426–428.

FOURTH VIEWPOINT: UNDERSTANDING ANIMAL WELFARE FROM A VETERINARIAN'S PERSPECTIVE

Gail C. Golab

INTRODUCTION

In the United States, veterinarians take an oath (AVMA, 2010) to provide for their animal patients, while ensuring that the interests of society are met through responsible animal use.

> Being admitted to the profession of veterinary medicine, I solemnly swear to use my scientific knowledge and skills for the benefit of society through the protection of animal health and welfare, the prevention and relief of animal suffering, the conservation of animal resources, the promotion of public health, and the advancement of medical knowledge.
>
> I will practice my profession conscientiously, with dignity, and in keeping with the principles of veterinary medical ethics.
>
> I accept as a lifelong obligation the continual improvement of my professional knowledge and competence.

Similar obligations exist and similar promises are made by veterinarians around the world (Hewson, 2006).

In serving both animals and society, veterinarians bring a unique skill set to the table. First, most veterinarians enter the profession because of their empathy for animals and their desire that they are cared for properly (Sprecher, 2004; Serpell, 2005). Empathy serves as a starting point in the examination of animal use and care. It leads to fundamental questions as to whether specific uses of animals are necessary and appropriate, and whether related animal care practices (e.g., genetic selection and manipulations, housing, handling, physical alterations) are important to facilitating that use. If that is so, are they being performed with due regard for the health and other welfare needs of individual animals and animal populations?

Second, during their training, veterinarians are provided with strong science-based knowledge about animal health and husbandry, and are schooled in the technical and practical application of that information. This combined skill set helps ensure that recommended approaches to animal care are likely to improve animal health and other aspects of animal welfare and can be realistically implemented.

Third, direct practitioner access to animals, the environments in which they are housed, and the people who own and care for them allows observation of what is actually occurring and provides a mechanism whereby veterinarians can actively encourage and demonstrate appropriate animal care. Veterinarians also interact regularly with the multiple individuals indirectly responsible for the welfare of those animals, including other scientists, policymakers in governmental agencies (local, state/territory, national, international), advocates in the animal agricultural industries and nongovernmental organizations, and the public.

Finally, veterinarians have tremendous credibility. A 2006 poll conducted in the United States on professional honesty and ethics ranked veterinarians third among 23 types of professionals (Gallup, 2006). Degree of credibility may vary by society, over time, and be affected by animal-related events; however, in general, veterinarians appear to be well respected. Credibility means that recommendations made by veterinarians are likely to be taken seriously.

Together, these attributes make veterinarians valuable advocates in assuring good animal welfare.

WHAT IS GOOD WELFARE?

There is general agreement that good welfare means satisfying an animal's needs, but when asked whether a particular situation or condition in which an animal finds itself is welfare-friendly, respondents, including veterinarians, may have different views.

Consider the question of whether the welfare of laying hens is better when they are kept in cages, barns, or allowed to range freely in a field (LayWel, 2006). In cages, hens have easy access to feed and water, individual birds are easily observed, aggressive interactions are infrequent and cannibalism is minimal, and their eggs are protected and easily collected. However, in conventional cages movement is restricted, and nest boxes and litter for dust bathing (both of which support the behavioral aspects of animal welfare) generally are not provided. Laying hens raised in barns most often have access to nest boxes and litter for dust bathing, but aggression, cannibalism, and flightiness are other behavioral characteristics of that environment, and feed and water are less easily monitored. Free-range systems allow great freedom of movement, usually include enclosures for sleeping and nesting, and natural substrates are readily available that provide multiple opportunities for expression of natural behaviors. On the other hand, laying hens in free-range systems have increased exposure to adverse weather conditions, pests, and predators (see Chapters 4 and 8 for further discussion).

Given these trade-offs, which of the three systems described does a veterinarian recommend to best ensure the hens' welfare? Would that veterinarian's colleague in the next town or state choose the same system? Are the veterinarian's recommendations likely to be consistent with client preferences? What about the expectations of the public (which may or may not be well-informed)? As health professionals, how veterinarians approach animal welfare will largely reflect their knowledge of the science behind animal care and use practices and their practical experience in applying that scientific knowledge; however, it will also depend upon their personal values, the needs and preferences of their clients, and various social influences. Veterinarians are challenged to assist in the decision-making process, while recognizing that even they are not immune to personal prejudices and external influences when making animal welfare decisions.

Personal Values

With respect to the laying hen example provided previously, many veterinarians are most comfortable with hens being kept in cages. That is because veterinarians (and many other biological scientists and producers) tend to emphasize measures of health, growth, and productivity in their evaluation of an animal's welfare. The veterinarian recognizes that keeping hens in cages allows better monitoring and control of disease, minimizes the risk of attack by the hen's conspecifics, protects the hen from predators, and ensures consistent provision of food and water. In other words, the veterinarian concludes that the hen is in a good state of welfare because its health, safety, and physical needs are met.

However, for others (including behavioral and social scientists, retailers, members of the public, and even colleagues of the veterinarians, scientists, and producers mentioned previously), the answer may not be so clear-cut. Fraser et al. (1997) suggested that views on animal welfare generally fall into three categories: Individuals who emphasize basic health and function of the body; those who are most concerned with how an animal "feels" (i.e., its psychological or affective states, such as pain, suffering, or contentment); and those who emphasize the animal's ability to lead a reasonably natural life and perform behaviors in which it might normally engage. None of these views can be classified as being inherently right or wrong, nor are they mutually exclusive. Rather, they represent different areas of focus or emphasis. Physical and health scientists are generally most comfortable with the functional view of animal welfare, animal behaviorists and psychologists tend to equate good animal welfare with positive affective states, and many members of the public, particularly those who rebel against what they perceive to be the wrongs of an industrialized society, look for components of natural living.

Sometimes the various views of what constitutes good animal welfare go hand-in-hand. For example, allowing a hen to nest may help it protect the integrity of its eggs (a functional criterion), may provide some comfort (an affective state criterion), and permit it to perform a natural behavior (a natural living criterion). Other times the various views conflict. For example, an owner feeding his or her dog treats on a regular basis may result in the dog having a positive psychological response and, depending on how the treats are provided, may meet its needs for exploratory or play behavior. However, too many treats can also cause the dog to become obese. In considering the welfare of animals, and through experience gained in practice, veterinarians soon learn the importance of balance in satisfying both their physical and psychological needs.

Experiences and Influences

While their patients are animals, veterinarians provide services for a human clientele. As such, what veterinarians recommend will be affected by social norms, and the relationship between people and their animals has changed dramatically over the past several decades.

Since the 1950s, there has been a shift in the American family unit from the nuclear family (represented by a mother, father, and children with extended family often living nearby) to families that may comprise younger or older couples with no children in the household, single parents with children, single persons, or same-sex partners, with or without children. Grandparents, parents, children, aunts, uncles, nieces, and nephews are often spread across the country. Both mothers and fathers often work outside the home, and latchkey children are the norm rather than the exception. Substantial traditional social support has been removed in the process and pets have filled the void as dependable companions. Higher per capita incomes have allowed owners to treat their animal companions more and more like the human companions they have replaced and to perceive such treatment as normal and appropriate. Almost simultaneously, direct experiences with animals as sources of food and fiber (i.e., functional animal uses) have been reduced. Since the 1950s, the United States has seen a dramatic trend toward urbanization (USDA, 1995) with fewer than 2% of the American public currently residing on farms. Together these factors put the American public in the position of viewing all animals and expectations for their care with the same spectacles they apply to the family dog, cat, or bird.

While the structure of families has changed, businesses have changed as well. After World War II, the United States saw a market-driven intensification of almost all industries, including those using animals (Colyer et al., 2001). Profit margins narrowed as production costs (especially wages) increased and prices dropped. Economies of scale and type were discovered and translated to animal production and care. A business culture emphasizing efficiency emerged, leading to increased specialization and economy of scale (e.g., farms became larger and shifted to a single species and, later, to a single phase of production), contract operations, and selection for animal characteristics (e.g., increased muscle mass, hardiness, susceptibility or resistance to particular diseases [as beneficial to their particular use]) that maximize return on food, housing, and care investments. Animal care interests correspondingly moved from a focus on the health of individual animals to an emphasis on the health of the herd and the quality and quantity of the final product.

Most members of the American public recognize, accept, and support the need to use animals as sources of food and fiber; however, the picture of animals as "commodities," with an emphasis on herd health and production, does present conflicts with their vision of animals as "family members," with its corresponding emphasis on the individual. Attempts to resolve this ideological conflict have resulted in (1) closer scrutiny of traditional animal use and care practices; (2) increasing prominence and public support of existing nongovernmental organizations focused on ensuring animal welfare, as well as the emergence of new ones; (3) retailers and their suppliers recognizing that members of the public can vote with their pocketbooks and acquiescing to their demands by creating business centers focused on issues of social responsibility, including animal welfare; and (4) governmental regulations and legal obligations directed toward aspects of animal use and abuse that the public finds most troubling. Because of their recognized scientific and practical expertise, as well as their regular contact with various stakeholders, veterinarians often find themselves in the challenging position of trying to bridge gaps between those with conflicting paradigms of animal use and care, while ensuring the needs of animals continue to be met. In the case of animal agriculture, veterinarians must protect the well-being of animals, assist farmers in producing sufficient product in a profitable way, and simultaneously respect the ethical norms of how society expects animals to be used and cared for.

Applying Science

Veterinarians want to believe that decisions about animal care primarily will be based on science. A look at the history of animal welfare decision-making, however, tells us otherwise. Science directed at the needs and wants of animals did not actually play a substantial role in animal welfare decision-making until the 1950s and 1960s, in concert with the publication of *The Principles of Humane Experimental Technique* by Russell and Burch (revised 1992; originally published in 1959) and the report of the Brambell Committee (1965). Concerns about animal welfare, however, have been raised since at least the time of Aristotle and it can be argued that mythological, cultural, and religious histories suggest an even earlier focus.

Science (and scientists) emerged as a player in the animal welfare debate when it was proposed as a possible way to help resolve conflicting perspectives. The strongest growth in animal welfare science has occurred since the mid-1980s, and the field is inherently inter- and multidisciplinary. Peer-reviewed information was initially published in journals of various established fields (e.g., animal science, laboratory animal science, animal behavior, veterinary medicine); more recently, animal welfare science-specific journals have been established.

Today's veterinarian who looks to use science in the evaluation of animal welfare includes multiple parameters to ensure a complete assessment. These parameters include the animal's biologic function (e.g., growth, reproduction, ability to maintain homeostasis), its health (e.g., absence/presence of disease or injury), and its behavior and social functions (e.g., adaptation, emotional states [distress, suffering], cognition/awareness, preferences). His or her assessment may look at what is provided for the animal (also referred to as inputs, resource-based criteria, or engineering criteria) or the effects of these inputs on welfare performance (also referred to as outputs, animal-based criteria, or performance criteria). More recently, animal welfare science and its proponents, including veterinarians, have shifted from an emphasis on easily measurable parameters (e.g., morbidity, mortality, production indices) to asking questions about the animal's perception of its own situation.

Interestingly, the basic parameters identified as being necessary components of a complete science-based animal welfare assessment mirror the views (i.e., function, affective states, natural living) discussed previously. The implication of this, of course, is that any data obtained may be differentially interpreted and emphasized based on these views. Therefore, a critical review and interpretation of the science demands the veterinarian be cognizant of the approach taken by the researcher involved, as well as his or her own views, and consider both during interpretation and during application. Science is almost never value-free or immune to experiential prejudice and animal welfare science and its applications are not exceptions to that truth.

Challenges for Veterinarians

What can pose the biggest challenges for veterinarians in successfully addressing animal welfare and the related concerns of other veterinarians, clients, businesses, policymakers, and the general public? To find out, the author asked 50 influential individuals that question. The individuals included veterinarians and non-veterinarians who worked in private practice, industry, not-for-profit organizations, and governmental service, and whose views on animal welfare were diverse. Their answers were amazingly consistent and relatively easily distilled into the following six challenges for the veterinary profession in addressing animal welfare questions.

Professional Homogeneity

Individuals attracted to veterinary school are generally science-focused, smart, conscientious, compassionate, and fascinated by animals, and are able to work under conditions that can be physically demanding (e.g., handling 1000+-lb cattle) and aesthetically (e.g., blood, animal pain or discomfort, feces/urine) difficult. Training in veterinary school instills knowledge about the various types, uses, and many of the practical realities of working with animals and acquaints these future veterinarians with a variety of owners and expectations. As students, veterinarians are taught to respect species differences and, as they mature in practice, they become very good at evaluating and predicting the responses of animals to various situations.

However, the attributes and training that allow veterinarians to become skilled practitioners can also create some separation from the experiences and expectations of the public. Most members of the public have a perspective reflecting their experience with mostly companion animals and they tend to apply that experience to everything animal-related. Veterinarians' experiences reflect a broader range of animals, uses, and owners, as well as a greater familiarity with animal pain and discomfort, its trade-offs with other stressors (e.g., handling), and the resulting choices that need to be made (e.g., restraint stress versus short-term pain). The result is that veterinarians working with agricultural animals can find themselves defending practices, and even their own activities, which their training and experience tells them are appropriate, but the public sees as questionable, based on analogies the public may draw from how veterinarians approach companion animals. Conversely, these same veterinarians may find themselves urging producers to change long-respected practices, based on new information about animals and their needs, the availability of new drugs and equipment, and the expectations of society for animal use and care. Disconnects in experiences, perspectives, and information are a significant challenge because veterinary medicine is a service industry and reaching satisfactory animal welfare conclusions (particularly for animals) requires that dialog and mutual understanding take place, not only between veterinarians and animal owners, but between veterinarians and a more encompassing public.

Veterinarians in the United States are currently largely Caucasian and middle to upper-middle class. This can create challenges in conveying animal welfare concerns and animal care needs to culturally diverse populations. Such failures in communication create a potential for animal suffering.

Professional Diversity in Service

Today's veterinarians are functionally diverse and different veterinary practice types carry different obligations. Companion animal practitioners focus on individual animals, and advanced medical and surgical procedures are common as pet owners seek health care for their pets that approximates what they seek for themselves. Companion animal owners expect a normal aging process for their pets, accompanied by interventions for treatable conditions, followed by as natural a death as possible. Care decisions are framed by owner attachment and ability to pay, and are less affected by the dollar value of the animal.

In contrast, veterinarians working with animals used to produce food and fiber most often focus on population health. Individual animals may need to be sacrificed for diagnostic purposes or the benefit of the herd or flock. Care decisions are framed by the goal of bringing a product to market and, in this context, a natural death is often a clear failure. Advanced procedures are limited by the market value of the animal, and some procedures traditionally performed by other types of veterinarians may be outsourced to non-veterinarian providers. Many farm animal species, while domesticated, are not as accustomed to handling as those species commonly kept as pets, and decisions made about animal care need to consider the impact of (and ways to ameliorate) that additional stressor, as well as inherent human safety risks associated with working with large, heavy animals.

Equine veterinarians deal with animals used for both pleasure and function. Care decisions are often framed by the horse's use, and return on investment can be a primary driver in the application of advanced procedures. Laboratory animal practitioners care for animals in the context of both individuals and groups. They may be faced with the additional challenge of research protocols that are purposely designed to affect the health and well-being of their patients.

While veterinarians are provided with a broad-based education and exposure to all of these areas of practice, concentrating their efforts in one segment or another will, over time, affect their perspectives and approach to animal care.

Veterinary clients are diverse as well. They may be individual owners (e.g., pet owners, small breeding facilities, or farms), companies or institutions (e.g., large food animal production facilities, research facilities, commercial breeders), governmental agencies (e.g., public health agencies, slaughterhouses, animal control, wildlife refuges), or nongovernmental agencies. Each of these clients has their own expectations for value in veterinary services and their definition of good (or even acceptable) animal welfare. Correspondingly, each may have less familiarity and comfort with the animal use and care paradigms embraced by others and may see different roles for veterinarians and owners in defining and assuring good animal care.

Professional Diversity in Demographics

Demographic changes occurring within the profession during the past 30 years (Brown and Silverman, 1999) have also substantively affected veterinary attitudes toward what is necessary for good animal welfare (Narver, 2007). Fewer students with rural roots are entering the profession (Prince, Andrus, and Gwinner, 2006; Andrus, Prince, and Gwinner, 2006), fewer students are choosing rural veterinary practice as a career (although modest increases appear to have resulted from recent recruitment efforts; Chieffo, Kelly, and Ferguson, 2008), and there has been an increase in the number of second-career entrants, particularly from non-science fields. In addition, the gender shift is dramatic. In 1950, there were 139 female veterinary graduates. By 1985, more than 50% of students attending veterinary schools in the United States were female, and it is estimated that women will comprise 67% of veterinary professionals by 2015 (Brown and Silverman, 1999).

These demographic changes have combined to create more interest in the affective and social components of agricultural animals' welfare. Data show that women focus more on social concerns and relationships (Heath and Lanyon, 1996; Paul and Podberscek, 2000; Hart and Melese-d'Hospital, 1989; Serpell, 2005; de Graaf, 2007) and animal welfare issues involve both. The perspectives of students and new graduates reflect their urban experiences, and the pace of demographic change has only served to increase the speed of the philosophical shift.

Functional and social diversity, not surprisingly, can create (and has created) conflicts among the various segments (e.g., practice types, generations) of the veterinary profession. If veterinarians in the various segments fail to consider the important insights that can be obtained from their colleagues, the result may be different recommendations as to what constitutes appropriate animal use and care. Inconsistent recommendations can give the impression of a profession that is unfocused, confused, and indecisive. In turn, this can reduce the trust of clients and the public and negatively affect the profession's ability to ensure that good decisions are made and appropriate care is delivered to animals.

Diversity of professional experience, however, has also been beneficial. When veterinarians with different experiences and perspectives collaborate, the result is a comprehensive look at animal welfare and recommendations that reflect a wealth of scientific and practical expertise and strike an appropriate balance between the needs of animals and people.

Understanding and Accepting the Role of Science

As mentioned previously, veterinarians are most comfortable when animal care decisions are science-based. Science can be of tremendous value in helping to inform and resolve disputes in animal welfare decision-making. However, science regarding the animal welfare implications of particular animal care practices can be of greater or lesser quality, may not always exist, may be ignored, or may be misrepresented and used selectively (by all sides) in related public policy debates.

While science can determine what type or degree of animal welfare risk exists with regard to a particular animal care practice, it cannot determine what type or degree of risk is acceptable. This social component of decision-making means that if the overwhelming perception is that a particular animal welfare risk is unacceptable (i.e., that doing something is "wrong"), then what the science says can become less relevant for those making the animal use/care decision. That science can be relegated to the back seat when animal welfare decisions are made is a reality that can be difficult for veterinarians to understand and accept.

As scientists, veterinarians are encouraged and trained to approach problems objectively. Unfortunately, veterinarians' efforts to be objective can sometimes give the appearance (or create the reality) of professional detachment. Such detachment is inconsistent with the aura of compassion that the public expects from those who serve as the protectors of animal welfare. As we strive toward science-based care decisions for agricultural animals, veterinarians cannot afford to forget that those looking for advice often do not care how much we know until they know how much we care.

Finally, our engagement with science is sometimes a love/hate relationship. As it was put by one respondent to my informal survey, "We're sometimes afraid to embrace the science because it may have implications for how we practice and what positions we may take as a profession. We make decisions based on their scientific merit, except when we don't like what the science says." More comfort with some animal welfare measures (e.g., physiologic and production indices, health status) than others (e.g., behavior) can be a source of conflict that exacerbates any tendency we may have to pick and choose. Incomplete application of the available science in animal welfare decision-making is in no way unique to veterinarians; there is ample evidence that other scientists, those in the industries and animal advocates, have all been guilty of selective appropriation. As trusted professionals, veterinarians must make a conscious effort to seek out, acknowledge, critically examine, and (if of good quality) embrace information from a multitude of disciplines to ensure that they continue to deliver the best possible recommendations for animal care.

The Value of Multidisciplinary Contributions

A veterinary degree plus compassion goes a long way toward ensuring animal welfare; unfortunately, it does not guarantee perfect knowledge of the subject matter nor does it mean we are the only individuals who can or should make valued contributions. Expertise in animal welfare is inherently multidisciplinary and a lot of specialized information contributes to the overall animal welfare knowledge base and the associated decision-making process. While some aspects of animal welfare (e.g., physical health, disease prevention, and treatment) are comprehensively addressed during veterinary medical education, other aspects (e.g., animal behavior, animal ethics) may not be. The goal should be continued assimilation and consolidation of as much information as possible so that the best decisions can be made.

A single animal welfare decision might take into account the animal's physiologic state, behavioral/social wellness, extent/absence of injury and disease, and adaptive potential, as well as ethical considerations and political and economic realities. Just as veterinarians work to develop special expertise in medicine, surgery, pathology, or epidemiology, they must work to develop expertise in the animal welfare field. Fortunately, courses on animal welfare science and ethics are becoming integral to veterinary curricula and opportunities for continuing education are expanding rapidly.

The Actuary, the Mechanic, or the Pediatrician?

Veterinarians may assume a variety of roles when it comes to animal welfare decision-making and choosing between those roles can be exceedingly difficult. We tend to vary between three approaches: The actuary, the mechanic, and the pediatrician (Rollin, 2006).

When behaving as actuaries, we try to base decisions on measurements and statistics and suggest that if we cannot measure it, we should refrain from making recommendations. Such an approach is clearly science-based and, accordingly, carries with it little outcome and professional risk. However, it also fails to take into account the social reality that if we do not see fit to make recommendations in the absence of irrefutable evidence, someone else will, and perhaps from a less knowledgeable and experienced perspective.

When we act like mechanics, we identify animal welfare problems and communicate our concerns and recommendations, but ultimately acquiesce to do what those "in charge" want, irrespective of what may be best for our patients. The ultimate risk resulting from this approach lies with the animal (or society, if the resulting animal care approach is deemed unacceptable), but we assign responsibility for that risk to another.

When we behave like pediatricians, we act like the knowledgeable advocate for the patient. This tends to result in the best animal welfare decision-making, at least from the perspective of the animal, but it also presents the greatest risk for the veterinarian because it makes him or her subject to owner and societal criticisms.

Some veterinarians believe their role should be limited to providing sound scientific information about what they can measure and that advocacy (and potentially decisions) should be left to others. Other (probably most) veterinarians agree that an appropriate role for the profession in animal welfare decision-making means that, most of the time, veterinarians need to behave like pediatricians.

Behaving like a pediatrician can be difficult when we are uncomfortable with the subject matter (due to either limitations in our knowledge base or the amorphous nature of the associated questions). We may also be put off by potentially aggressive dialog and criticism. For many veterinarians, the desired "James Herriot" image is not consistent with conflict in relationships. We are concerned about alienating other stakeholders, particularly our clients. Sometimes, like any other human being, we may simply not like being told (or be willing to acknowledge) that we might be wrong or what we should do.

ANIMAL WELFARE IN ANIMAL AGRICULTURE: OPPORTUNITIES AND OBLIGATIONS

Veterinarians in all types of practice within the agricultural sector have an opportunity and an obligation to help animal owners, caretakers, handlers, and policymakers improve animal welfare.

Those in private clinical practice provide direct-to-owner/caretaker assistance in ensuring good animal care. In addition to hands-on animal evaluations and care, they may raise awareness of animal welfare concerns, deliver training in best practices, and help animal owners and caretakers complete self-assessments of compliance with protocols and recommendations.

Consulting veterinarians may contribute to implementation of quality animal care by completing in-depth evaluations of facilities and using the results of those evaluations to recommend standard operating procedures and best practices. In so doing, they provide education for animal owners and caretakers as well as compliance assurance.

Veterinary educators have a critical role to play in schooling future generations of veterinarians in the scientific and ethical bases behind the development and adoption of appropriate animal care practices, as well as voluntary or regulatory structures related to compliance. Others contribute to the training of paraprofessionals, who may include members of the veterinary health care team in private practices, as well as other technicians and caretakers, who fulfill veterinary roles and perform veterinary tasks when veterinarians are not directly available to do so.

Those involved in veterinary research work to resolve welfare challenges associated with existing animal care systems and practices and, based on the results of their studies, can propose alternatives that may better accommodate animal needs. Basic researchers can help identify animal needs and possible approaches to meeting them in the laboratory. Applied researchers then take these proposed innovations into the field to evaluate their practical application.

Veterinarians employed in governmental and nongovernmental organizations are those most likely to help develop and certify animal care standards. Often they are assisted by multidisciplinary advisory bodies that may include veterinarians engaged in private or consulting practices. Such standards can then be embraced via market-driven (voluntary) or legislative/regulatory processes.

All veterinarians have an opportunity to provide education that can build industry, market, public, and governmental support for welfare-friendly animal care practices. In addition, veterinarians with specific animal welfare and species expertise can serve as highly qualified, independent auditors for assurance schemes. Veterinarians must not only work to implement existing standards, but must also contribute to ensuring continual improvement of those standards. Improvement typically comes through identification of gaps in maintaining good animal welfare and exploration of procedural changes and practice improvements that may help close those gaps. If such changes lead to demonstrable improvements in animal welfare, and are able to be implemented practically, then they are likely to become practices that will gain wide acceptance.

SUMMARY

Veterinarians serve both animals and society in unique ways, including empathy for animals and science-based knowledge of animal health and husbandry.

They have an inherent responsibility to help animal owners, the public, and other stakeholders understand the complexity and ramifications of animal care decisions. In addition to weighing effects on the animals involved, establishing and implementing good care for agricultural animals is a balancing act involving human needs (including occupational health and safety), environmental concerns, and economics.

REFERENCES

American Veterinary Medical Association. The veterinarian's oath. 2010. Available at: www.avma.org/about_avma/whoweare/oath.asp (Accessed March 30, 2011).

Andrus, D.M., Prince, J.B., and Gwinner, K. 2006. Work conditions, job preparation, and placement strategies for food-animal veterinarians. *J Vet Med Educ* 33: 509–516.

Brambell, F.W.R. 1965. Report of the technical committee to enquire into the welfare of animals kept under intensive livestock husbandry systems. London: Her Majesty's Stationery Office.

Brown, J.P. and Silverman, J.D. 1999. The current and future market for veterinarians and veterinary medical services in the United States. *J Am Vet Med Assoc* 215: 161–183.

Chieffo, C., Kelly, A.M., and Ferguson, J. 2008. Trends in gender, employment, salary, and debt of graduates of US veterinary medical schools and colleges. *J Am Vet Med Assoc* 233(6): 910–917.

Colyer, D., Kennedy, P.L., Amponsah, W.A. et al., Eds. 2001, *Competition in Agriculture: The United States in the World Market*. Binghamton, NY: Haworth Press.

de Graaf G. 2007. Veterinary students' views on animal patients and human clients, using Q-methodology. *J Vet Med Educ* 34(2): 127–138.

Economic Research Service, USDA. 1995. Understanding rural America. Agriculture Information Bulletin No. 710, Washington, DC. Available at: www.nal.usda.gov/ric/ricpubs/understd.htm (Accessed March 30, 2011).

Fraser, D., Weary, D.M., Pajor, E.A. et al. 1997. A scientific conception of animal welfare that reflects ethical concerns. *Anim Wel* 6: 187–205.

Gallup. 2006. The most honest and ethical professions. Poll conducted December 8–10, 2006. Available at: www.gallup.com/poll/25888/Nurses-Top-List-Most-Honest-Ethical-Professions.aspx (Accessed March 30, 2011).

Hart, L.A. and Melese-d'Hospital, P. 1989. The gender shift in the veterinary profession and attitudes toward animals: A survey and overview. *J Vet Med Educ* 16 :27–30.

Heath, T.J. and Lanyon, A. 1996. A longitudinal study of veterinary students and recent graduates. 4. Gender issues. *Aust Vet J* 74: 305–308.

Hewson, C.J. 2006. Veterinarians who swear: Animal welfare and the veterinary oath. *Can Vet J* 47(8): 807–811.

LayWel. Welfare implications of changes in production systems for laying hens. March 28, 2006. Available at: www.laywel.eu/web/pdf/final%20activity%20report.pdf (Accessed March 30, 2011).

Narver, H.L. 2007. Demographics, moral orientation, and veterinary shortages in food animal and laboratory animal medicine. *J Am Vet Med Assoc* 230(12): 1798–1804.

Paul, E.S. and Pdberscek, A.L. 2000. Veterinary education and students' attitudes towards animal welfare. *Vet Rec* 146: 269–272.

Prince, J.B., Andrus, D.M., and Gwinner, K.P. 2006. Future demand, probable shortages, and strategies for creating a better future in food supply veterinary medicine. *J Am Vet Med Assoc* 229: 57–69.

Rollin, B.E. 2006. *An Introduction to Veterinary Medical Ethics: Theory and Cases,* 2nd ed. Ames, IA: Blackwell Publishing.

Russell, W.M.S., Burch, R.L., and Hume, C.W. 1992. *The Principles of Humane Experimental Technique.* Hertfordshire, UK: Universities Federation for Animal Welfare (current edition, first published in 1959).

Serpell, J.A. 2005. Factors influencing veterinary students career choices and attitudes to animals. *J Vet Med Educ* 32(4): 491–496.

Sprecher, D.J. 2004. Insights into the future generation of veterinarians: Perspectives gained from the 13- and 14-year-olds who attended Michigan State University's veterinary camp and conclusions about our obligations. *J Vet Med Educ* 31(3): 199–202.

FIFTH VIEWPOINT: INDUSTRY PERSPECTIVE ON ANIMAL WELFARE

Charles Arnot

Virtually every sector of society has undergone significant change over the past 40 years and animal agriculture is no exception. Advancements in technology and structural changes in agriculture over the past two generations have radically altered how food animals are raised today. These changes have allowed Americans to enjoy a safe, nutritious, and remarkably affordable supply of meat, milk, and eggs. They have also raised questions about animal care on today's farms and animal agriculture needs to address those questions in a transparent and forthright manner.

Brent Sandidge is a third-generation central Missouri farmer specializing in pork production. His farm dates back to the 27 acres of land his family purchased in 1927. His father decided to get into the pig business in the mid-1950s.

"My father probably had 20 sows when he started," said Sandidge. "Almost everybody had some pigs, some cows, row crops, etc. Farmers were extremely diversified back then."

By the early 1960s, the Sandidge hog operation had grown to around 100 sows.

"That was considered big," said Sandidge. "We were definitely known as one of the largest pork producers in the state of Missouri."

Because animal agriculture is a low margin business, farmers have focused on reducing costs and increasing productivity to remain economically viable. According to USDA/AMS, from 1960 to 2005, the deflated average farm price of cattle declined by 49%, milk by 30%, hogs by 56%, eggs by 78%, chickens by 60%, and turkeys by 73% (Plain, 2010). Because the prices paid to farmers for these commodities did not keep up with inflation, farmers had four basic options: Increase the size of the operation to maintain the same basic income with more animals, live on less money year after year, find a specialty market to capture additional margin, or leave farming.

Sandidge recalls that when he returned to the farm after graduating from college in 1978, his county's pork producer association mailing list contained around 400 names. Today, only a handful would be considered traditional farrow to finish hog operations.

"What happened was, people who adopted the new technology continued to grow and thrive. For others, maybe the pig business wasn't their first love—they probably just didn't enjoy the pig business so they tended to leave it. They decided to concentrate on other things—corn and soybeans, for example."

While many left farming, the majority of those who remain have become more specialized and implemented technology to allow them to remain competitive. In 1950, 56% of farms had hogs, 67.8% had dairy cows, and 78.3% had chickens, according to the U.S. Census of Agriculture. By 2007, only 3.4% of farms had hogs, 3.2% had dairy cows, and 6.6% had chickens (Plain, 2010).

Specialization and adoption of technology allows larger operations to capture economies of scale and be more productive and more efficient than smaller operations.

The Sandidges steadily expanded their operation and built barns in which to feed pigs. In the early 1960s, they were building farrowing barns and within 10 years, the pigs were "completely off of the dirt," as Brent describes it, except for the gestating sows.

"We used to handle these sows in groups out on dirt. Some would get too fat—others too thin. They would fight and establish a pecking order. We would sort them into separate groups but then those groups would reestablish a pecking order. There would be a new group of sows that got too thin."

Sandidge recalls that after sows gave birth, they were moved to dirt lots when the piglets were around two weeks old.

"We fought all kinds of disease," he said. "The sows would lie on their pigs. The weather could be hard on them. We were doing well if we could get 50 or 60% of them to market."

"Today, we keep them in stalls through gestation and we can better manage their health and feed them exactly what they need for healthy growth. That has dramatically improved the health and productivity of our herd."

When the sows roamed freely, Sandidge recalls that a 70% piglet survivability rate was considered good. Today, 90% is not unusual.

Efficiencies and increased productivity allow U.S. consumers to enjoy more affordable meat, milk, and eggs than consumers in other countries. From 1960 to 2009, the average deflated retail price of beef decreased by 27%, pork by 31%, chicken by 58%, and turkey by 65% (Plain, 2010). As a result, consumers can afford more meat and poultry. According to the Livestock Marketing Information Center, in 1970, average Americans spent 4.2% of their income to buy 194 lb of meat and poultry. In 2005, average Americans spent 2.1% of their income to buy 221 lb of meat and poultry (Plain, 2010).

Many farmers have chosen contract production to minimize capital requirements and manage the extreme volatility of commodity markets. In contract production, the contractor or integrator owns the animals, and provides the feed, health supplies, and transportation. The grower or farmer is paid to care for the animals and generally gets to keep the manure to use as fertilizer. Today, 46% of U.S. hogs, 90% of chickens, and 75% of turkeys are raised on contract according to the University of Missouri (Plain, 2010).

The overwhelming majority of men and women involved in providing meat, milk, and eggs are committed to doing what's right, and while the size of today's farms and the use of technology have changed dramatically, the integrity and commitment of those in food production has not.

While the Sandidge farm has grown from 20 sows in the mid-1950s to 3000 sows today, Brent says he shares his father's commitment to do the right thing.

"If you're in the pig business, you've got to love pigs because it's a lot of hard work. I love raising pigs. I'm doing everything I can to improve their environment so they have less stress and they're more productive."

Less than 1% of the U.S. population listed their occupation as farming, forestry, or fishing in the 2000 Census (BLS, 2010). The remaining 99% of Americans are generationally and geographically removed from production agriculture. Many have a romanticized notion of what farming "should be" based on outdated information and a lack of education about today's production practices. While research proves that raising animals indoors protects them from weather extremes and predators and reduces disease (University of Missouri Extension, 2009), the integrated model of production is inconsistent with the nostalgic image of farming held by many. In qualitative consumer research conducted on behalf of the Center for Food Integrity, consumers indicated they have a high degree of trust and admiration for farmers, but they are not sure today's production methods should still be considered farming.

Consumers have a right to expect farmers, processors, restaurants, and food retailers to act responsibly and to hold accountable those who do not.

The change in size and structure of animal agriculture, the lack of public understanding of today's farming practices, and cultural confusion about the role and function of animals in developed countries requires those involved in animal agriculture not only to continue to produce safe, nutritious, and affordable meat, milk, and eggs, but also they must demonstrate their commitment to do so in a socially responsible manner to build and maintain public trust.

Historically, agriculture was perceived to be committed to the shared values of compassion, responsibility, respect, fairness, and truth. Farmers were granted a broad social license to operate because it was assumed they would "do the right thing." Today, some sectors of society are questioning that assumption.

Industry critics argue that today's systems put profits above principles. That is a primary tenet of the argument against today's animal agriculture and it is expressed in concerns about animal care, environmental practices, contribution to local communities, and employment practices.

When public trust is lost or violated, the social license to operate is replaced with social control in the form of legislation, regulation, market mandates, and litigation. If the public no longer believes those in animal agriculture will "do the right thing," they support laws and regulations to control what happens on the farm. Animal agriculture has seen an increase in social control related to animal care in the form of state legislation and ballot initiatives sponsored by activist groups.

Historically, those involved in animal agriculture have relied primarily on science to defend the increased use of technology and enhanced production systems. Research from Iowa State University (Sapp et al., 2009) shows that effectively communicating shared values is three to five times more important than demonstrating competency through science in building public trust, which protects the social license to operate.

To be successful today and in the future, animal agriculture needs to demonstrate a commitment to operating balanced systems that are ethically grounded, scientifically verified, and economically viable (Figure 5.2).

Those who focus on ethics want food system practices that are consistent with the shared values of compassion, responsibility, respect, fairness, and truth. They want to ensure that the increasingly sophisticated and technologically advanced food system does not put profits ahead of ethical principles and that science is not used as moral justification. When this side of the triangle is out of

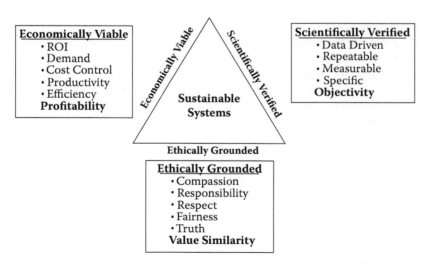

FIGURE 5.2 Balanced systems. (From CMA Consulting LLC©.)

balance, critics claim that there is no scientific basis for the claims being made and that the ethical demands will jeopardize the economic viability of the system.

Those with a primary interest in scientific verification are data driven. They want specific, measurable, and repeatable observations to provide the basis for their objective decisions. They believe science can provide the insight and guidance necessary to make reasonable determinations about how food systems should be managed. When this side of the triangle is out of balance, critics claim that the organization is relying on science while ignoring ethical considerations and that research may be done and recommendations made without consideration of the economic impact.

Those responsible for the bottom line are focused on profitability. They work every day to respond to demand, control costs, and increase efficiency to maximize the return on investment. They have to manage the increasingly complex demands of competing in a global marketplace with volatile commodity markets and ruthless competition. When this side of the triangle is out of balance, critics claim that profits outweigh ethical principles and that business decisions are made without the benefit of scientific verification, placing those decisions at risk when questioned by those who value validation.

If we cannot operate a balanced system that is ethically grounded, scientifically verified, and economically viable, it will collapse. That collapse may subject farmers, processors, restaurants, or retailers to undue pressure that includes consumer protests or boycotts, unfavorable shareholder resolutions, uninformed supply chain mandates, regulation, legislation, litigation, or bankruptcy.

There are some basic actions farmers and others in animal agriculture can take on the farm to build and maintain public trust in today's systems.

1. Do the right thing—above all else, make sure your farm meets or exceeds expectations for animal care and environmental stewardship.
2. Set codes of conduct for animal care—if you don't have them, establish animal care standards and ensure the standards are reviewed regularly and are consistently enforced. Require all workers who handle animals to sign the written code of conduct. This is important both for animal care protocol and to verify that all employees understand their shared obligation.
3. Hire the right people and provide ongoing training and consistent supervision—do background checks, establish clear expectations for animal care, and provide ongoing training in animal care and husbandry and consistent support and supervision.
4. Empower your workers—Let them know the critical role they play in providing animal care and assuring your care standards are met consistently throughout the farm. Create clear channels of communication for reporting concerns related to animal care.

Animal agriculture needs to communicate its genuine commitment to principles and shared values, not just because it is the right thing to do, but because it is good business. If animal agriculture fails to maintain a social license, it will be forced to comply with a more restrictive, higher cost, more bureaucratic system of social control.

Animal agriculture will be granted the greatest latitude in developing solutions and maintaining social license when farmers identify those issues that may challenge public trust and confidence in today's farming, and propose principle-driven solutions that maintain a sustainable balance of ethics, science, and profitability.

SUMMARY

Animal agriculture has changed significantly over the last 40 years, as has virtually every sector of society. Technological advances and structural changes in agriculture have allowed Americans to enjoy safe, nutritious, and very affordable food. Those changes have also raised questions about animal care on today's farms.

The economic reality is that prices paid to farmers for what they produce did not keep up with inflation, meaning they had to choose increasing the size of their operations, living on less money each year, finding a specialty market to capture additional margin, or leaving farming. The result is that there are fewer but larger farms in the United States today and new technology has allowed them to increase efficiency, productivity, and volume.

Research shows that modern production methods, such as raising animals indoors, is better for the animals in a number of ways but they are not consistent with the nostalgic image of farming held by many. Consumers have traditionally granted farmers a broad social license to operate because it was assumed they would "do the right thing." Since the public has little understanding of today's farming practices, farmers must demonstrate their commitment to produce food in a socially responsible manner to maintain the social license.

Historically, those involved in animal agriculture have relied on science to defend the increased use of technology. Research shows that effectively communicating shared values is three to five times more important than demonstrating competency through science in building public trust. To be successful, animal agriculture must demonstrate a commitment to operating balanced systems that are ethically grounded, scientifically verified, and economically viable. Failure to maintain this balance could subject the food system to undue pressure that includes consumer protests or boycotts, unfavorable shareholder relations, uninformed supply chain mandates, regulation, legislation, litigation, or bankruptcy.

References

BLS (Bureau of Labor Statistics). 2010. *Career Guide to Industries*, 2010–2011 edition, Agriculture, Forestry, and Fishing, www.bls.gov/oco/cg/cgs001.htm
Plain, R. 2010. Historical perspective of the integration of animal agriculture. CAST Food Animal Ag Symposium, June 8, 2010, Washington, DC.
Sapp, S.G., Arnot, C. et.al. 2009. Consumer trust in the U.S. food system: An examination of the recreancy theorem. *Rural Sociology* 74: (in press).
University of Missouri Extension. 2009. Study shows moving pigs inside has huge benefit. http://www.extension.org/pages/Study_Shows_Moving_Pigs_Inside_Has_Huge_Benefit

SIXTH VIEWPOINT: AN ACTIVIST'S PERSPECTIVE ON ANIMAL WELFARE

Paul Shapiro

America's animal agribusiness industry is being confronted with a new reality in the twenty-first century. For many decades, it cloaked itself in the protective mythology of Old MacDonald's Farm with images of contented cows and pampered pigs. However, that veneer is fading, as more and more Americans are learning how farm animals are really raised today.

When thinking about how farm animals are raised, it can be tempting to envision those young beef cattle we still see grazing in the countryside. The bucolic image is a powerful one, and one that even many involved in today's farming system seem to believe is the norm.

However, the beef industry, generally speaking, is the exception—not the norm—in animal agribusiness. Approximately 33 million beef cattle (USDA, 2010a) are slaughtered in the United States annually. Compare that to the nine *billion* chickens, turkeys, and pigs we consume (USDA, 2010b) and it becomes clear that if we are serious about discussing farm animal welfare, we need to be serious about conditions in the poultry and pig industries. Moreover, that is where a majority of the most pressing welfare concerns are found.

To put the disparity of scale in context, in just 36 hours the U.S. poultry industry slaughters more animals than the U.S. beef industry slaughters in an entire year.

Despite the U.S. animal protection movement's recent success in encouraging agribusiness to start moving away from some of its most extreme abuses, most of the billions of animals raised and killed each year still endure conditions that the majority of Americans would find simply appalling were they to actually witness them.

Animal science professor Peter Cheeke aptly describes this in his textbook, *Contemporary Issues in Animal Agriculture,* when he writes:

> One of the best things modern animal agriculture has going for it is that most people . . . haven't a clue how animals are raised and processed. . . . In my opinion, if most urban meat eaters were to visit an industrial broiler house, to see how the birds are raised, and could see the birds being "harvested" and then being "processed" in a poultry processing plant, they would not be impressed and some, perhaps many of them, would swear off eating chicken and perhaps all meat. For modern animal agriculture, the less the consumer knows about what's happening before the meat hits the plate, the better. (Cheeke, 1999)

Events in recent years give the impression that we are reaching a societal tipping point when it comes to establishing a better, more humane relationship with other animals. However, we need to balance that well-founded optimism with reality: In many ways, the treatment of the astronomical numbers of animals we raise and kill for food has grown steadily harsher in recent decades.

I don't anticipate that we'll soon reach societal agreement regarding the ethical permissibility (or lack thereof) of exploiting these animals. As interesting and worthwhile as that debate may be, it is a separate issue. We don't need to wait for such a broad discussion to conclude (or even to begin) before we can start making important animal welfare improvements that society already agrees on and that science and economics demonstrate are feasible. In short, it is incumbent upon us all to move forward on phasing out some standard practices that most of us *already agree* are simply unacceptable.

That is to say that there really is no excuse for failing to enact policies prohibiting many of the more egregious abuses animals face, and there are certainly plenty to go around. Such an effort would both reduce an enormous amount of unnecessary animal suffering and demonstrate that we are indeed capable of restraining ourselves when it comes to the virtually unlimited power we hold over farm animals.

Such progress is not intended to end the discussion about broader ethical questions, nor is its purpose to end all animal cruelty. The intent, simply put, is to allow our society to move in a positive direction by closing the gap between what Americans want for farm animals and what agribusiness is giving them.

WHERE DOES THE AMERICAN PUBLIC STAND?

The polling and the statewide votes regarding farm animal welfare are all fairly consistent.

A 2008 Gallup poll found that 64% percent of Americans support "passing strict laws concerning the treatment of farm animals" (Gallop, 2008). As well, a 2003 Zogby poll found that while a majority of Americans identify themselves as concerned about "the treatment of farm animals raised for food consumption," 82% agree that "there should be effective laws that protect farm animals against cruelty and abuse." The same poll found that 72% percent of Americans believe that "farms should be inspected by government inspectors to ensure that laws to protect animals from cruelty are being followed" (Zogby, 2003).

Even industry-funded polls show virtually identical results.

In 2007, the American Farm Bureau Federation paid Oklahoma State University to conduct a nationwide survey (Lusk, Norwood, and Prickett, 2007) on American attitudes toward farm animal protection. The results were revealing:

- 81% agree: Farm animals have roughly the same ability to feel pain and discomfort as humans.
- 75% agree: Would vote for a law in their state that would require farmers to treat their animals more humanely.
- 95% agree: It is important to me that animals on farms are well cared for.
- 68% agree: The government should take an active role in promoting farm animal welfare.
- 18% agree: Housing pregnant sows in crates is humane.

It could not be clearer: Americans believe farm animals have interests that matter (for example, not being confined in a virtually immobile state for months on end), and they believe those interests ought to be legally protected.

How Much Legal Protection Do Farm Animals Have Now?

If you spend any amount of time in agricultural circles, you would be hard-pressed to go for long without hearing complaints about a sea of regulation producers must endure. In reality, when it comes to how animals are actually treated, almost anything goes. It may be reassuring to pretend that animals on farms have significant legal protection from abuse, but that simply is untrue in most cases.

Animals used for food production have no federal legal protection whatsoever while they are on the farm. The federal Animal Welfare Act completely exempts animals used for food, and the Humane Methods of Slaughter Act (HMSA) only sets standards for the animals' final minutes—while they are at slaughter. Even worse, the U.S. Department of Agriculture (USDA) interprets the HMSA to exempt nearly all slaughtered animals (chickens, turkeys, rabbits, and several other species, which represent approximately 95% of the land animals who go through slaughter plants). Moreover, no person or company has ever been prosecuted under the HMSA for humane handling violations because USDA has no authority to do so; even in cases where the USDA has found repeated, blatant violations of the Act—such as an Iowa kosher cattle slaughterhouse that was documented repeatedly ripping the tracheas out of cows' throats while the animals were fully conscious (Eby, 2004)—violations go unprosecuted. Lastly, there is the federal 28-Hour Law, which regulates the transport of some farm animals, but which the USDA does not interpret to cover birds and is irrelevant as far as on-farm treatment is concerned (which is where the vast majority of farm animals' lives are spent).

At the state level, all 50 states have criminal anti-cruelty statutes, but most of them broadly exempt standard agricultural practices (which are often vaguely defined), essentially allowing any practice the industry chooses to widely utilize. Not surprisingly, a 2003 Zogby poll found that 66% percent of Americans find it "unacceptable" to exempt common agricultural practices from state cruelty laws (Zogby, 2003). However, even in states that do not exempt standard practices from their cruelty codes, animal abuse prosecutions against agribusiness operations are extremely rare.

The result of such regulatory laissez faire is that animals are left with very little protection, legally speaking, especially while they are on farms.

As American attitudes toward farm animals have grown increasingly sympathetic over the past few decades, some standard industry practices have gone in the opposite direction, especially in the poultry and pig industries. Poor farm animal welfare is not just a matter of a "few rotten eggs," but rather it is a case of some standard industry practices that most Americans find simply rotten.

This widening chasm between what Americans want for farm animals and what farm animals actually get is one of the most indefensible realities of our current animal agribusiness system. What many animal advocates are now proposing is simply that we narrow this gap by translating existing public support for animal welfare improvements into new policies that offer some semblance of protection to these animals. The following sections offer a few very brief concrete suggestions for such policies.

Cage Confinement of Laying Hens

More than 250 million U.S. egg-laying hens live in barren wire cages so restrictive that the animals can barely move for more than a year before they are slaughtered. With no opportunity to engage in many natural behaviors, including nesting, dust bathing, perching, and walking, these birds endure severe, chronic frustration. This near-immobilization takes a substantial toll on the animal's physical health. Deprived of exercise, the birds suffer from a weak skeletal system (Shipov et al., 2010), and combined with the commonly fed high-energy diet, they can suffer from "fatty liver hemorrhagic syndrome," a major cause of mortality in commercial flocks (Leeson, 2007).

If any system is emblematic of where the industry has gone far beyond what most Americans find acceptable, it is the cage confinement of laying hens. Even some in the meat industry seem uncomfortable with what happens in the egg industry. For example, consider what industry journalist and executive director of the Meat Industry Hall of Fame, Dan Murphy, has to say on the topic:

> Now, I don't know how many meat industry executives have spent any amount of time inside an egg production facility, but it's not a pleasant experience. In fact, I would argue that the egg industry is probably the sole exception to my conviction that producers and processors generally treat their livestock with care, if only to protect their investment. Egg producers operate from the principle of planned obsolescence. Since the hens are expendable, the goal is maximum production in the short time they are confined to their "living quarters"—if you can call the battery cage set-up anything that euphemistic. (Murphy, 2000)

Today's battery cage proponents frequently assert that the cages were invented for the welfare of the bird, an argument unsupported by much evidence. In fact, in 1971—long before animal welfare was a major topic in the industry—one poultry industry representative admitted:

> They can tell you all kinds of reasons why cages are good, but what they really did was to organize the hens in a production line where you can use more machinery, cut way down on labor, and allow just a few people to take care of a tremendous number of birds. (Sawyer, 1971, p. 216)

In other words, battery cages became popular because they made producing eggs cheaper, not because they were better for the birds.

Dr. Bernard Rollin of the Department of Animal Science at Colorado State University states that

> [v]irtually all aspects of hen behavior are thwarted by battery cages….The most obvious problem is lack of exercise and natural movement….Research has confirmed what common sense already knew—animals built to move must move. (Rollin, 1995, p. 120)

However, common sense does not always prevail, and basic movement is not an option for these animals.

When dealing with single facilities that confine hundreds of thousands—millions in many cases—of birds, individual inspection and veterinary care for each bird is impossible. The most that workers typically do for the birds is walk the aisles to remove the hundreds of newly-dead birds they find in cages each day (often, as numerous exposés have documented, the staff miss dead birds so frequently that carcasses become mummified in the cages).

The United Egg Producers (UEP) recommends that in a cage with multiple chickens, each laying hen get only 67 in.[2] of cage space (UEP, 2010). To put this in perspective, think about a letter-sized (8.5 × 11 in.) sheet of paper. That sheet of paper takes up 93.5 in.[2] of space. Now imagine folding the paper so that you hide almost a third of it, and then picture confining a 4-lb animal in that space for months on end. That is the plight of the modern egg-laying hen.

The extraordinarily restrictive amount of space is not the only major welfare assault for caged laying hens. Konrad Lorenz, the Nobel Prize-winning father of modern ethology, wrote that

> the worst torture to which a battery hen is exposed is the inability to retire somewhere for the laying act. For the person who knows something about animals it is truly heart-rending to watch how a chicken tries again and again to crawl beneath her fellow cagemates to search there in vain for cover. (Lorenz, 1980)

In fact, research has shown that laying hens will work as hard to gain access to an enclosed nesting area as they will to gain access to food after they have been starved for 27 hours (Follensbee, 1992). Such evidence makes it clear just how strongly these birds are motivated to nest.

The good news is that there is growing public opposition to the confinement of hens in cages, as evidenced by a flood of legislation, media attention, and corporate policies favoring cage-free production in recent years. For example:

- Several countries, such as Germany, Austria, and Switzerland, have already legislated against cages for laying hens and are presently phasing them out. Indeed, the entire European Union is phasing out barren battery cages (the kind that are standard in the United States) by 2012.
- California and Michigan—two large egg-producing states—have passed de facto bans (with phase-out periods) on cage confinement of hens.
- At the start of 2005, no major restaurant chains used any cage-free eggs; now, most do.

There is no question about the intersection of values that is driving change for laying hens. In the above-referenced American Farm Bureau poll, a majority of Americans thinks caging hens is inhumane, and a UEP-funded poll found that a plurality of Americans believe that caging hens is "not healthier nor safer."[1]

Animal scientist Dr. Michael Appleby sums it up well:

Battery cages present inherent animal welfare problems, most notably by their small size and barren conditions. Hens are unable to engage in many of their natural behaviors and endure high levels of stress and frustration. Cage-free egg production, while not perfect, does not entail such inherent animal welfare disadvantages and is a very good step in the right direction for the egg industry. (Appleby, 2006)

Commercial U.S. cage-free operations—which allow hens to walk, spread their wings, nest, perch, and more—are already raising millions of laying hens, and this number will likely increase as concerns about farm animal welfare grow stronger. The industry has a chance to embrace cage-free systems that better-accommodate both animal welfare and consumer desires.

GESTATION CRATE CONFINEMENT OF BREEDING PIGS

In 1968, after witnessing the economic results already achieved by the egg industry through confining increasing numbers of animals in small spaces, one pork industry analyst asked, "Why cannot greater efforts be made to introduce some of the economies of scale to hog production that have made the battery raising of chickens so efficient?" (Twedt, 1968).

So it began. There was indeed little to stop the pork industry from going in the same direction as the egg producers. This is especially so in the case of the female pigs who are used for breeding.

While most pigs used for pork production may have bleak lives living on concrete slatted floors with no bedding and little environmental enrichment, breeding sows are abused in ways so terrible, few people would support such treatment were they to see it firsthand.

Gestation crates are 2-ft-wide barren metal cages that confine impregnated pigs for months on end. They are unable even to turn around. Pigs confined in gestation crates suffer immensely, unable to exercise or engage in nearly any of their natural behaviors. The forced immobilization takes a serious physical and psychological toll, leading to both leg and joint problems along with psychosis resulting from extreme boredom and frustration.

Numerous animal scientists oppose these cruel crates. Colorado State University animal scientist Dr. Temple Grandin asserts, "Gestation crates for pigs are a real problem...Basically, you're asking a sow to live in an airline seat...I think it's something that needs to be phased out."[2]

Some in the pork industry still defend the use of gestation crates on the grounds that not only is it cheaper to pack pigs into the smallest spaces possible, but crating allegedly helps reduce sow aggression. Animal scientist and farm animal expert Dr. John Webster asserts that this defense "rests on the premise that it is acceptable to prevent an undesirable pattern of behaviour by restricting all forms of behaviour." Webster goes on to explain, "It would be as valid to claim that prisons would be much more manageable if all the inmates were kept permanently in solitary confinement" (Webster, 2005).

As well, the economic argument in favor of gestation crates isn't exactly strong. One need not look further than Iowa State University, where a 2-1/2-year-long study concluded that raising sows in groups in hoop housing rather than individual crates could cut the cost of production by 11% percent per weaned pig (Iowa State University, 2007).

As is the case with battery cages, the science seems to comport with the public's gut reaction against such extreme confinement. After the Scientific Veterinary Committee of the European Commission concluded, "Since overall welfare appears to be better when sows are not confined throughout gestation, sows should preferably be kept in groups" (Scientific Veterinary Committee, 1997), the entire European Union passed legislation phasing out gestation crates.

Seven U.S. states have passed legislation banning gestation crates. Even some parts of the industry, after years of defending such confinement, are beginning to see the light with major pork producers starting to move in the right direction.

In fact, a 2004 *National Hog Farmer* magazine article profiled Goldsboro Hog Farms, a major U.S. pork producer that has not used gestation crates for years (Miller, 2004). Cargill, a major pork producer, issued a press release in 2009 declaring that 50% of its sows are no longer in gestation crates (Cargill, 2009), and in 2010 the company's director of communications asserted that "Our plan is to ultimately move further away from gestation crates" (Forster, 2010). Smithfield Foods—the world's largest pork producer—has stated that its goal is to become gestation crate-free, although at present it doesn't have a timeline for achieving that aim.

The fact that many farms are using alternative systems is living proof of the unnecessary nature of gestation crates.

FORCED RAPID GROWTH OF BIRDS RAISED FOR MEAT

The 9 billion chickens and turkeys slaughtered in the United States each year are far removed in appearance from the wild animals that we originally domesticated. Unlike their fleet-footed ancestors, these animals are the products of intensive genetic selection for maximal weight gain with minimal feed consumption—as though animals could be transformed into meat-producing machines with enough human manipulation. Administration of growth-promoting antibiotics and other additives often help along the way, as do near-permanent lighting schedules that cause the birds to eat more than they would if they had a longer nighttime period of darkness.

In the 1950s, it took 84 days to raise a 5-lb chicken. Today, it takes an average of only 45 days, often even less (Havenstein, Ferket, and Qureshi, 2003). In 1947, just before this forced rapid growth of birds took off, the *Saturday Evening Post* described what the chicken industry was planning to do:

> No politician ever promised more than our poultrymen are now about to deliver. They expect to squelch that dream of two chickens in every pot by providing one bird chunky enough for the whole family—a chicken with breast meat so thick you can carve it into steaks, with drumsticks that contain a minimum of bone buried in layers of juicy dark meat, all costing less instead of more.[3]

They weren't really that far off.

Moreover, just as being morbidly overweight carries numerous health problems for humans, this forced rapid growth takes an enormous toll on the welfare of the birds. Poultry welfare expert Dr. Ian Duncan writes, "Without doubt, the biggest welfare problems for meat birds are those associated with fast growth" (Duncan, 2004).

Dr. Temple Grandin puts it more bluntly: "Today's poultry chicken has been bred to grow so rapidly that its legs can collapse under the weight of its ballooning body. It's awful" (Grandin and Johnson, 2005). Consequently, huge numbers of chickens raised for meat suffer from leg deformities and lameness. Studies consistently show that approximately 26 to 30% of broiler chickens suffer from gait defects severe enough to impair their walking ability (Knowles et al., 2008), and additional research strongly suggests that birds at this level of lameness are in pain (Danbury et al., 2000).

Additionally, rapid growth can lead to circulatory and pulmonary problems. "Sudden death syndrome" (SDS) is caused by acute heart failure and is common in broiler chickens (Riddell and Springer, 1985). Young birds die from SDS after sudden convulsions and wing-beating (Julian, 2004). Ascites is a condition in which rapidly growing broiler chickens do not have the heart and lung capacity needed to distribute oxygen throughout the body (Duncan, 2001) and is a leading cause of on-farm mortality as the birds reach market weight (Boersma, 2001).

Even though rapid growth increases mortality rates, it is not necessarily in producers' economic interests to improve the situation. Two University of Arkansas poultry industry researchers were straightforward in their assessment when they asked:

> Is it more profitable to grow the biggest bird possible and have increased mortality due to heart attacks, ascites and leg problems or should birds be grown slower so that birds are smaller, but have fewer heart, lung and skeletal problems?...A large portion of growers' pay is based on the pound of saleable meat produced, so simple calculations suggest that it is better to get the weight and ignore the mortality. (Tabler and Mendenhall, 2003, pp. 8–10)

But better for whom?

On the growth rate issue for meat-producing birds, animal scientist Dr. John Webster observes,

> On the balance of the evidence, we must conclude that approximately one quarter of the heavy strains of broiler chickens and turkeys are in chronic pain for approximately one third of their lives....This must constitute, in both magnitude and severity, the single most severe, systematic example of man's inhumanity to another sentient animal. (Webster, 1995, p. 156)

While slower-growing strains of birds do exist, they comprise an infinitesimal portion of the U.S. poultry market and are therefore not as easy for consumers to find. The companies that control nearly all poultry production have created the problem through intensive genetic selection for specific traits (mainly rapid growth and higher rates of feed conversion), and those same companies can instead select birds for health and welfare. In fact, nearly one-third of chickens raised for food in France are actually slow-growing, free-ranging birds, marketed as "Label Rouge" (Fanatico and Born, 2002).

Despite the enormity of the suffering forced rapid growth causes these animals, the costs associated with slowing these birds' growth rates are not as high as are those associated with some other important farm animal welfare improvements. The European Union's Scientific Committee on Animal Health and Animal Welfare found that slower growth would increase running costs principally by delaying the slaughter age, but that delaying slaughter age by only 10 days, while having a significant impact on welfare, would only cause approximately 5% higher costs than those of conventional breeds.[4]

Slowing today's astronomical growth rates would of course not address every form of suffering we inflict on the billions of birds we raise for food, but it would help improve their welfare in a meaningful way.

MOVING FORWARD TO A BETTER FUTURE FOR FARM ANIMALS

Concern about animal cruelty is far from the only consideration vying for the American public's attention, but the evidence is clear that our society considers it an important matter that warrants our serious attention. Farm animals are completely at our mercy, yet the abuses we force on them— including, but far from limited to, the three examples given in this chapter—are simply beyond the bounds of what our society considers ethically appropriate.

Some in the industry are consequently moving toward better systems and more realistic husbandry. Unfortunately, some trade groups that represent animal agribusinesses choose not to lead, but to fight the kinds of reforms outlined in this chapter, no matter how popular they may be with the American public.

As Nebraska cattle rancher Kevin Fulton writes,

> A lot of farmers I know don't support battery cages and gestation crates, but they fear being ostracized by the Farm Bureau and other trade groups if they speak out. I can't imagine anyone being proud to have to keep their animals locked up in tiny cages for their whole lives. Most farmers would rather use some husbandry than have to rely on such shortcuts, but they don't see a way out. If we had better leadership in our industries though, we could move in the right direction rather than being—correctly— perceived as hostile to any substantial animal welfare changes.[5]

The animal agribusiness industry has a chance to stop defending practices many Americans find indefensible and instead move toward systems that will better accommodate both animal welfare and consumer desires. Rather than trying to prevent change, these groups can and are beginning to seek incentives for producers to convert to higher welfare production methods.

REFERENCES

Appleby, M. 2006. "Clarification," letter to the editor published in *Minnesota Daily*, February 7, 2006.

Boersma, S. 2001. Managing rapid growth rate in broilers. *World Poultry* 17(8): 20–21.

Cargill. 2009. Cargill achieves eight critical animal welfare assurance goals, Cargill press release, April 15, 2009. Available at www.cargill.com/news-center/news-releases/2009/NA3011043.jsp

Cheeke, P. 1999. *Contemporary Issues in Animal Agriculture*, 2nd ed. Danville, IL: Interstate Publishers, p. 248.

Danbury, T.C., Weeks, C.A., Chambers, J.P., Waterman-Pearson, A.R., and Kestin, S.C. 2000. Self selection of the analgesic drug carprofen by lame broiler chickens. *The Veterinary Record* 146: 307–311.

Duncan, I.J.H. 2001. Animal welfare issues in the poultry industry: Is there a lesson to be learned? *Journal of Applied Animal Welfare Science* 4(3): 207–221.

Duncan, I.J.H. 2004. Welfare problems of poultry. In: *The Well-Being of Farm Animals*. Benson, J.B. and Rollin, B.E. Eds. Ames, IA: Blackwell, p. 310.

Eby, C. 2004. Ag Secretary Judge: Postville slaughter video is 'disturbing', December 7.www.globegazette.com/news/state-and-regional/article_3e20ddbd-ef9a-50cd-9202-daf25e5699aa.html

Fanatico, A. and Born, H. 2002. Label rouge: Pasture-based poultry production in France. National Sustainable Agriculture Information Service. www.attra.ncat.org/attra-pub/labelrouge.html

Follensbee, M. 1992. Quantifying the nesting motivation of domestic hens. Master's Thesis, University of Guelph, Ontario.

Forster, J. 2010. Humane Society buys ownership stake in Hormel. *St. Paul Pioneer Press*, September 13, 2010. Available at www.twincities.com/business/ci_16067517

Gallup. 2008. Post-Derby tragedy, 38% support banning animal racing. Gallup poll conducted May 8–11, 2008. Available at http://www.gallup.com/poll/107293/PostDerby-Tragedy-38-Support-Banning-Animal-Racing.aspx

Grandin, T. and Johnson, C. 2005. *Animals in Translation*. Harcourt Books, pp. 270–271.

Havenstein, G.B., Ferket, P.R., and Qureshi, M.A.. 2003. Growth, livability, and feed conversion of 1957 versus 2001 broilers when fed representative 1957 and 2001 broiler diets. *Poultry Science* 82: 1500–1508.

Iowa State University. 2007. Alternatives to Sow Gestation Stalls Researched at Iowa State. April 19, 2007. www.ag.iastate.edu/news/releases/319/

Julian, R.J. 2004. Evaluating the impact of metabolic disorders on the welfare of broilers. In: *Measuring and Auditing Broiler Welfare*. Weeks, C. and Butterworth, A., Eds. Wallingford, UK: CABI Publishing.

Knowles, T.G., Kestin, S.C., Haslam, S.M. et al. 2008. Leg disorders in broiler chickens: Prevalence, risk factors and prevention. *PLoS ONE* 3(2): e1545. doi:10.1371/journal.pone.0001545.

Leeson, S. 2007. Metabolic challenges: Past, present, and future. *Journal of Applied Poultry Research* 16: 121–125.

Lorenz, K. 1980. Animals are sentient beings: Konrad Lorenz on instinct and modern factory farming. *Der Spiegel*. 34(47): 264.

Lusk, J.L., Norwood, F.B., and Prickett, R.W. 2007. Consumer Preferences for Farm Animal Welfare: Results of a Nationwide Telephone Survey. Available at http://asp.okstate.edu/baileynorwood/AW2/InitialReporttoAFB.pdf

Miller, D. 2004. Sows flourish in gestation pens. *National Hog Farmer*, May 15, 2004. Also available at http://nationalhogfarmer.com/mag/farming_sows_flourish_pen/index.html

Murphy, D. 2000. Commentary: Fast-food chain proves service not limited to customers. *The Meatingplace*, August 25, 2000.

Riddell, C. and Springer, R. 1985. An epizootiological study of acute death syndrome and leg weakness in broiler chickens in Western Canada. *Avian Diseases* 29: 90–102.

Rollin, B.E. 1995. *Farm Animal Welfare: Social, Bioethical, and Research Issues*. Ames, IA: Iowa State Press, p. 120.

Sawyer, G. 1971. *The Agribusiness Poultry Industry: A History of Its Development*. Hicksville, NY: Exposition Press, p. 216.

Scientific Veterinary Committee, European Commission. 1997. The welfare of intensively kept pigs. Adopted September 30, 1997. http://ec.europa.eu/food/fs/sc/oldcomm4/out17_en.pdf (Accessed September 27, 2010).

Shipov, A., Sharir, A., Zelzer, E., Milgram, J., Monsonego-Ornan, E., and Shaher, R. 2010. The influence of severe prolonged exercise restriction on the mechanical and structural properties of bone in an avian model. *The Veterinary Journal* 183: 153–160.

Tabler, G.T. and Mendenhall, A.M. 2003. Broiler nutrition, feed intake and grower economics. *Avian Advice* 5(4): 8–10.

Twedt, D. 1968. General acceptance of pork. In: *The Pork Industry: Problems and Progress,* Topel, D.G., Ed. Ames, IA: Iowa State University Press, p. 7.

UEP (United Egg Producers). 2010. *Animal Husbandry Guidelines for U.S. Egg Laying Flocks*. 2010 edition. Available at www.uepcertified.com/media/pdf/UEP-Animal-Welfare-Guidelines.pdf

USDA (U.S. Department of Agriculture), National Agricultural Statistics Service. 2010a. Livestock Slaughter 2009 Summary, p 3. http://usda.mannlib.cornell.edu/usda/current/LiveSlauSu/LiveSlauSu-04-29-2010.pdf

USDA (U.S. Department of Agriculture), National Agricultural Statistics Service. 2010b. Poultry Slaughter 2009 Summary, p 2. http://usda.mannlib.cornell.edu/usda/current/PoulSlauSu/PoulSlauSu-02-25-2010.pdf

Webster, A.J.F. 1995. *Animal Welfare: A Cool Eye Towards Eden*. Oxford, UK: Blackwell, p. 156.

Webster, J. 2005. *Animal Welfare: Limping Towards Eden*. Oxford, UK: Blackwell Publishing, p. 112.

Zogby, J. 2003. Nationwide Views on the Treatment of Farm Animals. Zogby poll released on October 22, 2003. Available at www.civileats.com/wp-content/uploads/2009/09/AWT-final-poll-report-10-22.pdf

ENDNOTES

1. "Laying Out the Facts." Presentation by United Egg Producers spokesman Mitch Head, delivered to the American Meat Institute's "Animal Care and Handling Conference," Kansas City, MO, February 18, 2004.
2. Comments Temple Grandin made during a Q&A session on January 9, 2006 at Manhattan Columbus Circle, New York. They can be heard at: http://nycanimalrights.com/Temple%20Grandin%20Animals%20in%20Translation.htm
3. *Saturday Evening Post*, August 9, 1947. As cited in Sawyer, Gordon. 1971. *The Agribusiness Poultry Industry: A History of Its Development*. Hicksville, NY: Exposition Press, p. 116.
4. Scientific Committee on Animal Health and Animal Welfare. 2000. The welfare of chickens kept for meat production (broilers). For the European Commission; Health and Consumer Protection Directorate-General, March 21, 2000.
5. Personal email communication between Kevin Fulton and the author on September 29, 2010. Used with permission.

6 Contemporary Animal Agriculture

Rural Community Concerns in the United States

David Andrews

CONTENTS

INTRODUCTION

Industrialized agriculture can be defined as applying the values of simplification, specialization, and concentration to a manufacturing operation, including food production and farm animal production (Taylor, 1911; see also Kanagil, 1997). The methods of automobile production and food processing have been applied to livestock production beginning with poultry and moving to pigs, cattle, and dairy cows (Heinrichs and Welsh, 2003). These industrial farm animal production methods have had significant impacts on rural communities. The history of such impacts has been reviewed in much of academic literature in the fields of sociology, economics, and public health.

This chapter considers the community and social impacts of concentrated animal feeding operations (CAFOs). It will consider the impacts on the day-to-day lives of individuals, families, and communities. It will also provide as relevant global and structural considerations related to CAFO systems. Industrial production is of relatively recent vintage and a contemporary movement to consider issues of sustainability leads one to question the longevity of the industrial model for animal agriculture and CAFOs in particular.

HISTORICAL CONSIDERATIONS OF INDUSTRIAL PRODUCTION

Researchers have examined many aspects of industrialized animal production in rural communities and social groups. Linda Labao and Curtis Stofferahn, for example, are two established sociologists who have reviewed academic and public concern relative to the national food system, in topic areas such as agribusiness concentration, consumer health, food safety, and sustainability of the national eco-system. However, their chief concern has been to review the immediate effects of industrialized farming on the day-to-day lives of citizens residing in the places where CAFOs are located: rural communities. There is a long history of public, government, and academic concerns with the consequences of industrialized farming. CAFOs have significantly harmed rural communities by making enjoyable living conditions impossible to enjoy. The quality of life has suffered immensely and the concentration of many animals in one area has led to significant detrimental effects being reported, including the loss of quality of life, fresh clean air, and comfortable enjoyment of the outdoors, as well as significant health impacts.

Since the 1930s, government and academic researchers have investigated the extent to which large-scale, industrialized farms have adversely affected the communities in which they are located. One of the first studies was conducted by a sociologist, E.D. Tetreau (1938, 1940), who found that large-scale, hired labor-dependent farms were associated with poor social and economic well-being in rural Arizona communities.

In the early 1940s, the United States Department of Agriculture (USDA) sponsored a research project to assess the effects of industrialized farming using a matching-pair design of two California communities: Arvin, where large, absentee-owned, non-family farms were more numerous, and Dinuba, where locally owned, family-operated systems were more numerous. Walter Goldschmidt, a USDA anthropologist, prepared the report on this project. The purpose of the study was to assess the consequences of a California law with a provision placing acreage limitations on large farms located in California's Central Valley, to support family-sized farms in the region. Goldschmidt reported that the study "was designed to determine the social consequences that might be anticipated for rural communities if the established law was applied or rescinded" (Goldschmidt, 1978a, p. 458).

In his report, Goldschmidt (1978a) systematically documented the relationship between large-scale farming and its adverse consequences for a variety of community quality of life indicators. He found that, relative to the family farming community, Arvin's population had a small middle class and a high proportion of hired workers. Family incomes were lower and poverty was higher. Their schools were of poorer quality as were public services. There were also fewer churches, civic organizations, and retail establishments. Arvin's residents also had less local control over public decisions, or "lack of democratic decision-making," as local government was prone to be influenced by outside agribusiness interests. By contrast, the family farming city of Dinuba had a larger middle class, better socioeconomic conditions, higher community stability, and better civic participation. Goldschmidt's report was eventually published as Congressional testimony (Goldschmidt, 1968) and as a book (Goldschmidt, 1978a). His conclusion that large-scale industrialized farms create a variety of social problems for communities has been confirmed by many other studies.

California, in its Small Farm Viability Project (1977, pp. 229–230) affirmed Goldschmidt's findings by revisiting Arvin and Dinuba. They concluded:

> The disparity in local economic activity, civic participation, and quality of life between Arvin and Dinuba...remains today. In fact, the disparity is greater. The economic and social gaps have widened. There can be little doubt about the relative effects of farm size and farm ownership on the communities of Arvin and Dinuba.

Quality of life issues related to the structure and scale of agriculture have been examined since the early 1940s. MacCannell (1988) conducted a macro study that included family-farm and industrial agricultural communities in 98 industrial-farm counties in California, Arizona, Texas, and Florida. He found that farm size (in acres), gross farm sales, as well as high levels of mechanization "significantly predict declining community conditions not merely at the local agricultural community level, but in the entire county" (1988, p. 63). These studies of industrialized agricultural production are direct precedents for later studies of CAFOs, which engage in the specialization, concentration, and simplification of agricultural production systems, which are the hallmarks of industrialization.

Recent studies reveal tendencies of economic decline in communities with greater concentration of CAFOs, similar to Goldschmidt's thesis of greater rural community decline with greater industrialization of agriculture. The econometric analysis conducted by Gomez and Zhang (2000) over a decade revealed the negative impact of swine CAFOs on economic growth in rural Illinois counties, as indicated by sales receipts. They found that purchases from small businesses declined as concentrations of CAFOs intensified. In a Michigan study, Abeles-Allison and Connor (1990) found that local purchases of supplies for swine production decrease as CAFO concentrations increase. Local expenditures per hog were calculated at $67 for the small farms and $46 for the large farms. (Of interest here is the fact that one goal of CAFO is to reduce unit cost. Many more hogs would be sold and much more feed would be purchased in large swine CAFO.) The difference is largely due to bulk feed purchases from outside the community by the larger farms, but also is related to somewhat greater total expenditures per hog on the smaller farms. A significant conclusion is that rural community economic decline is related to vertical integration. Fewer inputs are purchased locally and businesses on Main Street tend to dry up as large-scale CAFOs purchase their services from a single related supplier outside the community where the CAFO is located.

TRENDS IN INDUSTRIAL AGRICULTURE DRIVEN BY CAFOs

There are significant negative consequences to rural communities due to the proliferation of CAFOs around the United States. Among these are a deterioration in the quality of life, an increase in income inequality, social disruption, the decline of community services and property values, public health, environmental racism (which recognizes differential impacts on minorities in environmental behaviors), negative attitudes toward living conditions, and decreased political vitality.

QUALITY OF LIFE

The social fabric of communities undergoes significant change as the industrialization of agriculture takes place and the many related undesirable effects occur. It has been shown recurrently through the research that there is a decline in local population size when family farms are replaced by industrialized farms; a smaller population is sustained by industrialized farms relative to family farms. Again, where capital-intensive agriculture relies more on technology than on labor in the efforts to simplify, standardize, and centralize operations, a distinct result is a decline in the social fabric of the community (Goldschmidt, 1978a; Heady and Sonka, 1974; Rodefeld, 1974; Wheelock, 1979).

An important aspect of the quality of life in a community is social capital, which includes mutual trust, reciprocity, and shared norms and identity. In general, communities with greater social capital provide greater quality of life (Flora and Flora, 1998). In addition, social capital emerges as an internal resource in instances of controversies. That is, where there is social conflict, there is a greater possibility of the division being enhanced by the grouping of activists on differing sides.

Wing and Wolf's (2000) study of 50 to 55 individuals from each of three North Carolina rural communities showed that quality of life was greatly diminished among residents near a 6000-head swine confinement operation, compared to residents near two intensive cattle operations or near an agricultural area without livestock operations that required liquid waste management. Quality of life was indicated by the number of times that neighbors could not open their windows or go outside due to odors from CAFOs. Of the respondents from around the hog CAFO, 30% (as compared to a maximum of 3% from the other two communities) indicated that these problems had occurred 12 or more times during the past six months prior to the survey. Many rural residents commented that it was difficult to plan social activities in their homes because of the uncertainty of whether the air would be tolerable for guests (see Wright et al., 2001, pp. 28–30, for similar health and social responses near Minnesota CAFOs). Such limitations on social relations with one's neighbors indicate a decline in community social capital.

Quality of life issues that related to agricultural structures are evident in eastern North Carolina. This region experienced a tremendous growth in the hog industry beginning in the 1980s that included both contract and corporate production facilities and meat packing plants. Many citizens there perceive that this has left them with a power structure in which the interests of large pork producers dominated those of local residents at all levels of government (McMillan and Schulman, 2003; Thu and Durrenberger, 1998).

The process of industrialization leads to the reduced enjoyment of property and deterioration in the landscape, especially when there is a recurrent odor problem in communities with hog CAFOs (Schiffman, Slattery-Miller, Suggs, and Graham, 1998; Wing and Wolf, 1999, 2000; Constance & Tuinstra, 2005; Kleiner, 2003; McMillan and Schulman, 2003).

Research reflects that various parts of the country have experienced a decline in quality of life and social capital related to CAFOs. Quality of life concerns incentivized citizen action such as the documented actions of anti-CAFO groups in the Texas panhandle. They focused on episodes of resistance carried out by local residents and environmental groups who were motivated mainly by human health and property value concerns. Corporate responses to community resistance primarily involved reconstruction of their corporate image as environmentally friendly. A decline in social capital is associated with swine CAFOs, according to rural residents of Iowa, North Carolina, Minnesota, Michigan, and Missouri who describe violations of core rural values of honesty, respect, and reciprocity, as reported in an interdisciplinary workshop held in Iowa on swine CAFOs. For example, CAFO neighbors often consider it a violation of respect when their concerns are labeled as emotional, perceptual, and subjective or are dismissed as invalid or unscientific. Findings that are more recent as presented by Kleiner, Rikoon, and Seipel (2000) indicate that in two northern Missouri counties where large-scale, corporately owned swine CAFOs are dominant, citizens expressed more negative attitudes regarding trust, neighborliness, community division, networks of acquaintanceship, democratic values, and community involvement. The county that was dominated by independently owned swine operations had the most positive attitudes regarding trust, neighborliness, community division, and networks of acquaintanceship. Quality of life issues reflecting the growth or decline of social capital are clearly reflected in these differing regions of the United States. Quality of life factors are emphasized in recent literature addressing the community impacts of CAFOs. In 2001, the state of Minnesota brought together the scientific and public policy communities to advise state government on how to address several CAFO issues, resulting in a Generic Environmental Impact Statement (GEIS) for animal agriculture. It suggests, "quality of life is related to perceptions of: (1) having alternatives in what one does on a daily or life cycle basis, and (2) being respected by family and communities of interest and place" (Flora et al., 1999, p. A24).

Wright et al. (2001) reported results from a six-county study in southern Minnesota regarding changes in animal agriculture. Over 100 producers, community leaders, and others were interviewed, either in roundtable discussions or individually. Three patterns reflect the decline of social capital that resulted from the siting of CAFOs in all six rural communities: (1) widening gaps between the farmers who produce livestock within CAFOs and their neighbors, including non-CAFO livestock producers; (2) harassment of vocal opponents of CAFOs; and

(3) perceptions by both CAFO supporters and opponents of hostility, neglect, or inattention by public institutions that resulted in perpetuation of an adversarial and inequitable community climate.

SOCIAL DISRUPTION

Social disruption is another category of negative impact of CAFOs on communities. These social disruptions are reflected in increases in crimes, lawsuits, police activity, stress, and social problems, which had not been experienced in the community prior to the arrival of CAFOs (North Central Regional Center for Rural Development, 1999). Research showed that the increase in local police activity was associated with CAFO laborers, especially when the operations relied upon large numbers of single men for their labor with relatively little social life integrated within the local community. A general increase in social conflict was evident from this in the research (Seipel, Hamed, Rikoon, and Kleiner, 1999) and included increased stress, social and psychological problems (Martinson, Wilkening, and Rodefeld, 1976; Schiffman et al., 1998), and teenage pregnancies (Labao, 1990). Research shows a deterioration of relationships among hog farmers and neighbors (Jackson-Smith and Gillespie, 2005; McMillan and Schulman, 2003) and more stressful, less neighborly relations (Constance and Tuinstra, 2005; Smithers, Johnson, and Joseph, 2004). The North Central Regional Center for Rural Development (1999) examined the dramatic increases in corporate hog production and meatpacking in a rural Oklahoma county. Social capital indicators measured mutual trust, reciprocity, and shared norms and identity. Individual security was measured in terms of crime, and community conflict was measured in terms of civil court cases. The overall crime rate increased dramatically between 1990 and 1997. Violent crimes increased 378% compared to an average 29% percent decrease in violent crimes over the same period in comparison farming-dependent counties with no dramatic changes in animal agriculture. Theft-related crimes also increased in the case county by 64% compared to a decrease of 11% in comparison counties. Civil court cases, indicating community conflict, increased in the case county by 7%, but decreased 11% in the comparison counties. This study reveals the dramatic costs of social disruption in counties experiencing rapid change due to the introduction of CAFOs.

CAFOs AND POVERTY

CAFOs seem to be located in communities' census blocks with high poverty and minority populations (Wilson, Howell, Wing, and Sobsey, 2002). There are lower incomes for certain segments of the community with a greater income inequality and poverty. There is evidence from research that shows communities differ depending upon the integration of a food production system within the local community and its economic life or if its operation is distinct, separated, and vertically integrated so that few of its inputs or activities are integrated fully with the local economic community. In industrialized agriculture, the "main street" is less active and the multiplier effect of interdependent economic activity is lost. This leads to a situation of greater income inequality and poverty in the area. The very recent research in the 1990s and later in 2000 and further on, as indicated in the following citations, reflect analysis that demonstrates that poverty and inequality are significant factors in CAFO areas (Tetreau, 1940; Goldschmidt, 1978a; Heady and Sonka, 1974; Rodefield, 1974; Flora, Brown, and Conby, 1977; Wheelock, 1979; Labao, 1990; Crowley, 1999; Deller, 2003; Crowley and Roscigno, 2004; Peters, 2002; Welsh and Lyson, 2001). There are higher unemployment rates in those CAFO communities where the industrialized system becomes more "efficient" through increased use of rendering technology rather than an increase in utilization of farm labor as the means to achieving successful economic operations (Skees and Swanson, 1988; Welsh and Lyson, 2001). Where industrialized agriculture comes to prevail, the social class structure becomes less desirable and poorer along with the increases in hired labor. In some areas, well-paid union labor gives way to migrant laborers who are willing to work for less and to rent apartments for a number of single males to occupy (Gilles and Dalecki, 1988; Goldschmidt, 1978a; Harris and Gilbert, 1982). Increased food stamp utilization is associated with industrialized hog production in Iowa, suggesting that industrial agriculture generates inequalities or that industrial agriculture thrives in counties with greater inequalities.

LOWER REAL ESTATE VALUES

Real estate values decline in residences closest to CAFOs. These homes experience declining values relative to those more distant from CAFOs (North Central Regional Center for Rural Development, 1999, p. 46; Seipel, Hamed, Rikoon, and Kleiner, 1998; Constance and Tuinstra, 2005; Wright et al., 2001). The decline in the values of homes has occurred in diverse parts of the country. The value of a house declines when a CAFO comes to the neighborhood.

ECONOMIC VITALITY DECLINES

When CAFOs arrive in an area, the local economy suffers. The retail trade is decreased and the community has fewer stores and a less diverse economic landscape (Goldschmidt, 1978a; Heady and Sonka, 1974; Rodefeld, 1974; Fujimoto, 1977; Marousek, 1979; Skees and Swanson, 1998; Gomez and Zhang, 2000). This has been documented throughout decades of research on industrialized livestock production including more recent studies of CAFOs.

CIVIC PARTICIPATION AND POLITICAL VITALITY SUFFERS

A reduction in civic participation leads to a deterioration in community organizations and less involvement in social life. This is a recurring factor in the literature about communities with CAFOs (Goldschmidt, 1978a; Heffernan and Lasley, 1978; Poole, 1981; Rodefeld, 1974; Smithers et al., 2004). People begin to withdraw and lose their impact on social interactions including the decline in the quality of local governance and less democratic political decision-making. The public becomes less involved as outside agribusiness interests increase their influence or control over local decision-making. This loss of democratic vitality has been discovered in industrial agriculture and in communities having CAFOs over the past several decades of reporting (Tetreau, 1940; Rodefeld, 1974; Goldschmidt, 1978a; McMillan and Schulman, 2003).

Community services decline with the industrialization of livestock production, leaving an area with fewer or poorer quality public services and fewer churches (Teatreau, 1940; Fujimoto, 1977; Goldschmidt, 1978a). This has been seen in many different places and times.

PUBLIC HEALTH CONCERNS GROW

There have been studies of neighbors of hog CAFOs that have developed health problems that include upper respiratory issues, digestive tract disorders, and vision and eye problems (Wing and Wolf, 1999; Constance and Tuinsra, 2005; Wing and Wolf, 2000; Kleiner, 2003). There have been more than 70 papers published on the adverse health effects of the CAFO confinement environment by authors in the United States, Canada, most European countries, and Australia. It is clear that at least 25% of confinement workers suffer from respiratory diseases including bronchitis, mucous membrane irritation, asthma-like problems, and acute respiratory distress syndrome. Recent findings substantiate anecdotal observations that a small proportion of workers experience acute respiratory symptoms early in their work history that may be sufficiently severe to cause their immediate withdrawal from the workplace (Dosmann et al., 2004). An additional acute respiratory condition, organic dust toxic syndrome, related to high concentrations of bio-aerosols in livestock buildings occurs episodically in more than 30% of workers in swine CAFOs.

The body of literature on adverse effects among residents living near swine operations has been increasing. Excessive respiratory symptoms in neighbors of large-scale CAFOs relative to comparison populations in low-density livestock-producing areas were documented. The pattern of development of these symptoms was similar among workers in Iowa, North Carolina, and Nebraska. Neighbors of confinement facilities reportedly experienced increased levels of mood disorders including anxiety, depression, and sleep disturbances attributable to malodorous compounds. Wheezing among schoolchildren has been identified as well as an increasing incidence of asthma among children living on these farms. Children's health has also been recognized as at risk from the effects of CAFOs (Thu and Durrenberger, 1998). Donham (2000) describes possible nontoxic mechanisms for CAFO odors to generate physical symptoms through complex interactions between nerves of the brain and somatic systems of the body. Shusterman (1992) describes some of these mechanisms in his review of the health effects of environmental odor and pollution on human health. There are well-researched linkages of physical symptoms to the uncontrolled variability of stressors, including environmental stressors, that may be applicable to CAFO odors. In addition, the variety of family, neighborhood, and community stressors sometimes associated with CAFOs may also generate stress-induced symptoms and illness.

The site of a swine confinement facility in Parma, Michigan, in the mid-1980s generated conflict when the firm established a five-unit CAFO with manure lagoons. Neighbors believed the three open-air 42-million-gallon lagoons compromised their health and quality of life. Local resistance culminated in the emergence of two grassroots organizations and a four-year litigation process. Consequences of this conflict were anger on the part of residents who believed that their environment and their integrity had been violated. This led to resentment toward public officials, polarization within the community, vandalism, alienation, and verbal threats and physical aggression by both sides. Although the opponents of the CAFO won the battle on the local level (the CAFO went bankrupt), when they were interviewed a few years later, the CAFO operator felt that the personal acrimony and divisions in the community resulting from the conflict over the smell from the lagoons was too high a price to pay.

Characteristics of the nearest CAFO and of the affected neighbor influence the latter's level of annoyance with CAFOs. In a study conducted in the early 1980s in British Columbia, Van Kleek and Bulley (1985) chose 14 swine farms, 14 beef farms, 11 laying hen farms, and 10 broiler farms located at least 800 m (somewhat less than a half mile) from any other livestock farm. At least 12 residents (non-producers of livestock) were within 800 m of each livestock farm. Those residents rated their perception of the livestock farm "as it relates to your living here" on a five-point scale with "no nuisance/very compatible" to "severe nuisance/incompatible."

The authors found that nuisance potential decreased with distance, but it decreased the least for hog farms. Larger farms were a greater nuisance than smaller ones, but the difference disappeared for residences that were at very close ranges to the livestock farms. Hog farms were considered the greatest nuisance, followed by cattle feedlots, and then poultry CAFOs. Odor represented 75% of the total nuisance, but the proportion differed according to the type of farm; for hog farms, 95% of the nuisance responses related to odor; for broilers, 75%; for layers, approximately 66%; and for feedlots, approximately 50%. People with rural backgrounds were less tolerant of livestock farms than were people who came from urban areas, while opinions of those with farm backgrounds did not differ from those without farm backgrounds. Lohr (1996) found that among neighbors of a swine farm, tenure of residence, previous contact with the farmer, and economic dependence on farming all correlated negatively with their perceived degree of odor annoyance.

All sides of controversies involving CAFOs tend to frame their issues and identities in terms of rights and entitlements, as described in McMillan and Schulman's (2003) research on the hog industry in North Carolina. For example, producers defend their property rights and right to earn a living from their land, while neighbors defend their right to enjoy their own property. DeLind (1995) documents that in response to local opposition to corporate CAFO or "hog hotels" in Parma, Michigan, the Farm Bureau, the Pork Producers Council, and other agricultural interest groups defended the right of "hog hotels" to exist without regulation by appealing to the right to farm.

An examination of local purchasing patterns of large and small dairy farms in Wisconsin found that the percentage of dairy feed purchased locally declined as herd size increased. Stronger indicators of local feed purchasing were the physical nearness and social attachment to the community. In Minnesota, Chism and Levins (1994) found that local spending was not related to gross sales volume on crop farms. However, local farm-related expenditures fell sharply when the scale of livestock operations increased.

ECONOMIC HEALTH

Economic concentration of agricultural operations tends to remove a higher percentage of money from rural communities than when the industry is dominated by smaller farm operations, which tend to circulate money within the community. A study by MacCannell (1988) of comparable types of communities researched in Goldschmidt's work found that the concentration and industrialization of agriculture were associated with economic and community decline locally and regionally. Results of studies in Illinois (Gomez and Zhang, 2000), Iowa (Durrenberger and Thu, 1996), Michigan (Abeles-Allison and Connor, 1990), and Wisconsin demonstrated decreases in tax receipts and declining local purchases with larger operations. A Minnesota study (Chism and Levins, 1994) found that the decline in local spending was related to enlargement in the scale of individual livestock operations rather than crop production. These findings consistently show that the social and economic well being of local rural communities benefits from increasing the number of farmers, not simply from increasing the volume of commodity produced (Osterberg and Wallinga, 2004).

MENTAL HEALTH

Living in proximity to large-scale CAFOs has been linked to symptoms of impaired mental health, as assessed by epidemiological measures. Greater self-reported depression and anxiety were reported among North Carolina residents living near CAFOs (Bullers, 2005; Schiffman et al., 1995).

Another example of mental health effects of CAFOs is that greater reporting of post-traumatic stress disorder (PTSD) cognitions have been reported among Iowans living in areas with a high concentration of CAFOs compared with Iowans living in areas of low concentrations of livestock production (Hodne, unpublished). The PTSD cognitions were consistent with multiple concerns of those persons interviewed about the decline in the quality of life and socioeconomic vitality caused by CAFOs in areas where increases in CAFOs resulted in concentration with declining family farm production.

SOCIAL HEALTH

One of the most significant social impacts of CAFOs is the disruption of the quality of life for neighboring residents. More than an unpleasant odor, the smell can have dramatic consequences for rural communities where lives are rooted in enjoying the outdoors. The encroachment of large-scale livestock facilities near homes is significantly disruptive to rural living. The highly cherished values of freedom and independence associated with life oriented toward the outdoors gives way to feelings of violation and infringement. Social gatherings where family and friends come together—backyard barbecues and visits by friends and family—are affected in practice or through disruption of routines that normally provide a sense of belonging and identity. Homes are no longer an extension of or a means for enjoying the outdoors. Rather, homes become a refuge to escape from the outdoors.

Studies evaluating the impacts of CAFOs on communities suggest that CAFOs generally attract controversy and often threaten community social capital (Kleiner et al., 2000). The rifts that develop among community members can be deep and long-standing (DeLind, 1998). Wright et al. (2001) conducted in an in-depth study of six counties in southern Minnesota and identified three patterns that reflect the decline of social capital that resulted from CAFOs. In all six rural communities, they studied: (1) widening gaps between CAFO and non-CAFO producers; (2) harassment of vocal opponents of CAFOs; and (3) perceptions by public institutions that resulted in perpetuation of an adversarial and inequitable community climate. Threats to CAFO neighbors have also been reported in North Carolina (Wing, 2002). Clearly, the community conflict often follows the siting or presence of a CAFO in a community.

Environmental Injustice

Disproportionate location of CAFOs in areas populated by people of color or people with low incomes is a form of environmental injustice that can have negative impacts on community health. In North Carolina, Wing, Cole, and Grant (2000) have found patterns of disproportionate siting of corporate CAFOs in rural, lower-income, and African-American communities. This places residents of those communities at disproportionate risk for health and socioeconomic problems. Results of several studies have shown that a disproportionate number of swine CAFOs are located in low-income and non-white areas (Ladd and Edward, 2002) and near predominantly low-income and non-white schools (Mirabelli, Wing, Marshall, and Wilcosky, 2006a, 2006b). These facilities and the hazardous agents and common problems associated with them are generally unwanted in local communities and are often thrust upon those sectors with the lowest levels of political influence. Low-income communities and populations that experience institutional discrimination based on race have higher susceptibilities to CAFO impacts due to poor health status and lack of access to medical care.

Failure of the Political Process

In 2005, the U.S. Government Accountability Office (GAO) issued a report on the effectiveness of the U.S. Environmental Protection Agency (EPA) in meeting its obligations to regulate CAFOs (U.S. Government Accountability Office, 2005). The report identified two major flaws: (1) allowing an estimated 60% of CAFOs in the United States to go unregulated; and (2) a lack of federal oversight of state governments to ensure that the CAFOs they are regulating are adequately implementing required federal regulations for CAFOs. Additionally, many states have not taken a proactive stance to comply with U.S. EPA regulations. Therefore, the concentration of livestock production, most noted by CAFO style production, has continued to expand in most states. This has resulted in many rural communities and individuals taking action on their own, through local ordinances or litigation, as they have not been able to find access through usual government channels.

Property Value Decrease

Several studies have found that property values decrease when CAFOs move into a community. Neighbors of CAFOs are interested in preventing loss of property value, loss of their homes and land, forced changes in their lifestyle, adverse changes in their communities, and threats to their health. However, the legislative process in many states has often been unresponsive to citizens' concerns regarding CAFOs. For example, 13 states have enacted laws that inhibit citizens from speaking freely about agriculture if remarks are deemed disparaging. All 50 states have some form of right-to-farm legislation. Nuisance suits have been the recourse of some residents. Most states have some form of environmental laws protecting the environment and require certain sizes of operations to apply for permits; however, there is little enforcement of these laws due to few provisions and little staff to provide such enforcement.

The following is derived from the part of the Farm Foundation study entitled: "Community and Labor" from the report "The Future of Animal Agriculture in North America, Farm Foundation" issued in 2004. The study links processing, slaughtering, and production, and includes the linkages. Results of research identified in this report document the significance of immigration to meat production most closely in all of the studies.

One significant outcome in the changing dimensions of animal agriculture is a change in the relationship between farms and rural communities. Production units have become larger and more technologically advanced, using supply chains and marketing channels to link to the economy at large. Much production has shifted from independent operators to vertically coordinated operations that largely bypass community linkages. New operations may bring new resources, opportunities, and economic growth to local economies. Large production systems, such as CAFOs, are linked to nearby processing plants that require a concentration of workers who may not be highly paid and who may have to be recruited from other locales. Such systems challenge the socioeconomic milieu of communities where those enterprises are located. New economic opportunities may affect the community's autonomy, norms, traditions, pace, culture, and control.

SUMMARY

This chapter reviews the documentation of and studies on the negative consequences of CAFOs in rural America. There is significant evidence of ongoing and cumulative negative effects. Unfortunately, there has been little political will to address these problems.

REFERENCES

Abeles-Allison, M. and Connor, L. 1990. An analysis of local benefits and costs of Michigan hog operations experiencing environmental conflicts. Department of Agricultural Economics, Michigan State University, East Lansing.

Banker, D.E. and MacDonald, J.M. 2005. Structural and financial characteristics of U.S. farms: 2004 family farm report. *Agriculture Information Bulletin* No. AIB979.

Bullers, S. 2005. Environmental stressors, perceived control and health: The case of residents near large scale hog farms in eastern North Carolina. *Human Ecology* 33: 1–16.

Chism, J.W. and Levins, R.A. 1994. Farm spending and local selling: How do they match up? *Minnesota Agricultural Economist* 676: 1–4.

Constance, D. and Tuinstra, R. 2005. Corporate chickens and community conflict in East Texas: Growers and neighbors' views on the impacts of industrial broiler production. *Culture and Agriculture* 27: 45–60.

Cox, J. 2007. *Industrial Animal Agriculture: Part of the Poverty Problem.* London: World Society for the Protection of Animals.

Craypo, C. 1994. Meatpacking: Industry restructuring and union decline. In: *Contemporary Collective Bargaining in the Private Sector*, Voos, P., Ed. Madison, WI: Industrial Relations Research Association, pp. 63–96.

Crowley, M.L. 1999. The impact of farm sector concentration on poverty and inequality: An analysis of north central U.S. counties. Master's Thesis, Department of Sociology, The Ohio State University, Columbus, OH.

Crowley, M.L. and Roscigno, V.J. 2004. Farm concentration, political economic process and stratification: The case of the North Central U.S. *Journal of Political and Military Sociology* 31: 133–155. Farming in Iowa: The applicability of Goldschmidt's findings fifty years later. *Human Organization* 55: 409–415.

DeLind, L.B. 1995. The state, hog hotels, and "the right to farm": A curious relationship. *Agriculture and Human Values* 12: 34–44.

DeLind, L.B. 1998. Parma, a story of hogs, hotels and local resistance. In: *Pigs, Profits, and Rural Communities*, Thu, K. and Durrenberger, E.P., Eds. Albany, NY: State University of New York Press, pp. 23–38.

Deller, S.C. 2003. Agriculture and rural economic growth. *Journal of Agricultural and Applied Economics* 35: 517–527.

Donham, K.J. 2000. The concentration of swine production: Effects on swine health, productivity, human health, and the environment. *Veterinary Clinics of North America: Food Animal Practice* 16: 559–597.

Donham, K.J., Wing, S., Osterberg, D., Flora, J.L., Hodne, C., Thu, K.M., and Thorne, P.S. 2007. Community health and socioeconomic issues surrounding concentrated animal feeding operations. *Environmental Health Perspectives* 115(2).

Dosman, J.A., Lawson, J.A., Kirychuck, S.P., Cormier, Y., Biem, J., and Koehcke, N. 2004. Occupational asthma in newly employed workers in intensive swine confinement facilities. *European Respiratory Journal* 24(4): 698–702.

Drabenstott, M. and Smith, T.R. 1996. The changing economy of the rural heartland. In: *Economic Forces Shaping the Rural Heartland*. Kansas City, KS: Federal Reserve Bank, pp. 1–11.

Flora, C.B. and Flora, J.L. 1988. Public policy, farm size and community well-being in farming dependent counties of the plains. In: *Agriculture and Community Change in the U.S.: The Congressional Research Reports,* Swanson, L.E., Ed. Boulder CO: Westview Press, pp. 76–129.

Flora, C.B., Carpenter, S., Hinrichs, C., Kroma, M., Lawrence, M., Pigg, K., Durgan, B., and Draeger, K. 1999. A summary of the literature related to social/community. Prepared for the Minnesota Environmental Quality Board by the University of Minnesota. A1–A68.

Flora, J.L., Brown, I., and Conby, J.L. 1977. Impact of type of agriculture on class structure, social well-being, and inequalities. Paper presented at the annual meeting of the Rural Sociological Society, Burlington, VT, August.

Flora, J.L. and Flora, C.B. 1986. Emerging agricultural technologies, farm size, public policy, and rural communities: The Great Plains and the West. In: *Technology, Public Policy and the Changing Structure of American Agriculture. Vol. 2, Background Papers, Part D: Rural Communities*. Washington, D.C.: Office of Technology Assessment, pp. 168–212.

Foltz, J.D., Jackson-Smith, D., and Chen, L. 2000. Do purchasing patterns differ between large and small dairy farms? Econometric evidence from three Wisconsin communities. Submitted to *Agriculture and Resource Economics Review.*

Fujimoto, I. 1977. The communities of the San Joaquin Valley: The relation between scale of farming, water use, and quality of life. In: *U.S. Congress, House of Representatives, Obstacles to Strengthening the Family Farm System. Hearings before the Subcommittee on Family Farms, Rural Development, and Special Studies of the Committee on Agriculture*, 95th Congress, first session. Washington, DC: U.S. Government Printing Office, pp. 480–500.

Gilles, J.L. and Dalecki, M. 1988. Rural well-being and agricultural change in two farming regions. *Rural Sociology* 53: 40–55.

Goldschmidt, W. 1968. Small business and the community: A study in the central valley of California on effects of scale of farm operations. In: *U.S. Congress, Senate, Corporation Farming, Hearings Before the Subcommittee on Monopoly of the Select Committee on Small Business,* U.S. Senate, 90th Congress, second session, May and July. Washington, DC: U.S. Government Printing Office, pp. 303–433.

Goldschmidt, W. 1978a. *As You Sow: Three Studies in the Social Consequences of Agribusiness*. Montclair, NJ: Allanheld, Osmun and Company.

Goldschmidt, W. 1978b. Large-scale farming and the rural social structure. *Rural Sociology* 43: 362–366.

Gomez, M.I. and Zhang, L. 2000. Impacts of concentration in hog production on economic growth in rural Illinois: An econometric analysis. American Agricultural Association Annual Meeting, Tampa, FL.

Green, G.P. 1985. Large-scale farming and the quality of life in rural communities: Further specification of the Goldschmidt hypothesis. *Rural Sociology* 50: 262–273.

Harris, C. and Gilbert, J. 1982 Large-scale farming, rural income and Goldschmidt's agrarian thesis. *Rural Sociology* 47: 449–458.

Heady, E.O. and Sonka, S.T. 1974. Farm size, rural community income, and consumer welfare. *American Journal of Agricultural Economics* 56: 534–542.

Heffernan, W.D. 1972. Sociological dimensions of agricultural structures in the United States. *Sociologia Ruralis* 12: 481–499.

Heffernan, W.D. 1974. Social consequences of vertical integration: A case study. *Proceedings of Southern Agricultural Scientists.*

Heffernan, W.D. and Lasley, P. 1978. Agricultural structure and interaction in the local community: A case study. *Rural Sociology* 43: 348–336.

Heffernan, W.D. and Hendrickson, M. 2007. University of Missouri at Columbia, MO. Update of Concentration Studies, Commissioned by the National Farmers Union, Denver, Colorado.

Heinrichs, C. and Welsch, R. 2003. The effects of the industrialization of U.S. livestock agriculture on promoting sustainable production practices. *Agriculture and Human Values* 20: 125–141.

Hodne, C. Unpublished research. Iowa State University, Ames, IA.

Humann, M.J., Donham, K.J., Jones, M.L., Achutan, C., and Smith, B.J. 2005. Occupational noise exposure assessment in intensive swine farrowing systems: Dosimetry, octave band, and specific task analysis. *Journal of Agromedicine* 10(1): 23–37.

Jackson-Smith, D. and Gillespie, Jr., G.W. 2005. Impact of farm structural change on farmers' social ties. *Society and Natural Resources* 18: 215–240.

Kanagil, R. 1997. *The One Best Way: Frederick Winslow Taylor and the Enigma of Efficiency.* New York: Penguin Books.

Kenney, M., Lobao, L.M., Curry, J., and Goe, W.R. 1989. Midwestern agriculture in U.S. fordism: From the new deal to economic restructuring. *Sociologia Ruralis* 29(2): 130–148.

Keystone Center. 2001. Keystone national policy dialogue on trends in agriculture. Keystone Center, Keystone, CO, March.

Kleiner, A.M., Rikoon, J.S., and Seipel, M. 2000. Pigs, participation, and the democratic process: The impacts of proximity to large-scale swine operations on elements of social capital in northern Missouri communities. Paper presented at the annual meeting of the Rural Sociological Society, Washington, DC.

Kleiner, A.M. 2003. Goldschmidt revisited: An extension of Lobao's work on units of analysis and quality of life. Paper presented at the annual meeting of the Rural Sociological Society. Montreal, Quebec, Canada, July.

Ladd A.E. and Edward, B. 2002. Corporate swine and capitalist pigs: A decade of environmental injustice and protest in North Carolina. *Social Justice* 29: 26–46.

Lobao, L. 1987. *Farm Structure, Industry Structure and Socioeconomic Condition.* Albany, NY: State University of New York Press.

Lobao, L. 1993. Forward. In: *Contemporary Sociology,* Barkley, D., Ed. 22(5): vii–ix.

Lobao, L.M. 1990. *Locality and Inequality: Farm and Industry, Structure and Socioeconomic Condition.* Albany, NY: State University of New York Press.

Lobao, L.M. 2000. Industrialized farming and its relationship to community well-being: Report prepared for the State of South Dakota. Pierre, SD: Office of the Attorney General.

Lohr, L., 1996. Perceptions of rural air quality: What will the neighbors think? *Journal of Agribusiness* 14.

Lyson, T.A., Torres, R.J., and Welsh, R. 2001. Scale of agricultural production, civic engagement, and community welfare. *Social Forces* 80: 311–327.

MacCannell, D. 1988. Industrial agriculture and rural community degradation. In: *Agriculture and Community Change in the U.S.: The Congressional Research Reports,* Swanson, L.E., Ed. Boulder CO: Westview Press, pp. 15–75.

MacCannell, D. and Dolber-Smith, E. 1986. Report on the structure of agriculture and impacts of new technologies on rural communities in Arizona, California, Florida, and Texas. In: *Technology, Public Policy and the Changing Structure of American Agriculture, Vol. 2, Background Paper, Part D: Rural.* pp. 19–167.

Marousek, G. 1979. Farm size and rural communities: Some economic relationships. *Southern Journal of Agricultural Economics* 11: 57–61.

Martinson, O.B., Wilkening, E.A., and Rodefeld, R.D. 1976. Feelings of powerlessness and social isolation among "large-scale" farm personnel. *Rural Sociology* 41: 452–472.

McCurdy, S.A. and Carroll, D.J. 2000. Agricultural injury. *American Journal of Industrial Medicine* 38(4): 463–480.

McMillan, M. and Schulman, M.D. 2003. Hogs and citizens: A report from the North Carolina front. In: *Communities of Work: Rural Restructuring in Local and Global Contexts,* Falk, W.W., Schulman, M.D., and Tickmayer, A.R., Eds. Athens, OH: Ohio University Press, pp. 219–239.

Minnesota Planning Agency. 2002. Environmental Quality Board, Final Animal Agriculture Generic Environmental Quality Impact Statement (GEIS), September 14.

Mirabelli, M.C., Wing, S., Marshall, S., and Wilcosky, T. 2006a Asthma symptoms among adolescents who attend public schools that are located near confined swine feeding operations. *Pediatrics* 118(1): e66–e75.

Mirabelli, M.C., Wing, S., Marshall, S., and Wilcosky, T. 2006b. Race, poverty, and potential exposure of middle school students to air emissions from confined swine feeding operations. *Environmental Health Perspectives* 114: 591–596.

North Central Regional Center for Rural Development (NCRCRD). 1999. The impact of recruiting vertically integrated hog production in agriculturally based counties of Oklahoma. Report to the Kerr Center for Sustainable Agriculture. Iowa State University, Ames, IA.

Osterberg, D. and Wallinga, D. 2004. Addressing externalities from swine production to reduce public health and environmental impacts. *American Journal of Public Health* 94(10): 1703–1708.

Pereira, F. and Goldsmith, P.D. 2005. From negative externalities to industrial illegitimacy, an empirical analysis of the Illinois livestock industry. Presented at the annual meeting of the Eastern Academy of Management Association, Springfield, MA.

Peters, D.J. 2002. Revisiting the Goldschmidt hypothesis: The effect of economic structure on socioeconomic conditions in the rural Midwest. Technical Paper P-0702-1, Missouri Department of Economic Development, Missouri Economic Research and Information Center, Jefferson City, MO.

President's Council on Sustainable Development. 1996. *Sustainable America: A New Consensus for the Prosperity, Opportunity and a Healthy Environment for the Future.* Washington, DC: U.S. Government Printing Office.

Poole, D.L. 1981. Farm scale, family life, and community participation. *Rural Sociology* 46: 112–127.

Rautiainen, R.H. and Reynolds, S.J. 2002. Agricultural Safety and Health Conference: Using Past and Present to Map Future Actions.

Rodefeld, R.D. 1974. The changing organization and occupational structure of farming and the implications for farm work force individuals, families and communities. Ph.D. dissertation, The University of Wisconsin, Madison.

Schiffman, S., Miller, E.A., Suggs, M.S., and Graham, B.G. 1995. The effect of environmental odors emanating from commercial swine operations on the mood of nearby residents. *Brain Research Bulletin* 37: 369–375.

Schiffman, S., Slattery-Miller, E.A., Suggs, M.S., and Graham, B.G. 1998. Mood changes experienced by persons living near commercial swine operations. In: *Pigs, Profits, and Rural Communities,* Thu, K.M. and Durrenberger, E.P., Eds. Albany, NY: The State University of New York Press, pp. 84–102.

Seipel, M., Hamed, M., Rikoon, J.S., and Kleiner, A.M. 1998. The impact of large-scale hog confinement facility sitings on rural property values. *Conference Proceedings: Agricultural Systems and the Environment,* pp. 415–418.

Seipel, M., Hamed, M., Rikoon, J.S., and Kleiner, A.M. 1999. Rural residents' attitudes toward increased regulation of large-scale swine production. Paper presented at the Annual Meetings of the Rural Sociological Society, August.

Shusterman, D. 1992. Critical review: The health significance of environmental odor pollution. *Archives of Environmental Health* 47: 76–87.

Skees, J.R. and Swanson, L.E. 1988. Farm structure and rural well-being in the south. In: *Agriculture and Community Change in the U.S.: The Congressional Research Reports,* Swanson, L.E., Ed. Boulder CO: Westview Press, pp. 238–321.

Skees, J.R. and Swanson, L.E. 1998. Agriculture in the South and the interaction between farm structure and well-being in rural areas. In: *Technology, Public Policy and the Changing Structure of American Agriculture. Vol. 2, Background Papers, Part D: Rural Communities.* Washington, D.C.: Office of Technology Assessment, pp. 373–495.

Small Farm Viability Project. 1977. *The Family Farm in California: Report of the Small Farm Viability Project.* Sacramento, CA: Employment Development, the Governor's Office of Planning and Research, the Department of Food and Agriculture in the Department of Housing and Community Development.

Smithers, J., Pal Johnson, P., and Joseph, A. 2004. The dynamics of family farming in North Huron County, Ontario. Part II: Farm-community interactions. *The Canadian Geographer* 48: 209–224.

Starmer, E., 2006. *Feeding the Factory Farm: Implicit Subsidies to the Broiler Chicken Industry.* Medford, MA: Global Development and Environment Institute, Tufts University.

Steinfeld, H., Gerber, P., Wassenaar, T., Castel, V., Rosales, M., and de Haan, C. 2006. *Livestock's Long Shadow.* Rome: Food and Agriculture Organization of the United Nations.

Stofferahn, C. 2006. *Industrialized Farming and Its Relationship to Community Well-Being: An Update of a 2000 Report by Linda Lobao.* Grand Forks, ND: University of North Dakota.

Stull, D., Broadway, M.J., and Griffith, D. 1995. *Any Way You Cut It: Meat Processing and Small-Town America.* Lawrence, KS: University Press of America.

Taylor, F.W. 1911. *Principles of Scientific Management.* New York.

Tetreau, E.D. 1938. The people of Arizona's irrigated areas. *Rural Sociology* 3: 177–187.

Tetreau, E.D. 1940. Social organization in Arizona's irrigated areas. *Rural Sociology* 5: 192–205.

Thompson, N. and Haskins, L. 1998. *Searching for Sound Science: A Critique of Three University Studies on the Economic Impacts of Large-Scale Hog Operations.* Walthill, NE: Center for Rural Affairs (article cited in the University of Minnesota Report).

Thu, K. and Durrenberger, P. Eds. 1998. *Pigs, Profits, and Rural Communities.* Albany, NY: State University of New York Press.

U.S. Government Accountability Office. 2005. Livestock market reporting: USDA has taken some steps to ensure quality, but additional efforts are needed. GAO-06-202, December 9, 2005. Available: http://www.gao.gov/new.items/d06202.pdf (accessed June 15, 2006).

Van Kleek, R.J. and Bulley, N.R. 1985. An assessment of separation distance as a tool for reducing farmer/neighbor conflict. *Proceedings of the Fifth International Symposium on Agricultural Wastes.* American Society of Agricultural Engineers, St. Joseph, MI.

Von Essen, S.G. and McCurdy, S.A. 1998. Health and safety risks in production agriculture. *Western Journal Medicine* 169(4): 214–220.

Welsh, R. and Lyson, T.A. 2001. Anti-corporate farming laws, the "Goldschmidt hypothesis" and rural community welfare. Paper presented at the annual meeting of the Rural Sociological Society, Albuquerque, NM.

Wheelock, G.C. 1979. Farm size, community structure and growth: Specification of a structural equation model. Paper presented at the annual meeting of the Rural Sociological Society, Burlington, Vermont, August.

Wilson, S.M., Howell, F., Wing, S., and Sobsey, M. 2002. Environmental injustice and the Mississippi hog industry. *Environmental Health Perspectives* 110(2): 195–201.

Wing, S. 2002. Social responsibility and research ethics in community driven studies of industrialized hog production. *Environmental Health Perspectives* 108: 225–231.

Wing, S., Cole, D., and Grant, G. 2000. Environmental injustice in North Carolina's hog industry. *Environmental Health Perspectives* 108: 225–231.

Wing, S. and Wolf, S. 1999. Intensive livestock operations, health and quality of life among Eastern North Carolina residents. Report to the North Carolina Department of Health and Human Services. Chapel Hill, NC: Department of Epidemiology, University of North Carolina.

Wing, S. and Wolf, S. 2000. Intensive livestock operations, health and quality of life among Eastern North Carolina residents. *Environmental Health Perspectives* 108(3): 233–238

Wise, T.A. and Starmer, E. 2007. Industrial livestock companies' gain from low feed prices, 1997–2005, Global Development and Environmental Institute, Tufts University, February.

Wright, W., Flora, C., Kremer, K., Goudy, W., Hinrichs, C., Lasley, P., Maney, A., Kronma, M., Brown, H., Pigg, K., Duncan, B., Coleman, J., and Morse, D. 2001. Technical work paper on social and community impacts. Prepared for the Generic Environmental Impact Statement on Animal Agriculture and the Minnesota Environmental Quality Board.

7 Implementing Effective Practices and Programs to Assess Animal Welfare

John J. McGlone and Temple Grandin

CONTENTS

INTRODUCTION

Farm animal welfare is important to livestock and poultry producers, governments, consumers, and retailers who sell animal products. In the United States, the Humane Methods of Slaughter Act regulates the humane slaughter of food animals and a 137-year-old law requires that pigs, cattle, and sheep not be transported more than 28 hours without feed and water, but the United States does not use government oversight to manage on-farm animal care. In contrast, the European Union has passed extensive laws and regulations that increase welfare standards for farm animals. There are no such national laws regulating humane treatment of animals on farms in the United States. Animal welfare concerns that may arise on farms range from the prevention of obvious abuse to ethical issues where scientific research alone cannot provide all the answers.

In the United States, the oversight of farm animal care falls to food retailers including grocery stores, restaurants, and intermediate suppliers. McDonald's, for instance, has taken the lead in auditing farm animal slaughter plants and some animal production facilities. Although McDonald's is recognized as an early adopter of humane oversight, virtually all of the major retailers have since developed some kind of animal welfare standards for the animals they purchase. This is primarily because corporations that sell products to consumers need to understand and manage their supply chains. It is in a company's interest to assure its customers that products have certain defined qualities, including that some humane standard has been met. In this way, the retailers are driving change in the U.S. animal agricultural industries.

The majority of animal products sold come from commodity producers. These farmers and processors produce relatively uniform animal products that meet defined minimum standards. Other "high welfare" animal products are also available for niche markets. Humane nongovernmental organizations (NGOs) have programs to certify animal products as having come from production systems that are viewed by the NGO as more humane. They use terms such as "natural," "free-range," "cage-free," "pasture-raised," and others to convey to their select consumer population that their products are from farms that meet certain animal welfare qualities, which shoppers deem important. Typically, these programs also include other non-welfare qualities of food production such as no antibiotic or hormone use, or that the animals were fed a vegetarian diet (no animal products in the feed). This chapter reviews major programs and practices for oversight of farm animal welfare for the majority of animal products provided for the commodity markets. Many programs are in development and are expected over time to be more inclusive of the entire production chain.

QUALITIES OF OVERSIGHT PROGRAMS

Any particular element of animal welfare is subjective, but people generally do not want animals to experience pain or distress, nor do they like the idea of animals being confined. Judging by two-thirds of the voters in Florida, Arizona, and California who voted to ban the use of gestation crates for pregnant sows, crates for veal calves, and cages for laying hens, one can assume that many Americans do not want farm animals kept in crates and cages. Importantly, most consumers are not familiar with production practices on commercial farms.

Livestock and poultry producers have focused on providing animal care in the form of a nutritious diet, a comfortable temperature, and sanitary conditions. They have not focused on, for instance, spacious environments. Economic pressures drive producers to provide the smallest space possible that still allows for normal animal growth and reproduction. Accommodating animal behavioral needs has also not been a worry of commodity producers. For example, veal and pork producers have not been concerned with the fact that animals cannot turn around in their stalls and crates. The consumer, however, was sufficiently unhappy about the inability to turn around that they voted by popular referendum to ban gestation crates for sows and battery crates for laying hens. A disconnect between consumer priorities and farming systems becomes apparent. Likewise, there is a disconnect between what animal producers and consumers want in on-farm welfare assessment tools. However, new farm animal welfare programs are being implemented at some level and we expect that they will evolve over time until the programs more directly reflect the issues consumers *and* producers think are important.

The role of science in the evolution of these industry programs is of interest and concern, as it is clearly and easily trumped by consumer/voter preferences. For example, two separate bodies of scientists concluded that sow productivity is similar or equivalent when sows are kept in well-managed pens or crates (McGlone et al., 2004; Rhodes et al., 2005). Consumers in Florida, Arizona, and California, however, made their decisions based on something other than scientific evidence. Both farm animal producers and consumers may accept science as helpful in making decisions but, in reality, each consumer uses a combination of science plus his or her own sensibilities or ethics to make a choice at the ballot box.

Economics also usually trump animal welfare. For example, pigs in the United States are castrated. This is a painful practice (McGlone and Hellman, 1988) done primarily to avoid an off-taste or odor in meat from boars. In the United Kingdom, they do not castrate, instead minimizing boar odor by processing the animals at a much younger age. In the United States, slaughter plant efficiencies drive up slaughter weights because it takes the same labor to process a 300-lb pig as it does a 200-lb pig. Thus, the cost of labor is reduced by getting more meat per worker from larger animals. In Brazil and other countries, some producers have eliminated castration altogether by using an injection that blocks male hormones (Pfizer is developing Improvac, which is a vaccine against GnRH that eliminates male sex hormones). Pork quality is preserved with this method (Gispert et al., 2010). Until this new product is approved in the United States, economic forces will favor castration as they favor the use of the gestation crates for commodity pork. The only way to solve problems such as these is through legislation or retailer requirements; for example, by regulating space requirements for sows in group housing.

Any effective animal welfare oversight program must be credible, workable, and affordable. Often, credible and affordable oversight programs are at cross-purposes. The most credible program would have inspectors or video surveillance at every farm all the time; however, this is not affordable at present. However, video auditing of slaughter plants by an outside auditor company is becoming more and more common. The Cargill Corporation conducts video audits by an independent auditing company in all their North American beef and pork plants. A balance must be struck to foster use of oversight systems that provide as much credibility as can be afforded. This often means that only a sample of animals can be assessed for short periods for compliance with program goals.

The Food Marketing Institute (FMI) and the National Council of Chain Restaurants (NCCR) attempted to establish on-farm animal welfare audits through an organization called Sustainable Environmental Solutions (SES), but the animal industries viewed this program as unacceptable and the program failed. The animal welfare oversight programs have been directed by the retailers and developed by each industry group. Retailers are asking for farm animal welfare oversight in the form of audits of farm animal care and acceptance of the best management practices. Most commercial farms have accepted that it is their responsibility to provide assurances to their customers (the food retailers) and animal welfare programs that are in development.

INDIVIDUAL ANIMAL INDUSTRY OVERSIGHT PROGRAMS

SWINE PROGRAM

The National Pork Board (NPB) is the national trade association for commercial pork producers. All producers who sell live pigs for slaughter are required to pay into the national check-off program. Therefore, essentially all commercial pork producers in the United States are part of the NPB. The association started its Swine Welfare Assurance Program (SWAP) some years ago, and its guidelines are detailed in the *Swine Care Handbook* (http://www.pork.org/filelibrary/AnimalWell-Being/swine%20care%20handbook%202003.pdf). At the same time, there was a program called the Pork Quality Assurance (PQA) program that had a small animal welfare component and a food safety/antibiotic control program. The two programs were merged to form the PQA+ program (http://www.pork.org/Certification/17/pqaPlusMaterials.aspx).

The PQA+ program focuses on 10 good production practices. These practices can be divided into topics of records, animal identification, education, medications, and animal well-being. The animal well-being component includes evaluation of the veterinary-client-patient relationship, animal records, animal observation, animal worker training, emergency back-up systems, and facility observations. The PQA+ program includes an educational program that ends in producer certification (that has to be renewed every 3 years), and an on-site assessment by an educator. A sample of animals on each farm is observed and documented. This program does not tell retailers or consumers which housing system or production practices are used (e.g., gestation crates or pens). Rather, it is meant to be an assessment of animal welfare in any production system. Most pork processors require their suppliers to participate in the PQA+ program.

The PQA+ program has enrolled over 50,000 sites in the United States. Another program is currently under development to perform a 9-point audit/verification that this program is, in fact, working. The verifiers are expected to visit 90 sites in 2011 to document whether the industry as a whole is complying. Assurance that individual farms are providing compliance with the PQA+ or other programs is expected in the future, but the cost is viewed as a barrier to industry-wide audits of on-farm animal welfare.

Dairy and Beef Cattle Programs

Some good sources of information on dairy cow well-being can be found at www.centerfordairyexcellence.org. Many states also have their own animal care programs published by their university extension services. For beef cattle, the national beef quality assurance program can be accessed at www. bqa.org. This website contains industry guidelines on transport, the care and handling of cattle, and a manual for trainers. The beef and dairy industries are not as far along as the pork industry. However, many individual or groups of dairy and beef feedlots have animal welfare programs of varying intensity (http://animalagalliance.org/current/home.cfm?Category=Animal_Welfare&Section=Beef).

Laying Hen Program

The primary national program for egg production by laying hens is the United Egg Producers (UEP). Not all laying hen producers are members of UEP; however, UEP members claim to sell approximately 90% of all U.S. eggs (http://www.uepcertified.com/about/). Most UEP members have hens in battery cages and UEP suggests 98% of hens are currently in battery cages (http://www.uepcertified.com/program/guidelines/categories/housing-space-feed-water). This means most hens are in cages with minimum space allowances (e.g., 67 in^2 per bird).

The UEP certified program provides on-farm animal welfare assessments. The program includes assurances that feed and water are provided, as well as management guidelines for molting, handling, transport, euthanasia, biosecurity, and animal health. The UEP also has guidelines for cage-free egg production.

The EU and California are phasing out the use of battery cages for egg production. It is unclear if the California egg producers will change their facilities, move out of state, or go out of business.

Meat Bird Programs (Chickens and Turkeys)

The National Chicken Council (NCC) is the organization that oversees commercial enterprises for production of meat chickens or broilers. The NCC has guidelines and an audit checklist for broilers and broiler-breeders (http://www.nationalchickencouncil.com/aboutIndustry/detail.cfm?id=19). There is no formal requirement to comply with this program, although most major chicken producers do follow these guidelines.

The NCC animal welfare program includes the categories of nutrition and feeding, comfort and shelter, health care, expression of normal behaviors, and best practices. The on-farm audit has nine pages with considerable detail. Chickens in the United States are typically kept in bedded large barns with thousands of other chickens. The broiler industry has not been under attack by activists or legislative action as in the case for the laying hen industry. Major issues of floor space, animal health, leg soundness, and euthanasia are included in the chicken audit.

The National Turkey Federation (NTF) is the national organization to which most turkey producers belong. They have published animal care best management practices (http://www.eatturkey. com/foodsrv/pdf/NTF_animal_care.pdf). The program includes guidelines and audits of the turkey hatchery, poults, and larger turkey barns. The 10-page audit instrument includes bird health and welfare, and handling and transport of turkeys, along with a biosecurity and facility audit.

Niche Market Programs

Many niche market programs are available that claim to certify that a higher level of on-farm animal welfare is provided. Often, these programs include other features such as no antibiotics or hormones. None of the programs are strictly animal welfare-based, although that is the main theme. Many audit companies will provide on-farm or in-processing plant audits, and will do so to any number of standards.

American Humane Certified has a program called The Humane Touch (http://thehumanetouch. org/). It has programs for producers of pork, chickens, turkeys, laying hens, buffalo, veal calves, dairy cows, and beef cattle. American Humane Certified conducts on-farm annual audits of its certified producers. Certified Humane Raised and Handled is an organization endorsed by 30 other humane organizations (including HSUS, ASPCA, and others) (http://www.certifiedhumane.org/). The Animal Welfare Institute (AWI) also provides an Animal Welfare Approved program (http:// www.awionline.org/ht/d/sp/i/11779/pid/11779). Besides animal welfare, it emphasizes production by family farmers. Organic livestock and poultry farms must comply with organic guidelines, which may include some animal welfare standards. Being certified organic, however, does not automatically mean compliance with any of the other humane organization programs.

OUTCOME VS. INPUT STANDARDS

Welfare standards can be divided into two main categories: (1) outcome based or animal based and (2) input based. Other names for input-based standards are engineering standards or resource-based standards. The modern trend in standards for welfare, the environment, and food safety is to use an outcome-based standard whenever possible (Wray, Main, Green, and Webster, 2003). An example of an outcome-based standard is scoring of swellings on the legs of dairy cows that are the outcome of many bad practices such as poor design of the cows' free-stalls, the wrong type of bedding, or poor maintenance of the bedding (Fulwider et al., 2007). An input standard would specify exactly how to design a free-stall for cows. It is not necessary to specify how a free-stall is designed because a wrong design will be detected by the outcome measure. The movement away from specific engineering design standards is good because a rigid requirement for a specific design can stifle innovation. Producers would be prevented from trying a new design that may improve animal welfare.

The desired types of input standards regulate prohibited practices and prevent severe welfare problems. Practices that may be prohibited may vary depending on the requirements of a niche market. For example, many of the guidelines for high welfare niche markets prohibit sow stalls (American Humane Certified, AWI, and Certified Humane). A common input standard that is used for many animal species is the maximum ammonia levels allowed for indoor housing. Using an outcome-based method to measure the bad effects of ammonia on animal health would be too difficult. A common standard for ammonia is 25 ppm maximum (NIOSH, www.cdc.gov/niosh/npgd0028. html), but 10 ppm is the goal (Jones, 2005).

When animals or birds are housed in enclosed indoor housing where the animal's life is dependent on a functional mechanical ventilation system, there must be backup systems for supplying ventilation when the electric power fails. Animals or birds can die from heat stress if the ventilation fails. Some common backup systems are automatic devices to open the sides of the building, automatic telephone dialers, diesel generators, or tractor-powered generators. During a welfare assessment, these backup devices are inspected. An input standard is essential because the outcome measure of a failure to provide ventilation could be many dead animals. Another common input standard is minimum space requirements for transport and housing. For some of the organic or welfare niche market programs, there may be additional input standards such as access to pasture or straw bedding for pigs.

MEAT PROCESSING PROGRAM

Successful outcome-based measures have been used by buyers of large restaurant companies since 1999 to avoid animal welfare issues in slaughter plants. The use of a simple numerical scoring audit system has resulted in great improvements (Grandin, 2000, 2005). Vague terms such as "proper" procedures or providing "adequate" space are avoided because one person's idea of a proper procedure may be very different from another person's idea. Numerical scoring is simple. Like traffic laws, a slaughter plant must achieve a specified level of performance to pass an audit from

McDonald's. Plant management knows exactly what is expected and there are five numerically scored outcome standards (Grandin, 1998a). The complete guideline is on www.animalhandling.org. To pass a McDonald's, Wendy's, or Whole Foods audit, the plant has to achieve a passing score on all five measures. Another advantage of numerical scoring is that a plant can also determine if it is improving or becoming worse. A plant will also fail an audit if an act of abuse occurs. Some examples of acts of abuse are dragging conscious, non-ambulatory animals, beating animals, or poking sensitive areas. The five numerically scored outcome measures are as follows:

1. Percentage of animals stunned effectively with one application of the stunner. For captive bolt stunning, the first shot must be effective on 95% of the animals. For electric stunning, the stunner must be placed correctly on 99% of animals. When the stunner is misapplied, it must be re-applied immediately before the animal is bled, hoisted, or cut. When this system was first started in 1999, only 30% of the beef plants could achieve this level (Grandin, 1998a). This was due to poor stunner maintenance (Grandin, 1998a, 2002). Now, over 90% of the plants can do this (Grandin, 2005, 2006).
2. Percentage of animals falling anywhere in the facility. For a passing score, the percentage of animals falling must be 1% or less. Falling is usually an outcome of either slick flooring or poor handling by people.
3. Percentage of animals vocalizing (bellowing, mooing, or squealing) during stunning or entry into the stunning area. Vocalization is scored on a per animal basis as either silent or vocal. To pass an animal welfare audit, vocalization must not exceed 5% of the pigs or 3% of the cattle. If a head holder is used on cattle, a score of 5% is allowed. Vocalization is correlated in both pigs and cattle with physiological measures of stress (Dunn, 1990; Lay et al., 1992; Warriss, Brown, and Adams, 1994; and White et al., 1995). In slaughter plants, vocalizations in cattle are associated with aversive events such as electric prods, missed stuns, or excessive pressure from a restraint device (Grandin, 1998b, 2003). Vocalization scoring cannot be used in sheep because they remain silent when they are stressed. Since the welfare audits started, the percentage of animals vocalizing has decreased greatly. In 1997, the worst plant had 32% of the cows vocalizing. More recent data obtained from restaurant audits shows that the worst plants are under 10% and the majority of plants have passing scores of 0% to 2% (www.grandin.com).
4. Insensibility. To pass a welfare audit, 100% of the animals must be unconscious and insensible before hoisting, bleeding, or skinning.
5. Percentage of animals moved with an electric prod. For pigs and cattle, an excellent score is 5% or less. A minimum passing score is 25% or less. A common question is: "Should electric prods be banned?" If the prods are totally banned, handlers may resort to more abusive practices such as poking a stick in the animal's rectum. However, the OIE (2006) banned electric prods on small calves, piglets, sheep, and horses. Before the audits were started, most animals were prodded multiple times with electric prods.

SIMPLE IMPROVEMENTS

Most slaughter plants were able to achieve passing scores by making simple improvements; they did not have to rebuild their entire facility. Some of these improvements are listed as follows:

1. Train employees in behavioral principles of animal handling (Grandin, 2007; Kilgour and Dalton, 1984).
2. Install non-slip flooring in unloading ramps and stunning boxes.
3. Install lamps on dark chute entrances because animals tend to move from a dark place to a brighter place (Grandin, 1982, 1996, 2007; Van Putten and Elshof, 1978).

4. Move lights to eliminate reflections on shiny surfaces or wet floors. Shadows and high contrasts of light and dark will cause animals to balk and refuse to move (Tanida, Miura, Tanaka, and Yoshimoto, 1996).
5. Install solid barriers to prevent animals from seeing moving people and equipment up ahead, or cover the sides of the races (Grandin, 2007; Kilgour, 1971).
6. Improve maintenance of stunners and store cartridges for captive bolt stunners in a dry location (Grandin, 2002).

BASIC OUTCOME MEASURES FOR ALL MAMMALS AND BIRDS ON FARMS

These basic outcome measures will help prevent severe welfare problems in agricultural animals that are raised for food, and can be easily used on farms:

1. Body condition score of breeding stock. Poor body condition may be an outcome of starvation, neglect, illness, inadequate diet, or tooth problems in older animals, and animals may be skinny or emaciated. Charts are available for body condition scoring of dairy cows (Wildman, Jones, Wagner, and Brown, 1982; University of Wisconsin, 2005), beef cows (http://www.cowbcs.info/), sows (NPB), and sheep (http://www.smallstock.info/tools/condscor/cs-sheep.htm). Animals that are emaciated and weak should be euthanized on the farm.
2. Lameness. Lame animals have poor welfare because lameness causes pain (Rushen, Pombourceq, and dePaisselle, 2006). It is also an important outcome measure, and can indicate a variety of problems. For example, foot diseases, poor flooring, or lack of hoof care can cause lameness in dairy cows. The best 10% of dairies had 5% or less lame cows. The national average is 24.6% (Espejo, Endres, and Salfer, 2006). Training materials to assess lameness in cattle can be found in Amstel and Shearer (2006) and online (http://www.csubeef.com/files/resources/Lameness-Rules_of_Thumbv2.pdf; http://www.merckvetmanual.com/mvm/index.jsp?cfile=htm/bc/90500.htm). For poultry, lameness (gait score) information can be found in Knowles et al. (2008).
3. Animal or bird cleanliness. Animals that are housed on poorly maintained litter or in a muddy feedlot may have manure or mud caked on their bodies. The most common cleanliness scoring systems use a 4-point scale, which ranges from a clean animal to an animal that has most of its body soiled. When pigs are housed outside with wallows the cleanliness score would be eliminated, but there must be an area of the pen where the pigs have access to clean, dry bedding or forage. In poultry, damage to the feet, hocks, and breasts due to wet or dirty litter can be easily assessed at the slaughter plant.
4. Mortality and morbidity. Health is an essential component of welfare, but it is not the only component (Fraser, 2008). Fulwider et al. (2007) found that high producing dairy cows have more leg lesions. Farms with high levels of sickness and death would have poor welfare. Some pigs are more susceptible to going down and becoming non-ambulatory during handling and transport. The use of beta-agonists to promote growth may make pigs weak and more difficult to handle (Marchant-Forde et al., 2003).
5. Sores and lesions. Each species and form of housing has specific types and patterns of lesions and injuries. Dairy cows housed in free-stalls often get swollen hocks (Fulwider et al., 2007) and sows housed in gestation stalls may develop ulcers (pressure sores) on their shoulders (Zurbrigg, 2006). Unpublished industry data has shown that sows on farms that measure rates of shoulder lesions have been able to greatly reduce them by repairing flooring and increasing sow body condition. Sows housed in groups may get severe wounds from fighting. In poultry, birds may inflict severe wounds on each other. Research has shown, in both pigs and chickens, that there are genetic differences in the tendency to fight or inflict wounds on other animals or birds (Muir and Craig, 1998; Løvendahl et al., 2005).

Improving the success of group-housed sows and cage-free chickens may require changes in animal genetics.

6. Coat or feather condition. Animals that have lice or other external parasites will often have bald spots or poor coat condition. In poultry, poor feather condition is associated with feather pecking from other birds, rubbing on the cage or other enclosure, or dirty litter. LayWel (http://www.laywel.eu/web/pdf/deliverables%2031-33%20health.pdf) has excellent feather condition scoring charts. One must remember that birds molt and most mammals will shed hair in the spring. Molting or shedding of hair must not be confused with poor coat or feather condition that is caused by external parasites or abrasions from housing.

7. Easily observed behavioral problems. Some examples are cribbing in horses, bar biting in sows housed without fibrous bedding, self-injurious behavior, urine sucking in calves, and tail biting in pigs. Obviously, a behavior that inflicts either self-injury or injury to another animal is detrimental to animal welfare. Other abnormal behaviors, such as tongue rolling in dairy cattle, do not cause damage to the animal. It is beyond the scope of this chapter to discuss all the welfare implications of abnormal behaviors or the scientific research on behavioral needs, although research clearly shows that some behaviors are highly motivated (Van der Weerd and Day, 2009; Duncan and Kite, 1989).

8. Animal handling practices. Handling of animals on the farm can be assessed using similar measures as the ones discussed previously for slaughter plants. The National Cattlemen's Beef Association has a scoring system for vocalization, falling, electric prod score, and exit speed from the squeeze chute. This can be used during routine handling procedures such as vaccination or pregnancy checking on the ranch or feedlot. The outcomes of poor handling can also be assessed. Rough handling of cattle or poultry during loading for transport significantly raises the level of bruising and broken wings (Grandin, 2010). Injuries from abusive handling, such as broken tails in dairy cows, can also be easily assessed when a welfare assessor walks through the herd.

9. Thermal stress. Heat stress both during transport and on the farm is becoming an increasingly important issue because both cattle and poultry are being grown to heavier weights. To prevent heat stress in chickens, growers have installed water-cooled ventilation systems. Modern, fast-growing pigs are also more susceptible to heat stress. Cattle or birds that are panting are severely heat stressed. Mader, Davis, and Brown-Branl (2005) developed a simple panting scoring system for cattle. Cattle that exhibit opened mouth panting are severely heat stressed and they are in danger of dying unless they are given relief. Cold stress is less of a thermal welfare problem overall in livestock and poultry enterprises unless frostbite is present.

DIRECTLY OBSERVABLE MEASURES

All the measures that have been discussed can be observed directly by an assessor who visits a slaughter plant or farm. They are not based on paperwork. There is an unfortunate tendency in some auditing systems to have an overemphasis on paperwork. There are often situations where all paperwork can be in order and the physical conditions of both the animals and the facility are bad.

THREE TYPES OR LEVELS OF ANIMAL WELFARE AUDITS

The most effective animal welfare and food safety auditing programs conducted by major retailers have three parts. They are: (1) internal audits done on the farm or plant; (2) outside audits done by a third party, independent of an auditing company; and (3) audits done by representatives of the retailer. In typical restaurant auditing programs, all of the slaughter plants are visited every year by a third-party auditor and a small percentage are visited by the retailer. Most large commodity companies that raise poultry or pigs have field staff who visit every farm monthly for an internal audit. Only a small portion of the farms is visited by a retailer or an auditing company. It would be too expensive for a retailer to visit all the commodity farms. Some specialized niche programs have either field staff or a third-party auditor visiting each farm on a fixed schedule (once per year or once per three years, for example). A retailer can easily audit 50 to 75 slaughter plants every year, but auditing hundreds of commodity farms annually becomes cost prohibitive.

COMMENTS MUST BE CLEAR

It is impossible to eliminate all subjectivity so, when something is in noncompliance with a published welfare standard, clear comments that describe what the assessor observed are essential. Vague comments such as "poor handling of pigs" provide little useful information. An example of a more clearly written comment would be "The handler kicked approximately 20 pigs as hard as he could." This would be a serious abusive event. A clear comment enables fairer judgment to be made on whether a farm or plant should be removed from an approved supplier list.

SUMMARY AND CONCLUSIONS

There is a disconnect between what retailers and consumers want and what the industries and niche market producers are providing. Commodity programs focus on providing basic needs of food, water, and shelter, but consumers also expect an ethical/emotional standard of animal welfare to be met (e.g., no battery cages or individual housing of sows or veal calves). Niche market animal welfare labels include more than just animal welfare in order to capture niche markets. However, it is unclear if a facility's adherence to an animal welfare niche market label automatically signifies all-around good animal welfare. Dairy cows raised under organic conditions have reduced lameness due to spending time on pasture, but internal and external parasites may be a greater problem due to restrictions in the use of pharmaceuticals (Rutherford et al., 2008). In addition, all niche market programs allow castration of pigs and cattle without anesthesia. Economics is still a major factor in animal welfare even within niche market programs.

For those consumers who want improved animal welfare on farms, it will come at some cost. That increased cost is not known in many cases and individual producers cannot provide those products and put themselves at an economic disadvantage with their peers. Still, we see incremental improvement in both farm animal care and oversight of animal welfare on commercial farms. This is driven mostly by retailers who seek such assurances, but they too are unwilling to pay much more for the same products that carry animal welfare assurances.

REFERENCES

Duncan, I.J.M. and Kite, V.G. (1989) Nest box selection and nest building behavior in the domestic hens, *Animal Behavior*, 37:215–231.

Dunn, C.S. (1990) Stress reaction in cattle undergoing ritual slaughter using two methods of restraint, *Veterinary Record*, 126:522–525.

Espejo, L.A., Endres, M.I., and Salfer, J.A. (2006) Prevalence of lameness in high-producing Holstein cows housed in freestall barns in Minnesota, *Journal of Dairy Science*, 89:3052–3058.

Fraser, D. (2008) *Understanding Animal Welfare*, West Sussex, UK: Wiley Blackwell.

Fulwider, W.K., Grandin, T., Garrick, D.J., Engle, T.E., Lamm, W.D., Dalsted, N.L., and Rollin, B.E. (2007) Influence of free-stall base on tarsal joint lesions and hygiene in dairy cows, *Journal of Dairy Science*, 90:3559–3566.

Gispert, M., Angeles, O.M., Velarde, A., Suarez, P., Perez, J., and Fontifurnois, M. (2010). Carcass and meat quality characteristics of immunocastrated male, surgically castrated male, and entire male and female pigs, *Meat Science*, 85:664–670.

Grandin, T. (1982) Pig behavior studies applied to slaughter plant design, *Applied Animal Ethology*, 9:141–151.

Grandin, T. (1996) Factors that impede animal movement in slaughter plants, *Journal of American Veterinary Medical Association*, 209:757–759.

Grandin, T. (1998a) Objective scoring of animal handling and stunning practices in slaughter plants, *Journal American Veterinary Medical Association*, 212:36–39.

Grandin, T. (1998b) The feasibility of using vocalization scoring as an indicator of poor welfare during slaughter, *Applied Animal Behavior Science*, 56:121–128.

Grandin, T. (2000) Effect of animal welfare audits of slaughter plants by a major fast food company on cattle handling and stunning practices, *Journal of the American Veterinary Medical Association*, 216:848–851.

Grandin, T. (2002) Return to sensibility problems after penetrating captive bolt stunning of cattle in commercial slaughter plants, *Journal of the American Veterinary Medical Association*, 221:1258–1261.

Grandin, T. (2003) Cattle vocalizations are associated with handling and equipment problems at beef slaughter plants, *Applied Animal Behavior Science*, 71:191–201.

Grandin, T. (2005) Maintenance of good animal welfare standards in beef slaughter plants by use of auditing programs, *Journal of American Veterinary Medical Association*, 226:370–373.

Grandin, T. (2006) Progress and challenges in animal handling and slaughter in the U.S., *Applied Animal Behaviour Science*, 100:129–139.

Grandin, T. (2007) *Livestock Handling and Transport,* 3rd ed. Wallingford, Oxfordshire, UK: CABI International.

Grandin, T. (2010) *Improving Animal Welfare: A Practical Approach,* Wallingford, Oxfordshire, UK: CABI International.

Jones, E.K.M. (2005) Avoidance of atmospheric ammonia by domestic fowl and the effect of early experience, *Applied Animal Behavior Science,* 90:293–308.

Kilgour, R. 1971. Animal handling in works, pertinent behaviour studies. 13th Meat Industry Research Conference. Hamiltion, New Zealand, pp. 9–12.

Kilgour, R. and Dalton, D.C. 1984. *Livestock Behaviour: A Practical Guide.* Collins Technical Books. Glasgow, UK.

Knowles, T.G., Kestin, S.C., Hasslam, S.M. et al. (2008) Leg disorders in broiler chickens: Prevalence, risk factors and prevention, *PLOS One,* 3(2):e1545.doi:10.1371/journal.pone.0001545

Lay, D.C., Friend, T.H., Randel, R.D., Bowers, C.L., Grissom, K.K., and Jenkins, O.C. (1992) Behavioral and physiological effects of freeze and hot iron branding on crossbred cattle, *Journal of Animal Science*, 70:330–336.

Løvendahl, P.L., Darngaard, L.H., Nielsen, B.L., Thodberg, K., Su, G., and Rydhmer, L. (2005) Aggressive behavior in sows at mixing and maternal behavior are heritable and genetically correlated traits, *Livestock Production Science*, 93:73–85.

Mader, T.L., Davis M.S., and Brown-Branl, T. (2005) Environmental factors influencing heat stress in feedlot cattle, *Journal of Animal Science*, 84:712–719.

Marchant-Forde, J.N., Lay, D.C., Pajor, J.A., Richert, B.T., and Schinckel, A.P. (2003) The effects of ractopamine on the behavior and physiology of finishing pigs, *Journal of Animal Science*, 81:416–422.

McGlone, J.J. and Hellman, J.M. (1988) Local and general anesthetic effects on the behavior and performance of two and seven week-old castrated and uncastrated piglets, *Journal of Animal Science*, 66:3049–3058.

McGlone, J.J., von Borell, E., Deen, J., Johnson, A.K., Levis, D.G., Meunier-Salaun, M., Morrow, J., Reeves, D., Salak-Johnson, J.L., and Sundberg, P.L. (2004) Review: Compilation of the scientific literature comparing housing systems for gestating sows and gilts using measures of physiology, behavior, performance and health, *Professional Animal Scientist*, 20:105–117.

Muir, W.M. and Craig, J.V. 1998. Improving animal well-being through genetic selection. *Poultry Science*, 77:1781–1788.

OIE. (2006) Introduction to the recommendations for animal welfare, terrestrial animal health code, World Animal Health Organization, Paris, France.

Rhodes, T.R., Appleby, M.C., Chinn, K., Douglas, L., Firkins, L.D., Houpt, K.A., McGlone, J.J., Sundberg, P., Tokach, L., and Wills, R.W. 2005 A comprehensive review of housing for pregnant sows, *Journal of American Veterinary Medical Association (JAVMA)*, 227(10):1580–1590.

Rushen, J., Pombourceq, E., and dePaisselle, A.M. (2006) Validation of two measures of lameness in dairy cows, *Applied Animal Behaviour Science*, 106:173–177.

Rutherford, K.M., Langford, F.M., Jack, M.C., Sherwood, L., Lawrence, A.B., and Haskell, M.J. (2008) Lameness prevalence and risk factors in organic and non-organic dairy herds in the United Kingdom, *Veterinary Journal*, May 5, 2009.

Tanida, H., Miura, A., Tanaka, T., and Yoshimoto, T. (1996) Behavioral responses of piglets to darkness and shadows, *Applied Animal Behavior Science*, 49:173–183.

University of Wisconsin. 2005. Body condition score (dairy cattle). http://dairynutrient.wisc.edu/302/page. php?id=36 Accessed 9-12-2011.

van Amstel, S.R. and Shearer, J. 2006. *Manual for the treatment and control of lameness in cattle.* Blackwell Publishing. Oxford, UK.

Van de Weerd, H.A. and Day, J.E.L. (2009) A review of environmental enrichment for pigs housed in intensive housing systems, *Applied Animal Behaviour Science*, 116:1–20.

van Putten, G. and Elshof, W.J. 1978. Observations on the effect of transport on the well being and lean quality of slaughter pigs. *Animal Regulation Studies*, 1:247–271.

Warriss, P.D., Brown, S.N., and Adams, S.I.M. (1994) Relationship between subjective and objective assessment stress at slaughter and meat quality in pigs, *Meat Science*, 38:329–340.

White, R.G., DeShazer, I.A., Tressler, C.J., Borcher, G.M., Davey, S., Waninge, A., Parkhurst, A.M., Milanuk, M.J., and Clems, E.T. (1995) Vocalizations and physiological response of pigs during castration with and without anesthetic, *Journal of Animal Science*, 73:381–386.

Wray, H.R., Main, D.C.J., Green, L.E., and Webster, A.J.F. (2003) Assessment of welfare of dairy cattle using animal based measurements, direct observations, and investigation of farm records, *Veterinary Record*, 153:197–202.

Wildman, E.F., Jones, G.M., Wagner, P.E., and Brown, R.L. (1982) A dairy cow body condition scoring system and its relationship to selected production characteristics, *Journal of Dairy Science*, 65:495–501.

Zurbrigg, K. (2006) Sow shoulder lesions: Risk factors and treatment effects on an Ontario farm, *Journal of Animal Science*, 84:2509–2514.

8 Animal Welfare

Synthesizing Contemporary Animal Agriculture/Engineering and Animal Comfort and Social Responsibility

Bernard E. Rollin, John J. McGlone, Judith L. Capper, Kenneth Anderson, and Terry Engle

CONTENTS

PROLOGUE

Bernard E. Rollin

In earlier chapters, we discussed the way in which animal agriculture moved from being based on animal husbandry to an industrial approach. We also discussed the ever-increasing societal demand in Europe, North America, Australia, and New Zealand for the restoration of the kind of respect for animals' biological and psychological needs and natures that was historically presuppositional to good husbandry and thereby more or less guaranteed good welfare for agricultural animals.

It is evident that the sort of extensive, pastoral agriculture that was historically dominant would be difficult if not impossible to achieve in the current socioeconomic milieu. Much prime grazing land has been transmuted into urban and suburban commercial property and housing, with less than 1% of the American public engaged in animal agriculture as a vocation. Both the escalating value of land and the major population increases experienced in most Western countries makes the restoration of pastoral animal agriculture a practical impossibility. Only Western cattle ranching, and to a lesser extent sheep ranching, have protected large tracts of private land from rapacious development, and such preservation requires increasing ingenuity. How, then, is society to meet the major demand for animal products, while still acknowledging and respecting ever-increasing demands for animal welfare based on the model of animal husbandry?

The answer to this very difficult conundrum lies in reassessing the conceptual basis for industrialized agriculture. As was pointed out earlier, those who developed these systems assumed that if animals were economically productive, they were necessarily experiencing positive welfare. We have pointed out that while such an argument was quite legitimate under husbandry conditions, it had considerably less validity for industrialization, where animals could have various aspects of their welfare requirements ignored, yet still be economically productive. For example, numerous scholars have argued that sows maintained in gestation crates experience compromised welfare by virtue of their inability to engage in normal, species-specific behaviors and because of their marked limitations in movement.

At the same time, the field of animal behavior and animal welfare science has provided us with tools for understanding animal natures and needs. One very valuable approach to this issue has been the development of animal preference testing as a way of determining the sorts of accommodations, feedstuffs, light cycles, ambient temperatures, and so on, that animals will consistently choose. Marion Dawkins pointed out one important caveat regarding this approach when she cautioned that animals, as much as people, may have preferences that are not in their best interest.

What is needed, then, is the redesign of these confinement systems with a conscious demand for incorporating into them accommodation to animals' needs and natures, what I, following Aristotle, have called animal *telos*—the "pigness" of the pig, the "cowness" of the cow. Much spadework has already been done in this area in Great Britain and in Europe, where the demand for welfare-friendly reform of confinement systems developed considerably earlier than in the United States. Thus, there is no need to start completely cold; various modified systems have been developed and tested in other countries. In the United States, a variety of people have worked on such systems. Particularly notable is Temple Grandin's work on incorporating knowledge of cattle behavior into feedlot pen design, slaughterhouse systems, transport systems, restraint devices, and so on. Grandin has demonstrated that welfare-friendly systems are not only congenial to the animals, but also they are easier to implement, and in many cases, increase profit.

In this chapter, individual animal scientists specializing in swine (McGlone), dairy cattle (Capper), poultry (Anderson), and beef cattle (Engle), address progress and concerns prominent in meeting animal welfare needs and goals.

SWINE

John J. McGlone

INTRODUCTION

To produce animal products in sufficient quantities and of the quality desired, humans keep animals in production systems. The driving force for development of production systems in the last century has been to develop systems that produce quality products at the lowest possible cost. In real current dollars, animal meats are less expensive today than they were years ago. The cost of production, for example, for pigs and poultry was higher in 1950 than it is today. The cost of labor is less; in 1945, it required 5 hours per 100 pounds of broilers, but by 1980, the labor requirement was far less than 15 minutes per 100 pounds of broiler produced. Feed efficiency (pound of feed consumed per pound of weight gain) was cut in half in both pigs and poultry. Pigs weaned per litter increased from 7 to over 9 pigs per sow. Chicken went from 40 cents per pound in 1960 (in 1960 dollars) to approximately 80 cents per pound—without the technological advances, chicken would have been over $1.60 per pound in 1995. When retail broiler prices are deflated, the retail cost has lowered from $2.20 per pound to less than $1 per pound from 1955 to 1979; pork has similarly declined in consumer cost by 30% (Martinez, 1999). Vertical integration and the wonders of science have brought us what we wanted—less expensive animal products. But at what cost?

Often in society, driving toward a single goal results in unintended consequences. Older animal producers who can still remember older production systems will be frustrated at society's concerns about production methods. They may say:

"They wanted inexpensive meat. We gave it to them. Then they wanted meat with less fat. We gave it to them. Then they wanted it to taste good, not pollute the environment, be safe to eat, and be good for the animal's welfare."

The drive toward low-cost, intensive animal production systems has resulted in a laundry list of unintended consequences. Among them are real or perceived concerns:

- Food safety, including bacterial contamination, especially antibiotic-resistant bacteria
- Movement of viruses among animals and people, potentially creating new, more virulent strains
- Healthfulness of animal products containing fat in unacceptable quantities and qualities
- Environmental pollution of water, air, and soil
- Worker health and safety issues including injuries and respiratory problems
- Animal welfare concerns

It is important when trying to focus on animal welfare concerns that we not be distracted by other important societal concerns. More importantly, if we are to make progress on animal welfare concerns, it is important to do so without losing ground on other societal issues (McGlone, 2001). We simply must develop sustainable production systems—those that produce wholesome products with positive impacts on society's concerns.

The philosophers have not been much help in defining animal rights and welfare. Peter Singer argues for animal liberation by taking a utilitarian cost-benefit approach while dismissing the notion of animal rights. Tom Regan dismisses the utilitarian approach in favor of an animal rights approach. Matthew Scully dismisses both the utilitarian approach and the rights approach in favor of a theology-based care for animals that serve humans as a quality food. If no clear, logical, defensible theme surfaces for animal rights/liberation/theology, then the concept is of little use. Instead, we must provide for the best animal welfare that is possible in a practical way. Providing for sound animal welfare may be more easily said than done.

Once one morally accepts that animals may be humanely killed for our palate, then animal welfare can focus on how best to breed, grow, transport, and kill animals so that their welfare is accommodated. This may not be completely possible. For example, animals must be transported to be "processed." The first transport experience (and often the last) for most growing food animals is a novel, frightening experience. While it is possible to ameliorate that negative experience, this will have to be a research and development goal—it is simply not possible at this time.

Other practices involve pain. Everyone, including farmers, agree that causing pain to farm animals is not good for their welfare. However, the marketplace makes demands on the production system. Animals are castrated to avoid behavioral problems and bad odor, and to increase meat quality (more marbling). Castration hurts. Some countries have banned castration of pigs (e.g., Norway), some do not castrate (Denmark), and some require anesthesia for the procedure (Belgium and the Netherlands), but most male pigs in the world are castrated. And in the United States most beef cattle fed in feedlots are castrated without anesthesia or analgesia. More will be discussed on this topic in the species sections.

Besides the seemingly inevitable distress and pain associated with production systems, society has other animal welfare issues. They might prefer that animals have more space, that social animals not be housed individually, and that there be some relief to the interminable boredom, which, to the casual observer, must be present when large numbers of animals are kept in relatively barren environments with feed, water, and a comfortable thermal environment meeting their physical and physiological needs. Resolving the long list of animal welfare concerns of intensive animal agriculture will take some time and a great deal of effort.

In manufacturing (a negative term when referring to animal systems), one strives for continuous improvement in the production process. Will those caring consumers be satisfied with continuous improvement in animal systems or must our systems be abruptly overhauled? Moreover, if we are to change our modern systems, do we have systems to move toward that improve animal welfare while not losing ground on other societal issues? How much time do we have?

PIG AND PORK PRODUCTION SYSTEMS

The modern pork production system is a marvel of efficiency. Many farms produce 25 pigs per sow per year. A 280-lb pig is grown from a 3-lb piglet in 6 months with less than 3-lb of feed to produce each pound of body weight gain. These would have been unthinkable numbers 50 years ago, even 20 years ago. The wonders of genetic selection, improvements in animal health products, a better understanding of nutrition, and use of environmentally controlled barns has allowed animal scientists, veterinarians, and engineers to create these improvements.

Pigs were kept outdoors for most of the last few millennia. Just as feral pigs in the southern United States today have parasites and zoonotic diseases, the outdoor pigs of the past were largely unhealthy. Diseases like hog cholera and foot and mouth were common in U.S. pigs (they are now eradicated). Most farm pigs had internal and external parasites. Although they had mud to wallow in and earth to root, they had significant health problems. How were those problems to be solved?

The simple answer was that they were moved indoors—not indoors on earthen floors, but on sanitizable metal, plastic, and concrete slatted flooring. It is indeed a rarity to find internal or external parasites on farmed pigs in the United States today. Bacterial diseases are largely under control. Of course, "new" pig viruses have emerged such as PRRS and PCV (porcine reproductive and respiratory syndrome virus and porcine circovirus)—viruses that either were not present or were not detected when pigs were mostly outdoors. Still, the modern pork producer can bring in healthy pigs, hold them inside buildings in isolation from other pigs, and have a good chance of keeping out these pathogens.

The pig industry in the United States, the American pork producers, has not actually conducted a survey to know the percentage of pigs kept indoors. Maybe they are afraid to know. The percentage of pigs indoors is well over 90%. Of those, most are in standardized systems that result in very high levels of productivity. In 2008, just 20 pork producers had approximately 3 million sows of the 5 million in the United States (The Pig Site, 2008). Continued vertical integration and consolidation are likely as profit margins shrink for commodity pork production.

The most common production system for pigs in the United States is to have pregnant sows in gestation crates, lactating sows and piglets in farrowing crates, boars in all-male boar studs producing semen for artificial insemination, and weaned pigs on cleanable slotted flooring, all in mechanically ventilated buildings.

Animal welfare issues that are real or perceived in modern pork production units include:

- Crating of pregnant sows
- Crating of lactating sows
- Tail docking without anesthesia
- Castration without anesthesia
- Slatted, non-rootable flooring
- Transportation
- Euthanasia of ill or injured pigs
- Boredom
- Lack of sufficient space
- Injury to bone and lameness
- Wounds, scratches, and abscesses
- Health problems, usually respiratory and enteric
- Rough handling of pigs by people

It is now illegal (or soon will be illegal) to keep pregnant sows in gestation crates in some American states. This issue is frustrating for mainstream pig industry folks because they truly believe that pregnant sows are fine or even better off in gestation crates than in group housing. Then we have the AVMA and leading animal scientists writing reviews of the scientific literature and reaching the same conclusion: The welfare of sows in gestation crates is no better or worse than sows in group housing systems (Rhodes et al., 2005; McGlone et al., 2004). However, the public clearly perceives the gestation crate as an animal welfare problem.

One of the concerns about sows in individual gestation crates is the expression of stereotyped behaviors. Sows in gestation crates bite the bars, and they chew, root, and lick the bars, floor, fencing, and feeder. For an individually kept sow, it appears that the sow is experiencing behaviors associated with frustration or boredom. However, these stereotyped behaviors are caused by mechanisms that are yet undefined. We are not sure they are frustrated.

Many of the studies of gestating sows where the crate and group systems are compared do not have equal penning materials, sow previous experiences, or management (flooring, feeding, etc.). When all else is held constant, sows express about the same level of stereotyped behaviors when kept in indoor group pens and crates. Hulbert and McGlone (2006) found pregnant sows in drop-fed crates showed 2.23% of their time in stereotyped oral-nasal-facial (ONF) behaviors, while pregnant sows in groups showed 2.24% of their time engaged in this behavior. It could be that (1) stereotyped behavior is not indicative of a problem (or stress), (2) that both the crate and pen similarly lack enrichment and thus cause issues, or (3) neither the crate nor the pen induces a particular behavioral problem when well managed. Showing 2.2% of their time in ONF stereotyped behavior might not seem like a lot of time; however, sows spent 96% of their time lying down, so in effect, sows are spending about half of their active moments engaged in ONF stereotyped behaviors. That an animal spends half its active time engaged in a behavior that we do not understand is worthy of study.

Sows in gestation crates do not move much. This has been reported to cause problems with their feet and legs. However, the work that has attempted to show this was inadequately controlled and so objective information to support this observation is lacking.

Think about a pregnant sow that is well cared for in an individual crate. She is hungry because we have bred pigs to grow fast and large and, to prevent obesity, we must limit their feed—so they are often hungry. She has a comfortable thermal environment, but for a dominant sow at least, she may experience a problem because she does not have the ability to interact socially. For the submissive sows, they may have a problem with group housing. The industry has traded one welfare concern (social stress) for another (boredom and social isolation) largely to accommodate economic pressures. To provide an individual environment is positive for welfare. However, why does the space have to be so small? It is so you can get more sows in each expensive building if they have the absolutely least amount of space with which they can get by. Group-housed sows need approximately 20 sq. ft per sow; crated sows need only 14 sq. ft each.

One could keep pregnant sows individually in larger pens (rather than crates). This would allow them to turn around, but not fully socially interact (including not wound each other through aggressive social interactions). Why is this not done? Economics. Once you decide to give sows more space, then you remove penning/fencing materials to lower the cost of the building. If a new building were built, one could have individual pens with more penning materials or group housing with less penning materials. Once the decision is made to give them more space, then economics drives the decision toward group housing.

Therefore, the gestation crate imposes two constraints on the sow that are problems perceived by the public: Too little space and lack of social interaction. American state laws lack detail to distinguish between these two. Moreover, a pork producer could keep group-housed sows in spaces of less than 14 sq. ft per sow, which would introduce a crowding-induced welfare problem of increased skin lesions (Salak-Johnson et al., 2007).

In visiting a reported "high welfare" farm in Sweden, McGlone (2006) found very high levels of wounds and scratches on straw-bedded, group-housed pregnant sows and high piglet mortality in loose-housed lactating sows. So even when the public perceives a welfare problem and takes regulatory action, they may, in fact, cause the use of systems that introduce new stressors.

What would be a welfare-friendly system for pregnant sows? Clearly, if one applies all the features that are important to the public (more space, straw, social interaction), the sows can still experience considerable social stress and lack of reproductive performance (McGlone, 2006). To reduce the stress of social groups requires more space. In the outdoor system, aggression-related scratches and wounds are virtually non-existent. However, the zoonotic and pig disease challenges (including, in particular, influenza, internal parasites, and *Salmonella*) and uncontrolled manure flows offset the welfare advantages for the outdoor system. What we are left with is finding indoor systems that accommodate the individual sow's needs for social protection and access to feed. This ideal system would have an enriched, spacious indoor environment, with individual, safe feeding stations. Such a system, when it is eventually designed and tested, may not be affordable given today's economic pressures in pork production systems.

Gestation crates cause arguable welfare problems. What about other issues? Nobody has argued that castration and tail docking are good for the welfare of pigs. Both procedures intentionally cause pain, distress, and behavioral disruptions for economic reasons. We castrate pigs primarily because of an off-flavor and odor in boar meat. However, they do not castrate in the U.K. and Denmark. Denmark is a major exporter of pork. How is this possible? It is because they market pigs at a lighter weight and they test for boar odor.

Economics favors "processing" pigs at heavier weights. It is the same labor to turn the live pig into a carcass if the live pig is 300 lb or 220 lb. You get more meat per worker and per physical plant investment. Castration—the induction of intentional pain and distress—is performed entirely for economic reasons in the pig industry.

Tail docking is the same story. Pigs tail bite when they are housed indoors on concrete floors (solid or slatted). When they have bedding, they bite less often. When they are outdoors, they still may tail bite, but at a very low level. The housing system—introduced to lower diseases and improve pig performance—causes this behavioral problem. More expensive buildings (with more space and straw bedding, for example) or outdoor rearing would significantly reduce tail biting to the point that tail docking would not be needed. However, several studies report that pigs kept outdoors or on straw have a higher prevalence of *Salmonella* (van der Wolf et al., 2001; Calloway et al., 2005). Most people would not want to choose between a food safety risk and an animal welfare benefit. People would probably want an assurance that their meat is not contaminated and that the animals experienced a healthy life, including their behavioral health.

The animal welfare problem of small spaces and barren, boring environments is a concern throughout pig production (lactating sows, weaned and growing pigs, pregnant sows, and boars). We cannot say pigs under our care are entitled to enrichment; at least thus far, this has only been legislated for laboratory non-human primates and dogs, but not farm animals. However, enrichment in barren environments is recommended for animals in teaching and research protocols, but not on commercial farms at this time.

If one were to ask a pork producer or an active pig welfare scientist, they would say that the major issues of pig welfare revolve around (1) pig health including not just infectious diseases, but also wounds, injury, and behavioral problems and (2) procedures we are forced to perform (castration, tail docking, transport) for economic reasons. The public has only learned about close confinement of sows from activist political activities. The public is very disconnected from commercial pork production—they don't know what happens on the farm. Because so few people actually raise pigs, they cannot afford to educate the public. Nor would producers be willing to put themselves at an economic disadvantage by providing more quality space, enriched environments, or marketing to avoid pig castration.

Alternative pork products are available that meet some animal welfare standards; particularly things like no gestation crates (which may be the only appreciable difference between some animal welfare certified products and conventional products) or access to outdoors (free range). In the absence of legislative action, and given the economic realities of animal production, any change in the name of animal welfare will be made based on market forces. If people start buying alternative products, more will be produced. In the mean time, activists will continue to chip away at conventional pork production to make changes that the public may perceive as positive for animal welfare.

REFERENCES

Callaway, T.R., J.L. Morrow, A.K. Johnson, J.W. Dailey, F.M. Wallace, E.A. Wagstrom, J.J. McGlone, A.R. Lewis, S.E. Dowd, T.L. Poole, T.S. Edrington, R.C. Anderson, K.J. Genovese, J.A. Byrd, R.B. Harvey, and D.J. Nisbet. 2005. Environmental prevalence and persistence of *Salmonella* spp. in outdoor swine wallows. *Foodborne Pathogens and Disease.* 2(3): 263–273.

Hulbert, L.E., and J.J. McGlone. 2006. Evaluation of drop versus trickle-feeding systems for crated or group-penned gestating sows. *Journal of Animal Science.* 84: 1004–1014.

Martinez, S.W. 1999. Verticial coordination in the pork and broiler industries: Implications for pork and chicken products. Agricultural Economic Report No. 777. http://ddr.nal.usda.gov/bitstream/10113/33407/1/CAT10872618.pdf

McGlone, J.J. 2001. Farm animal welfare in the context of other society issues: Toward sustainable systems. *Livestock Production Science.* 72: 75–81.

McGlone, J.J. 2006. Comparison of sow welfare in the Swedish deep-bedded system and the US crated-sow system. *Journal of the American Veterinary Medical Association* 229: 1377–1380.

McGlone, J.J., E. von Borell, J. Deen, A.K. Johnson, D.G. Levis, M. Meunier-Salaun, J. Morrow, D. Reeves, J.L. Salak-Johnson, and P.L. Sundberg. 2004. Review: Compilation of the scientific literature comparing housing systems for gestating sows and gilts using measures of physiology, behavior, performance and health. *Professional Animal Scientist.* 20: 105–117.

The Pig Site. 2008. Pork powerhouse. http://www.thepigsite.com/swinenews/18994/pork-commentary-pork-powerhouse

Rhodes, T.R., M.C. Appleby, K. Chinn, L. Douglas, L.D. Firkins, K.A. Houpt, C., J.J. McGlone, P. Sundberg, L. Tokach, and R.W. Wills. 2005 A comprehensive review of housing for pregnant sows. *Journal of the American Veterinary Medical Association (JAVMA)* 227(10): 1580–1590.

Salak-Johnson, J.L., S.R. Niekamp, S.L. Rodriguiez-Zas, M. Ellis, and S.E. Cirtis. 2007. Space allowance for
    dry, pregnant sows in pens: Body condition, skin lesions, and performance. *Journal of Animal Science.*
    85: 1758–1769.
Van der Wolf, P.J., A.R.W. Elbers, H.M.J.F. van der Heijden, F.W. van Schie, W.A. Hunneman, and M.J.M.
    Tielen. 2001. *Salmonella* seroprevelance at the population and herd level in pigs in the Netherlands.
    Veterinary Microbiology. 80: 171–184.

DAIRY CATTLE

Judith L. Capper

INTRODUCTION

Animal welfare concerns usually center around three areas of focus—productivity, ability to express "natural" behaviors, and the absence of pain or suffering (Fraser et al., 1997). Nonetheless, it can be argued that dairy cattle welfare is a function of the three aforementioned criteria, with notable interconnections between each issue. The degree to which husbandry systems satisfy the mental and physical needs of dairy cattle is somewhat difficult to assess. Traditionally, animal productivity has been accepted as an indicator of animal welfare—with higher productivity (milk yield, fertility, growth rate) implying that the animal's needs are met to a satisfactory degree. There can be no doubt that in the case of the lactating dairy cow, sustained high productivity cannot be achieved in the absence of good welfare. Nonetheless, other parameters such as physiological data (circulating hormone and enzyme concentrations, heart rate, immunosuppression), measures of morbidity and mortality, and behavioral adaptations that suggest compromised welfare or adoption of coping strategies provide indicators by which we can benchmark the effects of differing management practices or husbandry systems.

UNIQUE ASPECTS OF DAIRY PRODUCTION IN ANIMAL WELFARE ISSUES

Animal welfare is often related to the animal's ability to express natural behaviors (von Keyserlingk et al., 2009). Concern exists that animals kept under conditions considered abnormal may suffer, although abnormality is difficult to define in modern livestock. The issue of natural behavior expression may be overtaken by emotive language propagated by those who are opposed to animal agriculture and wish, for example, for "pigs to express their pigginess." Such groups neglect to acknowledge the role of animal agriculture in providing high-quality protein to the growing population, and fail to acknowledge animals' contributions to human life in terms of clothing, land maintenance and diversity, by-products for industrial manufacture, etc. When directed at the dairy industry, emotive language serves to further promote the popular consumer perception that the small-scale production systems present in the 1940s and 1950s had considerably higher welfare standards than current production systems. This is an entirely disingenuous suggestion—few people would suggest that standards of human welfare (health, nutrition, behavior) were significantly better in the 1940s, where the average life expectancy was 62.9 years (compared to 77.8 for 2005; National Center for Health Statistics, 2006). The U.S. industrial revolution demonstrated the short-term improvements in productivity gained by running factories for 24 hours per day. However, this short-term increase in productivity was at a considerable cost to human welfare— poorly ventilated, cramped working conditions without adequate time allowances for breaks or meals and no health care provision led to increased disease, reduced morale, and a long-term productivity decline (Brezina, 2005). To take this example further, factories still run on a 24-hour cycle in many industries; however, with considerably improved working conditions, scheduled breaks and vacation, and provision of health care and benefits, productivity has improved considerably. It has become clear that maximum short- and long-term productivity is gained through improving worker health and welfare, allowing the human components of the system to perform at the optimum level. The same approach may be applied to animal production—turning the "high productivity = high welfare" suggestion on its head, one can suggest that "high welfare = high productivity." There is no doubt that early innovations demonstrated to improve dairy productivity had undesirable consequences when taken to extremes. However, improved knowledge and understanding of dairy cow nutrition and metabolism has led to a system, which allows for improved animal welfare and productivity when applied appropriately.

The bucolic image of small-scale, extensive dairy systems often leads to the characterization of modern large-scale agriculture as "factory farms," thereby implying that these systems have an extremely low level of concern for animal welfare. Nonetheless, examination of the characteristics of mid-1940s dairy farms shows that the agrarian idyll may not be an appropriate image. Dairy production in 1944 was characterized by extensive pasture-based systems with an average herd size of approximately six cows (Capper, Cady, and Bauman, 2009). Dairy cow nutrition was reliant on homegrown forages with few purchased concentrate feeds (Woodward, 1939) and with only a basic understanding of the nutritional and metabolic interactions between animal nutrition and productivity. Perhaps the most striking aspect of this so-called high animal welfare system was the low productivity—the average dairy cow in 1944 yielded only 2074 kg/year. Since this time, the milk yield per cow has increased at an average of 136 kg/year, of which half to two-thirds of the increase has been attributed to improved genetics (Shook, 2006). However, the remaining component can be attributed to improved understanding of nutrition, management, and welfare, thus allowing the modern dairy cow to produce more than 9333 kg of milk per year (USDA/NASS, 2010). Nonetheless, efficiency within modern production systems is sometimes perceived by the consumer as being undesirable or to occur at the expense of optimum animal welfare and well-being.

The sustainability of any dairy system depends upon balancing economic and environmental sustainability while maintaining the social license to operate. Average dairy product consumption has steadily risen over the past 20 years, with a decline in fluid milk consumption more than compensated for by an increase in consumption of cheese and other dairy products. Although milk is still considered a staple food, competition from other beverages and concern over the portrayal of dairy management practices by media and activist groups may threaten social license, particularly when animal welfare is the issue under discussion. This is exacerbated by anthropomorphic views of animal welfare and the perception that the modern dairy cow has been "removed" from its natural environment. In contrast to the dairy population in the 1940s, which comprised a mixture of small (Jersey, Guernsey) and large breeds (Holstein, Ayrshire, Shorthorn), the modern U.S. dairy population is distinctly more homogenous, containing over 90% Holsteins, approximately 5% Jerseys, and 5% other breeds (Majeskie, 1993). The modern dairy cow may therefore be considered to be a human creation—selection pressure augmented by the introduction of technologies including artificial insemination, embryo transfer, genetic evaluation, and genome mapping has allowed for animals that have significantly higher milk yields, yet these come with their own management challenges that must be met for productivity and animal welfare to be optimized. It appears that selection for high milk production may confer a higher susceptibility to stress and therefore a greater risk of behavioral, physiological, and immune problems (Rauw et al., 1998) than demonstrated by lower producing cows. It should be noted that milk production per se does not confer an increase in cortisol or stress-related behaviors—it is the very absence of stress that allows dairy cattle to perform to their genetic potential for lactation. Improvements in management practices that result in a system more conducive to dairy cow welfare therefore have demonstrable effects upon performance. Major contributors to animal welfare and productivity include the physical environment, disease prevention and treatment, and nutrition, all of which should be considered both as singular effects and as interacting factors.

Physical Environment

To maximize productivity and animal welfare, dairy management systems should be founded upon the behavioral routines of the animal. This does not necessarily extend to a situation where animals are allowed to forage on pastureland and to run in traditional herds containing both female and male animals, without human intervention, as might be suggested by some of the more extreme anti-animal agriculture groups. Nonetheless, the behavioral needs and routines of the cow must be considered when designing a dairy system that is effective in optimizing animal welfare. According to Grant and Albright (2001), dairy cows spend 3 to 5 h/d eating, thus consuming 9 to 14 meals per day. In addition, they ruminate for 7 to 10 h/d, spend approximately 30 min/d drinking, and require approximately 10 h/d of lying or resting time. This only leaves a minor period free for daily management practices including milking. Compromising the cow's ability to perform these activities has negative effects on productivity and may increase stress levels.

Groups of dairy cattle quickly establish a dominance hierarchy, which is maintained according to age, body weight, and social status within the population (Friend and Polan, 1974). Research demonstrates that when maintained in groups containing greater than 100 animals, dairy cattle may lose the ability to recognize individuals and assess their relative position within the hierarchy (Albright, 1978). This would appear to favor small-scale dairy production systems; however, it can easily be achieved within larger dairies, which, for ease of management, group cows according to stage of lactation or parity. However, significant stress behaviors are often exhibited as a result of moving animals between established groups, for example, from a "far-off" (60 to 30 days pre-partum) to a "close-up" (30 days pre-partum to parturition) dry cow group. Abnormal feeding behaviors and an increased incidence of metabolic disorders have been exhibited by cows subjected to abrupt environmental or social changes during the peri-parturient period (Bazeley and Pinsent, 1984) with consequent effects on productivity. This may be alleviated by moving large numbers of cows at a time, in order to minimize individual animal stress from handling and to reduce social disruption (Grant and Albright, 2001) but this practice is again better suited to a large facility.

Grant and Albright (2001) note that optimal grouping strategies minimize negative social interactions and encourage positive interactions, with an overall aim of maximizing cow comfort and productivity. Fighting within the group is an obvious stressor and may reduce productivity—although conflict is thought to be reduced by the maintenance of a stable dominance hierarchy, it is not eliminated and can only be minimized. Competition for feed is an inevitable consequence of modern dairy production systems unless animals are confined to tie-stalls (which are associated with a different group of welfare issues). For example, the increase in dry matter intake during the first few weeks of lactation occurs at a faster rate in older cows than in heifers (Kertz, Reutzel, and Thomas, 1991) and may lead to negative interactions at the feed bunk. This provides a rationale for grouping cows according to parity during early lactation. Fox (1983) suggests that the welfare of cows within small- and medium-scale production systems is higher than in other farm animal species. However, it is interesting to note that grouping cows is more suited to a medium- or large-scale dairy than a small-scale dairy, despite their generally negative image with consumers.

Anecdotal evidence from the U.S. dairy industry suggests that when herd sizes were reduced in California in an attempt to decrease milk supply, milk production per facility increased because of improved dry matter intake (DMI) and extra feeding space per cow. Despite the potential for hierarchal conflicts within large groups, it appears that these may be mediated though the provision of adequate feeding space and supplies of fresh feed (Grant and Albright, 2001). The ideal group size is difficult to define, but is a function of competition for feed and water, space in the lot and holding area, stall use, and time diverted from productive behaviors (eating, drinking, resting, and ruminating).

Over time, greater knowledge of cow behavioral requirements has led to the understanding that provision of comfortable stalls has a direct effect upon productivity. Tremendous evolution has occurred from original wooden stalls that did not allow adequate forward or side space for animals to lunge forward in a natural manner but facilitated free movement within the pen, to modern free-stalls with sand bedding and ample space to extend their front legs and lunge forward or sideways, while still allowing for natural herd behavior within the pen. Poorly designed stalls that are too short or that have inadequate bedding material reduce occupancy of free-stalls, thus reducing the proportion of time spent lying or resting and increasing the chance of injury and lameness.

The debate as to whether cattle should be confined, grazed on pasture, or kept within a system that makes use of both practices continues to rage. Critics of confinement systems claim that they stifle natural behaviors, yet given the increase in human population size that is predicted to occur within the next 40 years, the intensity of competition for land use is likely to increase. Assuming that dairy consumption per capita stays stable, an industry-scale move to grazing systems is not a feasible alternative simply based upon the lower productivity in grazing herds (USDA, 2007) and thus the increase in land requirements per unit of milk (Capper et al., 2008). Grazing systems are often perceived to be more welfare-friendly than are confinement systems; nonetheless, the welfare issues associated with grazing may have different symptoms, but are equally detrimental to dairy productivity and well-being. There is little evidence that cows within these grazing systems have higher overall welfare than animals in a well-managed confinement system, especially given the relative lack of control over environmental factors such as temperature, humidity, and ventilation. Indeed, over time, conventional dairy systems have progressed from extensive pasture-based systems, through completely enclosed tie-stall and stanchion barns to modern open side-walled barns with ventilation fans or cross-ventilated barns, which create an environment that allows animals to remain within their thermo-neutral zone without expending excess energy on heat generation or dispersion. Where a market or sufficient resources are present to allow for grazing systems to prosper, it is essential to match the animal characteristics to the system. This is exemplified by the results observed when U.S. Holstein genetics were imported into New Zealand: Initially milk production was increased compared to the New Zealand Holstein, but the grass-based system is nutritionally insufficient to support high milk production and leads to lower survival rates as cows fail to cycle or become pregnant and are culled as a result of the demands of the seasonal antipodean calving system (Lucy, 2001).

Arguably, one of the most significant advances in both dairy and beef cattle has been the development of handling systems that minimize stress and maximize productivity. Researchers such as Dr. Temple Grandin at Colorado State University have designed and implemented movement systems that allow the animal's natural flight zone to be manipulated to facilitate handling with reduced animal stress and thus greater ease and efficiency of management (Grandin, 2007). Cattle that have a positive relationship with their handlers tend to move more smoothly, are less nervous within the milking parlor or handling systems, and acclimatize more easily to changes in routine, for example, when moving groups or during initial introduction to the milking process. Fox (1983) states that maximum biological efficiency is achieved through a close human–cow bond, lack of fear, zero flight distance, and selection for docility; nonetheless, these characteristics do not compensate for low genetic merit for milk yield or poor management within the herd.

Disease Prevention and Treatment

The introduction of antibiotics for animal use was a major step forward in improving dairy welfare and productivity. Modern animal production is often criticized for the extent to which antibiotics are used, with ongoing debate as to whether antibiotic use within agriculture has contributed to the rise of antibiotic resistance and related human health issues. Given that one of the cornerstones of animal welfare according to the "five freedoms" first originated by Brambell (1965) is the ability to be "free from pain, injury, and disease," promotion of a dairy system whereby antibiotic use is prohibited seems counter to the suggestion that animal welfare and productivity should be maximized. If it is accepted that animal welfare is paramount within production systems, the increasing popularity of extensive or low-input systems that make marketing claims based upon non-use of therapeutic antibiotics should be questioned. Groups opposed to animal agriculture often suggest that modern-day conventional dairy producers are motivated simply by profit, with little regard for animal welfare or well-being (Sustainable Table, 2009). However, this suggestion is inappropriate as productivity is negatively affected by suboptimal animal welfare or increased morbidity and mortality. Any management practice or system that negatively affects morbidity or mortality rates is neither economically viable nor practicable.

Within any system analysis, it is vital to consider the scientific basis behind the livestock production practices rather than allowing decisions to be made based on emotional or philosophic arguments (Pretty, 2007). This is exemplified by animal welfare legislation that is coming into play across the United States and the rest of the world. For example, restricting the use of individual housing for calves after eight weeks of age in Europe facilitates social interactions and allows the development of natural herd behaviors (von Keyserlingk et al., 2009), but also increases the potential for disease transmission through direct contact, with a concomitant risk of increased morbidity and mortality. The conflict between public perception, scientific evidence, and traditional production methods is perhaps best exemplified by the current discussion relating to tail docking in dairy cattle. Proponents of tail docking suggest that it promotes cleanliness within the herd, reduces tail-related injuries (predominantly eye infections) in workers, and reduces the incidence of mastitis. There is little scientific evidence to support these claims either from an animal or human welfare perspective and as the practice is not supported by the major animal welfare or wellness organizations, nor the general public as a whole, it appears that it may soon be legislated against. It is impossible to justify production practices for which no scientific data exist to demonstrate either a lack of negative effects or an improvement in welfare—this underlines the importance of devoting further resources to welfare issues in future research protocols.

Dr. Temple Grandin, a pioneer in the field of animal behavior and movement, often refers to the concept of "bad becoming normal," which may be defined as a situation that is detrimental, yet is seen so often that it becomes commonplace (Grandin and Johnson, 2006). Dr. Grandin applies this principle to the relatively high incidence of lameness within the dairy industry—an issue that is cited by consumers as a particular welfare issue. There is some debate as to whether an increased incidence of lameness is an inevitable consequence of industrialization within the dairy industry: Certainly lameness reduces productivity (Green et al., 2002) and is undesirable both from an economic and welfare perspective. However, milk yield itself has not been shown to be a contributing factor (Haskell et al., 2006). In addition, there was no association between herd size and lameness incidence in the study of Espejo and Endres (2007), although the authors noted that studies in England had found differing results. The frequency of hoof-trimming, time spent away from the pen (without access to stalls, food, or water), and cow-comfort quotient were reported to have significant effects upon lameness (Espejo and Endres, 2007). Matching stall size and design to cow size and weight was also cited as a major factor in lameness incidence by both Haskell et al. (2006) and Espejo and Endres (2007). This is often seen in older facilities where average cow size has increased over time, without a corresponding increase in stall size or change in design. It is somewhat comforting to know that these management factors can be controlled or changed in most farm situations; therefore, significant potential exists to reduce lameness and improve overall animal welfare, provided that the producer has sufficient incentive to do so. The increasing number of certification schemes that include animal welfare as a major component and provide a market advantage may achieve this.

Mastitis is arguably one of the most significant issues within the dairy industry, with potential production losses of 135 kg milk in the first lactation or 270 kg milk in the second lactation per unit increase in average log somatic cell count (Raubertas and Shook, 1982). Mastitis's nature as an inflammatory condition causing pain and loss of production is by definition a welfare issue. The severity of this issue is highlighted by the fact that producers report 16.5% of animals suffering from the condition, and udder or mastitis problems rank second in the list of producer-reported reasons for culling (USDA, 2007). There appears to be an association between milk yield and mastitis incidence (Phipps, 1989), yet there is some discussion as to whether this is a direct cause-effect relationship, for which there seems to be little biological foundation, or whether it results from greater time spent in the milking parlor with associated potential for infectious transfer, as a consequence of increased yield. For example, the biotechnological tool recombinant bovine somatotropin (rbST) increases milk yield by approximately 4.5 kg/d if sufficient feed is supplied to support milk yield (Capper et al., 2008). The FDA-approved label for rbST includes a warning that cows injected with the product are at an increased risk for mastitis, which groups opposed to biotechnology have taken as evidence that rbST use causes mastitis. However, a 1300-cow study undertaken by Poulet (1982) demonstrated no correlation between the relative incidence of mastitis and the use of rbST. As demonstrated by the U.S. dairy industry over the past century, greater intensification, including an increase in herd size, is an inevitable consequence of the need to produce more milk to feed the increasing population using fewer animals and non-renewable resources. However, mastitis incidence is not linked to herd size (USDA, 2007) and its control is dependent upon the implementation of best management practices including milking parlor hygiene, use of teat disinfectants, and clean bedding materials. It is worth noting that there are few studies relating to mastitis incidence in organic herds in which antibiotic use is not permitted (Hamilton et al., 2006; Ruegg, 2009). Anecdotal evidence suggests that many large organic herds also maintain a conventional herd into which animals may be moved if antibiotic treatment becomes necessary, or these animals may simply be sold. Given that milk yields in organic dairy herds are generally 20 to 40% lower (Zwald et al., 2004; Rotz et al., 2007) than those of conventional herds, any demonstrable reduction in mastitis may simply result from lower productivity. It appears that there is little to be gained from adopting management practices characteristic of organic or extensive production in preventing and controlling mastitis, but implementing best management practices as exhibited by the most productive and efficient farms currently within the industry paves the pathway to improving animal welfare.

Increases in milk production over the past 30 years have been associated with a reduction in fertility (Lucy, 2002). It is debatable as to whether this is an animal welfare issue per se. Reduced fertility may be taken as an indicator of underlying health issues, but it may also be argued that achieving pregnancy after milk production peaks and the cow is able to attain a positive energy balance is more desirable for the animal and is more likely to result in a successful pregnancy. Drying-off high-yielding cows that continue to yield 30 or 35 kg of milk per day at 365 days into lactation is undesirable and may lead to problems in the subsequent lactation (Church et al., 2008). Nonetheless, infertility is a major reason for culling with a producer-reported 26.3% of animals being removed from the herd due to reproductive problems (USDA, 2007). A recent report from the Farm Animal Welfare Council (2009) suggested that the average lifespan of 3.3 lactations for U.K.-based cows is an indicator of suboptimal welfare given that cattle can live to 12 years or older. If we set aside the previously discussed effects of genetic merit upon productivity and the market forces in place that favor replacing older cattle with heifers within the current dairy herd, improving fertility would be expected to have positive effects upon lifespan and welfare. It should be noted that dairy cow fertility is not an objective measure—pregnancy rate (defined as the proportions of cows that become pregnant divided by the total number of cows eligible to become pregnant within a specific time frame) is significantly affected by the ability of herders to detect heat. Indeed, Coleman (1993) reported that 90% of low estrus detection rates could be attributed to herders versus 10% to the cow herself. This does not necessarily account for the increase in non-behavioral estrus ("silent" heats) exhibited by high-producing animals under thermal or other stresses (Her et al., 1988), but demonstrates the value of heat detection methods such as tail chalking in improving fertility. The current average U.S. pregnancy rate ranges from 16% to 20%. Nonetheless, the author is personally aware of more than one U.S. dairy herd averaging over 41 kg of milk per day with a pregnancy rate of 29%—an example of a production facility whose management practices should be emulated both now and in future.

The relatively high incidence of culling within the U.S. dairy herd is often cited as evidence of poor animal welfare compared to less intensive systems. Holstein cows spend an average of 2.54 lactations within the herd (DairyMetrics™ database, Dairy Records Management Systems, Raleigh, NC; accessed November 13, 2009) before being sold or diverted to the beef market (culling). Just as any dairy production system has to function as a fiscally efficient business to be economically sustainable, it can be argued that the concept of "involuntary culling," that is, culling that is not under the producer's control, can be restricted to only two occasions—animal death or theft. Other incidences of culling due to low yields, poor fertility, or disease are an economic decision—if the cost invested in rectifying the issue or the return gained by keeping the animal in the herd outweighs the cost of replacing the animal with a freshly calved heifer, and providing such a heifer is available, it is inherently logical to replace the cow. It should be noted that the movement of cows from the dairy herd to the beef supply should not be considered "wastage"—approximately 7% of animals slaughtered for beef production in 2009 originated from the dairy herd, allowing sufficient beef to be produced without having to increase the size of the national beef herd. Although the majority of dairy bulls are diverted into beef and veal production systems, dairy heifers comprise only 1.4% of animals within beef feedlots (USDA, 2000), reflecting their relative value as dairy versus beef animals. On an idealistic basis, it is tempting to suggest that cattle would perform to their genetic merit and only leave the herd when they have completed their natural lifespan; however, this situation may not be best-placed to fulfill the needs and constraints of the modern dairy industry, especially given that a cow necessitates the production of a calf in order to lactate, and approximately half of the calves born are heifers. Discussion is occurring as to the potential effects of increasing sexed semen use within the dairy industry—it is possible that the future U.S. dairy industry will only use female-sexed semen upon the highest genetic merit cows, with the remainder being bred to a beef bull, or inseminated with male-bearing sperm.

Nutrition

Nutrition is the foundation upon which dairy cow productivity and welfare is built. Multifaceted links exist between the three pillars of animal welfare, yet without an adequate high-quality feed provision to supply the nutrients required to support maintenance, lactation, pregnancy over the long-term, productivity, efficiency, and health and welfare suffer. As previously discussed, adoption of the credo that high productivity goes hand-in-hand with optimal animal welfare carries the inherent assumption that nutritional strategies that encourage high production also ensure that animal welfare is maintained. Provision of sufficient time and physical space for feeding behavior to occur is a key to maintaining productivity—Grant and Albright (2001) suggest that feeding is the predominant behavior in dairy cattle until requirements are satisfied, with rumination taking precedence only when its feed has been abnormally restricted. From a physiological aspect, disturbances in rumen function or nutrient digestion lead to reduced productivity; for example, the early discovery that supplementing ruminants with highly fermentable grain (e.g., corn) also led to a considerable increase in mortality until correct feeding levels were established. Once these were in place, the next issue to become known was the fluctuations in ruminal pH and subsequent acidosis conferred by feeding forage separately from concentrate feeds. Over time, the adoption of total mixed rations (TMRs) within conventional dairy production has increased from 35.6% in 1996 to 51.5% in 2007, with 70.1% of herds with a rolling herd average of over 9072 kg/y (slightly below the average annual milk yield for the United States in 2007) feeding a TMR. Feeding a diet that is balanced to maintain energy and protein supply and that reduces adverse changes in ruminal or intestinal digestion has demonstrably improved digestibility, productivity, and welfare. These are only two brief examples of the interaction between nutrition, health, and physical environment, but there are many more. An in-depth discussion of the effects of inadequate or inappropriate nutrition upon welfare is beyond the scope of this review, yet the subject should be considered in any welfare discussion.

Conclusion

Animal welfare, productivity, and efficiency are keys to the continued sustainability of the dairy industry. Rather than focusing on individual practices from conventional or alternative production systems, best progress can be made by highlighting the management principles that maximize all three components of animal welfare, thus indicating that productivity and welfare are intrinsically linked. Within the current industry, this means examining the systems employed by the top 20% of producers, shifting the bell-shaped curve from the current average to a better average, and gaining momentum for future change in the process. Early adopters of innovation within any industry make the fastest progress, with the difference between early and late adopters being demonstrated by product quality—in this case milk production and indicators of animal welfare. Ideally, proactive adoption of best management practices will improve productivity and welfare—if adoption is so low that regulation or legislation is required to bring the lowest performers up to average performance, it should be questioned as to whether those producers will remain competitive within an industry that is increasingly reliant on social license to operate. Ultimately, one of the biggest threats the dairy industry faces concerning animal welfare is the presence of producers who fail to value the interaction between animal welfare and productivity and who are inevitably the subject of exposés by anti-animal agriculture groups. The importance of animal welfare and productivity in maintaining the socioeconomic sustainability of the dairy industry cannot and should not be underestimated.

References

Albright, J.L. 1978. Social considerations in grouping cows. In: *Large Dairy Herd Management*. C.J. Wilcox and H.H. Van Horn, Eds. Gainesville, FL: University Press of Florida.

Bazely, K., and P.J.N. Pinset. 1984. Preliminary observations in a series of outbreaks of acute laminitis in dairy cattle. *Veterinary Record* 115: 619–622.

Brambell, F.W.R. 1965. *Report of the Technical Committee to Enquire into the Welfare of Animals Kept under Intensive Livestock Husbandry Systems*. London, UK: HMSO.

Brezina, C. 2005. *The Industrial Revolution in America: A Primary Source History of America's Transformation into an Industrial Society*. New York: The Rosen Publishing Group.

Capper, J.L., E. Castañeda-Gutiérrez, R.A. Cady, and D.E. Bauman. 2008. The environmental impact of recombinant bovine somatotropin (rbST) use in dairy production. *Proceedings of the National Academy of Sciences USA* 105: 9668–9673.

Capper, J.L., R.A. Cady, and D.E. Bauman. 2009. The environmental impact of dairy production: 1944 compared with 2007. *Journal of Animal Science* 87: 2160–2167.

Church, G.T., L.K. Fox, C.T. Gaskins, D.D. Hancock, and J.M. Gay. 2008. The effect of a shortened dry period on intramammary infections during the subsequent lactation. *Journal of Dairy Science* 91: 4219–4225.

Coleman, D.A. 1993. Detecting estrus in dairy cattle. USA National Dairy Database, Reproduction Collection, University of Maryland, College Park, MD. Text is now only available on a CD rom from the University of Maryland.

Espejo, L.A., and M.I. Endres. 2007. Herd-level risk factors for lameness in high-producing Holstein cows housed in freestall barns. *Journal of Dairy Science* 90: 306–314.

Farm Animal Welfare Council. 2009. *Opinion of the Welfare of the Dairy Cow*. London, UK: Farm Animal Welfare Council.

Fox, M.W. 1983. Animal welfare and the dairy industry. *Journal of Dairy Science* 66: 2221.

Fraser, D., D.M. Weary, E.A. Pajor, and B.N. Milligan. 1997. A scientific conception of animal welfare that reflects ethical concerns. *Animal Welfare* 6: 187–205.

Friend, T.H., and C.E. Polan. 1974. Social rank, feeding behavior, and free stall utilization by dairy cattle. *Journal of Dairy Science* 57: 1214–1222.

Grandin, T. 2007. *Livestock Handling and Transport*. Wallingford, UK: CABI.

Grandin, T., and C. Johnson. 2006. *Animals in Translation: Using the Mysteries of Autism to Decode Animal Behavior*. New York: Mariner Books.

Grant, R.J., and J.L. Albright. 2001. Effect of animal grouping on feeding behavior and intake of dairy cattle. *Journal of Dairy Science* (E. Suppl.): E156–163.

Green, L.E., V.J. Hodges, Y.H. Schukken, R.W. Blowey, and A.J. Packington. 2002. The impact of clinical lameness on the milk yield of dairy cows. *Journal of Dairy Science* 85: 2250–2256.

Hamilton, C., U. Emanuelson, K. Forslund, I. Hansson, and T. Elkman. 2006. Mastitis and related management factors in certified organic dairy herds in Sweden. *Acta Veterinaria Scandinavica* 48: 11.

Haskell, M.J., L.J. Rennie, V.A. Bowell, M.J. Bell, and A.B. Lawrence. 2006. Housing system, milk production, and zero-grazing effects on lameness and leg injury in dairy cows. *Journal of Dairy Science* 89: 4259–4266.

Her, E., D. Wolfenson, I. Flamenbaum, Y. Folman, M. Kaim, and A. Berman. 1988. Thermal, productive, and reproductive responses of high yielding cows exposed to short-term cooling in summer. *Journal of Dairy Science* 71: 1085–1092.

Kertz, A.F., L.F. Reutzel, and G.R. Thomas. 1991. Dry matter intake from parturition to midlactation. *Journal of Dairy Science* 74: 2290–2295.

Lucy, M. 2001. Reproductive physiology and management of high-yielding dairy cattle. In: *Proceedings of the 61st Conference of the New Zealand Society of Animal Production*, Christchurch, New Zealand.

Lucy, M. 2002. The future of dairy reproductive management. *Advances in Dairy Technology* 14: 161–173.

Majeskie, L.J. 1993. Status of United States Dairy Cattle. Dairy Herd Improvement Collection. University of Maryland, College Park, MD.

National Center for Health Statistics. 2006. National Vital Statistics Reports. 54: 19. June 28, 2006. http://www.cdc.gov/nchs

Phipps, R.H. 1989. A review of the influence of somatotropin on health, reproduction and welfare in lactating dairy cows. In: *Use of Somatotropin in Livestock Production*. K. Sejrsen, M. Vestergaard, and A. Neimann-Sorensen, Eds. New York: Elsevier.

Poutrel, B. 1982. Susceptibility to mastitis: A review of factors related to the cow. *Annals of Veterinary Research* 13: 85.

Pretty, J. 2007. Agricultural sustainability: Concepts, principles and evidence. *Philosophical Transactions of the Royal Society B* 363: 447–465.

Raubertas, R.F., and G.E. Shook. 1982. Relationship between lactation measures of somatic cell concentration and milk yield. *Journal of Dairy Science* 65: 419–425.

Rauw, W.M., E. Kanis, E.N. Noordhuizen-Stassen, and F.J. Grommers. 1998. Undesirable side effects of selection for high production efficiency in farm animals: A review. *Livestock Production Science* 56: 15–33.

Rotz, C.A., G.H. Kamphuis, H.D. Karsten, and R.D. Weaver. 2007. Organic dairy production systems in Pennsylvania: A case study evaluation. *Journal of Dairy Science* 90: 3961–3979.

Ruegg, P.L. 2009. Management of mastitis on organic and conventional dairy farms. *Journal of Dairy Science* 87: 43–55.

Shook, G.E. 2006. Major advances in determining appropriate selection goals. *Journal of Dairy Science* 89: 1349–1361.

Sustainable Table. 2009. http://www.sustainabletable.org/issues/dairy/ (accessed 09/09/2011).

USDA. 2000. *Part I: Baseline Reference of Feedlot Management Practices, 1999*. Fort Collins, CO: USDA.

USDA. 2007. *Dairy 2007 Part I: Reference of Dairy Cattle Health and Management Practices in the United States, 2007*. Washington, DC: USDA.

USDA/NASS. 2010. Statistics by Subject. http://www.nass.usda.gov/Statistics_by_Subject/index.php (Accessed November 2010).

von Keyserlingk, M.A.G., J. Rushen, A.M. de Passille, and D.M. Weary. 2009. The welfare of dairy cattle — Key concepts and the role of science. *Journal of Dairy Science* 92: 4101–4111.

Woodward, T.E. 1939. Figuring the rations of dairy cows. In: *Yearbook of Agriculture*. Washington, DC: USDA, p. 592.

Zwald, A.G., P.L. Ruegg, J.B. Kaneene, L.D. Warnick, S.J. Wells, C. Fossler, and L.W. Halbert. 2004. Management practices and reported antimicrobial usage on conventional and organic dairy farms. *Journal of Dairy Science* 87: 191–201.

POULTRY

Kenneth Anderson

POULTRY AND POULTRY PRODUCTION SYSTEMS

Over the last 100 years, the poultry industry has developed into three highly efficient systems made up of the commercial egg, broiler, and turkey segments. Back in the early 1900s when small self-sustaining farms were everywhere in the United States, free-range chickens for eggs as well as meat were a standard commodity on most every farm (Dryden, 1918). By the 1930s, free range was the main form of egg production being utilized, but farmers needed a more economical way to produce eggs year round for market and to get away from diseases caused by having the chickens on the floor. Thus, a battery system of caging chickens began to be developed in the early 1950s (Jull, 1951). Cages resulted in farmers being able to decrease the cost of production and increase the bird-to-space ratio, which made egg and meat production more profitable. Battery systems for eggs and litter systems for meat have been the standard now for decades, but entering into the twenty-first century there is a huge push from animal rights activists as well as a segment of the consumer market to get birds out of cages, back on the floor, and provide outdoor access. It is ironic how the industry is making a huge circle right back to where it all began. Today, hens on many of the poultry farms produce 489 eggs in 110 weeks (Anderson, 2007), 6.4-lb broilers in 42 days with 1.58 lb of feed per pound of gain (Havenstein, Ferket, and Qureshi, 2003), and 50-lb turkey males in 22 weeks with a feed conversion of 2.7 lb of feed per pound of gain (Krueger, 2008). These performance numbers were undreamed of 60 years ago, and even 20 years ago, layers were only producing 380 eggs in 110 weeks (Anderson, 1991). These advances in performance are the result of genetic selection, better understanding of disease and vaccines, nutrition, and environmental management. Within each of these sectors, there are subsectors made up of the breeders, hatcheries, broiler growers, egg production, transport, and processing. Currently, broilers and turkeys are predominantly reared on litter floor operations where the birds are contained in a large building with deep litter. Commercial layers are predominantly housed in some type of cage environment, with approximately 80% of the U.S. laying flock housed in cages, 10% housed in environmentally enriched production environments, and approximately 9% in a cage-free range system. Because of the extensive use of cages, the layer industry has been a primary target of organizations to end the use of battery cages in the United States. This criticism and activism is coming primarily from external coalitions of animal rights organizations, environmentalists, vegetarians, individuals within the animal research community, and the consumer (Anderson, 2009c). As a result, state ballot initiatives and state agreements targeting the layer industry have emerged, resulting in the affected industries rapidly changing to meet the imposed requirements. The organizations sponsoring these initiatives have become very astute at manipulating the public perception and influencing regulations.

The poultry industry is being criticized from all sides for its management of facilities, husbandry practices, disease prevention, and environmental management. There are a number of practices within the poultry industry that can be misconstrued as deleterious to the welfare of animals. However, these practices have been researched and are constantly being examined by the industry for their benefit to welfare and quality of the product produced. In a number of instances, practices have been abandoned in commercial operations because of their potential negative impact on the bird and lack of benefit to the commercial producer or product quality. Part of this may be a result of the efforts of poultry breeders to select for behavior traits that benefit the birds in a more intensive setting (Craig and Muir, 1996). Issues in the poultry industry that have been noted as affecting animal well-being are discussed in the following sections.

HATCHERY

The handling of newly hatched chicks, poults, or ducklings has been associated with a number of animal welfare concerns regarding hatcheries and the movement of hatchlings through the hatchery system (Agriculture Canada, 1989). Growing concerns are focused on the way the neonatal chick or poult is handled once it is removed from the incubator. The keys to humane handling of these young animals are related to gentle handling of chicks from the hatching tray, separating them from hatch residue and piped embryos, and ensuring that they are not dropped from high places. Chicks experience short drops of a few inches during processing and have no changes in their livability in the growing house. Hatchery processes begin with the chicks, broken shells, and unhatched embryos in the hatching trays being gently tipped onto the chick and eggshell separator, which allows the chicks to fall through the rollers onto a rod conveyor. This separates the chicks from the large shell components and the small shell particles. The chicks then slide into a chick-go-round. From this carousel, the chicks can easily be handled for sorting, sexing, and vaccinating (Bell and Weaver, 2002). The chicks are then placed in chick boxes for transport to the rearing facilities.

   Cull or non-salable hatchlings that do not enter production such as males (layers) and chicks with defects or injuries are humanely euthanized immediately after hatch. Three methods are used for euthanasia in hatcheries. They include immediate mechanical destruction (maceration), vacuum with impact plate, and modified atmospheric gas (asphyxiation). The Humane Slaughter Association (2002) recommended the use of two methods: maceration and modified atmosphere gas euthanasia (Raj and Whittington, 1995). The key to each of these methods is the immediate death of the chick with no excessive pain or struggling. All of these methods of euthanasia are acceptable if they are done according to standard operating procedure and the equipment is maintained and functioning properly. The result of this process should be evaluated rigorously because the animal welfare concerns are very high. The same can be said for methods for the disposal of unhatched embryos. Live pips and the embryos that have not hatched are now treated in the same manner as cull chicks. As such, they should be disposed of in a similar manner with constant checking of the results to ensure that no live embryos survive. Two additional methods, rapid cooling and freezing, are also acceptable means of euthanizing unhatched embryos. Most hatcheries utilize some form of maceration as their primary euthanasia method, which results in immediate death (Beckman, 2010). In other circumstances or in an emergency, euthanasia may be accomplished using CO_2 for large groups and, for individual chicks, cervical dislocation can be used by properly trained individuals.

BEAK TRIMMING, DUBBING/DE-SNOODING, AND TOE TRIMMING

These are morphological alterations in a number of different ways, including elective surgery, amputations, or mutilations. These descriptors vary depending upon who is describing them. If these procedures are utilized, one must ensure that the equipment used to carry out these procedures is working properly, and that the personnel involved in carrying out these procedures are adequately trained. If these procedures are not needed, they should be eliminated from chick processing practices. Breeders are selecting for behavioral patterns that diminish the need for these practices (Craig and Muir, 1996).

   Beak trimming was developed to curtail the development of abnormal behaviors such as cannibalism or excessive feather pecking. In these cases, the hen's welfare was enhanced with beak trimming. Beak trimming continues to be the method of choice worldwide for the control of cannibalism and general improvements in performance and livability. When performed at the proper age using the hot blade (HB), infrared (IR), or scalable continuous wave lasers (SL), there are few long-term negative effects. There are advantages and disadvantages with each method. Some of the advantages of all methods are reduced mortality (Craig and Lee, 1989, 1990), lower feed consumption, improved feed efficiency (Lee, 1980), and improved egg production (Kuo, Craig, and Muir, 1991). Some disadvantages associated with beak trimming of older birds or severe trimming are delayed sexual maturity (Carey, 1990), potential neuroma formation, and chronic stress in the trimmed pullets. Indications are that beak trimming likely results in pain to the bird due to the mechanoreceptor and thermoreceptor cells present in the beak (Gentle and Breward, 1985; Gottschaldt et al., 1982). However, the length of time that the pain may endure appears to be related to quality of the trim (Gentle, 1986a, 1986b), trimming age, and severity of the trim (Davis, Anderson, and Jones, 2004). Davis showed that corticosterone levels in birds trimmed at 6 days returned to the same as non-trimmed flock mates within 24 h, while hens trimmed at 11 weeks of age had elevated corticosterone levels at 5 weeks after the trim. Regardless of the methods, the negative aspects of beak trimming that may occur in the pullet phase appear to be offset by the positive aspects in the layer phase with enhanced performance and improved livability of the flock. These changes are in part due to changes in behavioral patterns, which result from beak trimming (Craig and Lee, 1989) that includes increased feeding activity, increased resting pattern, and a reduction in pecking by cage mates. The chickens adapt quickly to the beak alteration and there does not seem to be a long-term negative effect on the birds.

   Dubbing is a procedure to remove the comb from the head of the bird at hatch in an attempt to limit later damage by injury, freezing, or cannibalism. Dubbing roosters and hens is a practice that has not persisted in the layer industry due to increased climate control of the production houses (Hester, 2005). Dubbing is still used for special cases that include research facilities where the combs of roosters may become caught or injured due to caging for selective artificial insemination practices; however, hens are no longer dubbed. The comb of breeding males in cages can become so large they become a potential entrapment component or may restrict access to the feed trough. Dubbing eliminates this impediment and, when done properly at hatching, results in a reduction in comb size of 50 to 75%. This is only used in strains with large combs such as egg-type strains. The second reason is to minimize the comb's exposure to cold temperatures. Full-size combs have a greater potential of freezing in cold climates and dubbed hens perform better than their non-dubbed counterparts do in cold weather (Cole and Hutt, 1954). However, as the poultry industry is forced to revert to extensive production systems in cooler climates, the use of dubbing may be revived to help the birds cope with cold or freezing temperatures in the winter. In this case, the producers are balancing one husbandry practice with another. Whether the practice is dubbing or housing chickens in a confined space, each has welfare considerations, which will improve the overall welfare of the bird in one instance, but may not improve welfare of birds in another. If necessary, dubbing is best completed at hatching due to the lack of vascularization of the comb at that age (Cole and Hutt, 1954) although it can be done through 8 weeks of age with special care to prevent bleeding. The comb is removed at its base using surgical scissors.

De-snooding is the removal of the snood (dewbill) to prevent head injuries from picking or fighting in a growing flock (TNAU, 2010). The snood is removed at hatch by pinching the snood off between the thumbnail and forefinger or using a small clipper. It can also be removed with scissors at 3 weeks of age. As with many practices in poultry, this practice has alternative names and meanings especially in the way they are presented to the public. One case in point is the Wales Statutory Instruments 2007 No. 1029 (W.96) regulation entitled "The Mutilations (Permitted Procedures) (Wales) Regulations 2007." With this type of title, de-snooding would not be a very welcomed procedure even if the benefits to the bird were significant. However, recent research has shown that the snood may enhance heat loss in males (Buchholz, 1996) and that, behaviorally, de-snooding does not appear to result in overt aggression in the rearing environment. In support of discontinuing de-snooding, growers have found that there is no advantage to the male turkey and that the snood may help the turkey dissipate body heat. Therefore, in discussions with experts, it was concluded that de-snooding be abandoned as unnecessary for the welfare of birds in the turkey industry.

Toe clipping is only used in the turkey industry for females grown for roasting and in the broiler industry for male breeders to reduce the incidence of injuries to the other birds in the flock from scratches to the back, breast, and legs. This practice was shown to diminish the nervousness of the flock and to reduce body injury to flock mates from moving and fighting as the birds reach maturity (McEwen and Barbut, 1992). However, advances in genetic selection, husbandry, and nutrition have minimized the need to use this practice. Toe trimming is typically done at the hatchery using a hot blade, infrared, or microwave (Honaker and Ruszler, 2004). Broiler breeder females are no longer trimmed and the males typically only have the dewclaw removed (Bell and Weaver, 2002). Ouart, Russell, and Wilson (1989) indicated that trimming of multiple toes might contribute to decreased mating efficiency and fertility. When toe trimming is done in the hatcheries, the infrared method is preferred to minimize pain and stress (Wang et al., 2008) associated with older methods. This practice does reduce the incidence of injuries to other birds; however, the question of whether the procedure results in long-term pain or discomfort to the animal has not been resolved. One report indicated that removal of one toe in breeder chicks did not appear to cause chronic pain (Gentle and Hunter, 1988). Esthetically the procedure is not pleasant to observe, but neither turkeys nor broilers appear to suffer any long-term negative consequences.

Chick transport from the hatchery is another area of concern for animal welfare groups. Items that need to be monitored include the cleanliness of the chick boxes and pads, handling of the chick boxes, temperature of the transport truck, ventilation in the transport truck, exposure to excessive stress and noise, and the duration of the delivery trip. If these components are monitored and maintained, then both good chick quality and bird welfare are ensured. Mitchell and Kettlewell (2004) indicated that a transport time of 12 h is acceptable if conditions such as temperature, humidity, and ventilation within the transport vehicle are well controlled and monitored to ensure chick well-being.

HUSBANDRY PRACTICES

Poultry housing issues have focused on space and housing for laying hens in cages and it is probably the most controversial issue facing the poultry industry today. It is by far the most pressing issue in the commercial egg industry, but less pressing in other sectors of the poultry industry in which birds are reared on the floor in litter facilities (Bell and Weaver, 2002; Hester 2005). Housing density, the amount of space provided to the hens, is a combination of two factors—the amount of floor space allocated to each bird and group size. In a cage-house setting, both of these factors can have a negative impact on production and behavior of the flock (Adams and Craig, 1985; Anderson, 1996; Anderson, 2009b). As space per hen is diminished and as group size increases, productivity declines and mortality increases. These impacts are present even when the population is held constant with decreasing space and when the population is increased with a constant density (Anderson, 1996). However, is it correct to interpret this response as being due to diminished well-being? Bogner et al. (1979) determined that Leghorns need between 458 and 581 cm^2 in order to accommodate behaviors of preening and comfort movements. Lagadic and Faure (1987) taught hens that if they performed a task, pecked a specific button, a portion of the cage would move to increase the determined floor space available. With this type of testing, they determined that hens selected floor space of between 400 to 619 cm^2. Currently, the egg industry is providing 432 cm^2 (67 in.^{2}) for white egg strains and 490 cm^2 (76 in.^{2}) for brown egg layers (United Egg Producers, 2010). These amounts of floor space for the hen, as well as the physical structures within the environment, promote the display of comfort movements from a more natural behavioral repertoire. There is a transition within the egg industry toward housing birds in more extensive systems that include environmentally enriched housing systems (Tauson, 2000), cage-free space or aviaries (Gibson et al., 1989) and free-range facilities (Hughes and Dunn, 1986; Appleby and Hughes, 1991). Spaces within these facilities range from 929 cm^2 (1 ft^2) for slat/litter houses and aviaries to 1393 cm^2 (1.5 ft^2) in all litter and free-range operations (United Egg Producers, 2010; Anderson, 2009a). These systems provide roosts, nest boxes, litter areas, and, in the case of free-range operations, the opportunity for hens to access the outdoors (Anderson, 2009a). In these environments, adequate space for roosting (13 to 15 cm per bird), nesting (1 nest per 5 to 8 hens), feeding (3.8 to 5.1 cm per bird and the hen should not have to move more than 7.9 m), and watering (1 to 2.54 cm per hen depending on device configuration or 1 nipple per 10 hens) are important. These extensive systems provide a more enriched and stimulating environment that allows hens to exhibit a complete behavioral repertoire. However, there are negative aspects associated with extensive systems such as sternum deformities, bone fractures from falls, exposure to inclement weather, increased risk of disease and parasitism, and increased risk of predation.

Broilers, broiler breeders, turkeys, and turkey breeders are housed in floor facilities that contain litter areas, feeders, waterers, and nest boxes; therefore, these segments of the poultry industry have not had the level of scrutiny focused on the layer segment of the poultry industry. However, as with all commercial poultry operations, the primary concerns are related to the housing and maintenance of such flocks. These concerns are associated with bird density and adequate space allocations for the resources of feed and water.

Broiler breeder density allocations recommended for litter and slat/litter houses are 3 and 2 ft^2 per bird, respectively, and for commercial broilers the desired density is 0.8 to 1.0 ft^2 per bird depending on the final body weight desired (Bell and Weaver, 2002). Bird density, whether excessive or not, can and will affect growth, feed conversion, and behavior of birds, which can negatively affect their welfare. The undesirable behaviors in breeder flocks are cannibalism, excessive feather pecking, and fear-related behaviors such as avoidance and escape responses or flock hysteria. Many of these behaviors are readily observable by producers and, if noted, measures should be taken to rectify them. Space at the feeder should be adequate for all birds in a pen to eat at once, as this is especially important in breeder flocks. In skip-a-day feed restriction programs, this may be especially important. If space is not adequate, there may be observable increases in aggressive behaviors. Inadequate feeder space will not necessarily result in injury to the subordinate animals, but will influence the subordinate bird's ability to obtain adequate nutrition, and will result in non-uniform body weights and poor productivity. In many instances, it may only be a single bird dominating a feeder. The birds in a flock utilize water space differently and aggressive behaviors associated with water consumption are not an issue in facilities with adequate space. As long as watering space does not limit water consumption, watering space is not an area that needs to be controlled. Hens will typically stand around a cup or nipple drinker and take turns drinking. Nesting space is important in breeder operations and should provide 1 nest per 4 to 5 hens or 1 m of community nest per 35 to 40 hens. If this space is inadequate, there will be an increased number of eggs laid on the floor. Inadequate nesting space can also lead to increases in breakage and eating of eggs. The height of the nests from the floor (>20 in.) is also thought to increase the potential for the development of hysteria. In floor production systems, hens should be kept out of nests at night and early morning, and then the rests should be opened for egg laying in the morning. This keeps the nests cleaner and allows free access to the nests when eggs are being laid.

Feed and water restriction programs are used to control body weight in fast-growing, high-feed consuming breeder birds and water restriction keeps them from over-drinking after the feed has been consumed (Bell and Weaver, 2002). Such programs go hand in hand, one to restrict feed intake, and the other to limit growth rate. Water restriction is also used to prevent birds from consuming excessive amounts of water in an attempt to satisfy their desire for more food. Water restriction also helps maintain better litter conditions. Thus, monitoring of behavior with regard to feed and water consumption can provide insight into the well-being of hens.

Commercial turkey breeder hens are maintained in facilities separate from the breeding toms. Due to the size of the males, natural mating is no longer used, and lighting and feeding programs are different for the two populations. The recommended space is 0.3 m^2 per hen and 0.4 m^2 per tom. If the space is not adequate, feather picking, cannibalism, and other health problems can ensue (Spratt, 1993).

Molting is used extensively in the layer industry to extend the productive life of laying hens (Bell and Weaver, 2002; Anderson and Havenstein, 2007). It is also used in the broiler breeder, turkey (Lilburn et al., 1993), and duck segments of the poultry industry to extend egg production (Rolon, Buhr, and Cunningham, 1993; Hurwitz et al., 1995, 1998). The molting procedures result in the initiation of a natural process in which the hen enters into a phase of reproductive quiescence that allows her to replace her feather coat and replenish her body systems before entering into another reproductive cycle. The stimulus for entering into this phase consists of environmental stimuli, such as reducing lighting, temperature, and some level of anorexia. In the avian species, molt inducement has been accomplished by limiting the nutrient intake of all or selected nutrients as a commercial husbandry practice. The methods used to induce molt in laying hens are stressful and have been condemned as inhumane husbandry practices. There are times when wild birds do not eat in spite of having food readily available, for example, during molting, breeding, and egg incubation.

Stevens (1996) indicates the importance birds place upon seasonal breeding and other activities. He indicated that fasting is especially pronounced in geese that may be anorexic for 2.5 months and king penguins that fast for 4 to 6 months. It must be remembered that stress is not something that can be avoided throughout the course of life and there is stress that is actually beneficial to the animal. By definition, the absence of stress is death (Selye, 1973). Fasting can also be the result of an alteration in the endocrinology of the hen (Swanson and Bell, 1974a). In wild birds, hormonal changes are typically associated with molting and broodiness, and seasonal changes result in limited food supply, so the husbandry practice of molting in the commercial egg and breeder industries is based on those principles. The hen is capable of coping with and compensating for changing conditions in its environment to maintain physiological homeostasis (Clarenburg, 1986; Freeman, 1987). The hen responds by using physical, chemical, anatomical, and physiological mechanisms to maintain this homeostasis. The hen has functions that are constitutive or always functioning, and others that are adaptive, that is, they are used as the need arises to maintain the homeostatic state.

The following are some of the physiological mechanisms, both constitutive and adaptive, that are used to respond to limited or total restriction of food that occurs postprandial, between meals, and during a fast, as determining when one mechanism starts and another begins is arbitrary (Clarenburg, 1986). The metabolism of chickens readily evokes these physiological processes throughout the course of a regular day. Upon prolonged absence of food, other essential nutrients are depleted (for example, vitamins, minerals, essential amino and fatty acids, lipotropic factors, and carbohydrates), which can be life threatening. Starvation triggers a collapse of homeostasis as basal metabolic rate declines and the hen minimizes all energy expenditures in order to survive. This response does not occur in anorexia associated with animal husbandry practices. Rice (1905) and Rice, Nixon, and Rogers (1908) were the first to report on fasting in laying chickens to induce molting of hens in commercial layer flocks. However, during eras of depressed financial returns on egg production, research on molting experienced renewed interest as a means of extending the productive life of the hen (King and Trollope, 1934; Frasier, 1948; Swanson and Bell, 1974a). Modified photoperiods combined with withdrawal of feed and water were used in the 1940s and research interest in induced molting has continued. Several types of induced anorexia and durations of anorexia have been widely examined in chickens based on total feed restriction (Frasier, 1948; Marble, 1963; Bierer and Eleazer, 1966; Noles, 1966; Bell, 1970, 1984; Swanson and Bell, 1970, 1974a, 1974b, 1974c, 1974d; Summers and Leeson, 1977; Brake, Thaxton, and Benton, 1979; Brake and Thaxton, 1979a, 1979b; Washburn, Peavey, and Renwick, 1980; Lee, 1982, 1984; Rowland and Brake, 1982; van Kempen, 1983; Brake and Carey, 1983; Garlich et al., 1984; Zimmerman, Andrews, and McGinnis, 1987; Kuney and Bell, 1987; Carey and Brake, 1989; Savage, 1992; Koelkebeck, Parsons, and Leeper, 1993; Brake, 1994; Bell et al., 1995; Hurwitz et al., 1995; Anderson, 1998, 2000; Davis, Anderson, and Carrol, 2000). Other areas of research have included limited feeding, altering the mineral content of the diet, such as excessive dietary magnesium (Shippee et al., 1979), excessive dietary iodide (Arrington et al., 1967), excessive dietary zinc (Shippee et al., 1979; Bell, Swanson, and Kuney, 1980; Berry and Brake, 1985; Goodman, Norton, and Diambra, 1986; Berry, Gildersleeve, and Brake, 1987; Breeding, 1991), dietary calcium restriction (Douglas, Harms, and Wilson, 1972), and dietary sodium restriction (Whitehead and Shannon, 1974; Hughes and Whitehead, 1974; Whitehead and Sharp, 1976; Nesbeth, Douglas, and Harms, 1976a, 1976b; Wakeling, 1978; Said et al., 1984; Berry and Brake, 1985). However, all of these methods resulted in a forced anorexic state and a significant loss in body weight. Water deprivation was also employed, but Palafox (1976) and Swanson, Bell, and Kuney (1978) reported no beneficial effects and, in fact, found undesired post-molt effects on performance of laying hens. Thus, water deprivation during the molt was abandoned. New molting methods have been reviewed and developed as non-anorexic methods have been adopted by the layer hen industry (Anderson and Havenstein, 2007; Biggs et al., 2003, 2004; Anderson, 2002). All concurred that the birds produce an equivalent total number of eggs and a greater egg income. They further suggested that economically feasible alternatives to the more traditional molting methods resulted in better performance of hens compared to that for hens not induced to molt.

EUTHANASIA

Euthanasia is the act of inducing humane death in an animal. Ultimately, this means that the animal should be exposed to minimal stress and anxiety brought on by the pain that the animal might perceive before unconsciousness and death. The poultry industries are faced with two needs in this area. There is a need for euthanasia of individual birds that become sick or injured during the course of the production period and a need for mass euthanasia of whole houses of birds in instances such as infectious disease outbreaks (Benson et al., 2009). The use of gas (CO_2) and cervical dislocation are two methods that work well for immediate euthanasia of sick or injured birds.

The Canadian Council on Animal Care (2010) defines the use of CO_2 as conditionally acceptable with emphasis on proper methods if used. Carbon dioxide would normally be used as emergency backup on small populations of poultry. A proper chamber must be used, and proper precautions must be taken to protect workers involved. Compressed CO_2 gas in cylinders should be used to allow inflow into the chamber to be regulated precisely. With an animal in the chamber, an optimal flow rate should displace at least 20% of the chamber volume per minute. It is important to verify that an animal is dead before removing it from the chamber. Chambers for exposing poultry to CO_2 must have a view port to allow verification that the birds are down for at least 2 min before being removed from the chamber. A clear plastic bag is suitable for administering CO_2 to very young poultry, generally less than 10 days of age, or for live piped embryos, which are still in the shell. A sealed box with the ability to maintain a 60 to 70% concentration of CO_2 gas as it is gradually increased at a rate of 20 to 30% per minute, exhaust, and view ports is acceptable for older birds as long as the CO_2 atmosphere within the chamber is sufficient to euthanize the bird (AVMA, 2007). Loss of consciousness is caused within 10 to 15 sec and death is typically induced within 5 min of exposure. Death should be verified by extending the exposure time of the bird to the CO_2 atmosphere for an additional 10 min.

Cervical dislocation by hand is a second method that can be used for smaller birds, but the Burdizzo Emasculator Apparatus is used for larger birds. The procedures for cervical dislocation by hand begin by restraining the bird by both legs at the hock joint. Then the head is grasped by placing the index finger or thumb at the occipital crest just above the neck at the junction of the atlas and caudal vertebra and the other finger being placed under the lower mandible (Chamberlin, 1943). Then with one quick motion, the neck is stretched and the head rotated backward, simultaneously by pinching it between the thumb and forefinger. The vertebrae between the atlas and caudal vertebra are dislocated simultaneously, which severs the spinal cord and tears the jugular vein and carotid artery. The procedures for cervical dislocation using the Burdizzo Emasculator Apparatus begin with restraining the bird's legs and/or wings (depending on body size) using an appropriate device or having one person hold the bird by both legs at the shanks, resting the bird with its breast on a table or on the floor. The neck of the bird is placed between the jaws of the Burdizzo Apparatus at the junction of the atlas and caudal vertebra and the jaws are closed quickly by pulling the handles together until the handles of the Burdizzo Apparatus lock together. The bird is released after all reflexes cease.

Govrin-Lippmann and Devor (1978) and Jensen et al. (1985) indicated that injury resulting from discharges of peripheral nerves subside within seconds and that all afferent activity ceases. This response causes activity of the muscles in poultry immediately after the severing of the spinal cord. Hughes and Gentle (1995) and Gentle (1991) provided physiological evidence that there is no peripheral neural input immediately after severing of the nerves of the spinal cord, indicating a pain-free period immediately after the severing of the spinal cord. This indicates, in the case of cervical dislocation and decapitation, that when the burst of nerve discharge occurs, there is no cerebral receptor site functioning to perceive the nerve impulses sent to the brain. Therefore, the brain of the animal does not sense the burst of neural activity through cervical dislocation or decapitation. The EEG recordings made from severed heads are merely recording the random firing of neurons that are not indicative of pain (Scadding, 1981). Chapman et al. (1985) indicated that animals have responses to neural stimulation that differ from humans. This makes it difficult to draw strong, clinically relevant conclusions from experimental observations on animals. Cervical dislocation is one of the primary and easiest methods of euthanasia. Mass euthanasia because of diseases or natural disasters is relatively new to the industry, but the need became apparent because of diseases such as avian influenza in Southeast Asia and natural disasters like Hurricane Floyd in North Carolina. Methods using water-based foams, used in fire suppression, have been developed for emergencies where large numbers of birds must be euthanized at once. These methods were conditionally approved by USDA-APHIS in 2006 for meat-type chickens. This process has been verified as effective in a number of other species (Benson et al., 2009).

Stunning prior to euthanasia for processing is now done by two methods: electrical and modified atmosphere (Raj, 1998). The issue associated with electrical stunning is that birds may not be stunned properly and may recover their somatosensory evoked potentials in the brain, which is a significant welfare concern. New electrical stunning methods appear to have minimized this problem (Prinz et al., 2010). Modified atmosphere stunning has been developed and used successfully in the European community (Poole and Fletcher, 1998). Both methods are acceptable and, depending on the gasses used and timing of the euthanasia sequence in the processing plant, have a similar disadvantage of somatosensory recovery if euthanasia is not done promptly.

TRANSPORT AND CATCHING

The transport of older birds requires catching them for transport, which is followed by movement of the birds on trucks from the rearing facilities to the production unit and later to the processing plant (Lacy and Czarick, 1998; Scott, Connell, and Lambe, 1998; Kannan et al., 1997). Catching and transport are novel experiences for birds and they are equally stressful regardless of rearing environment. The key in all of these processes is gentle handling of the birds to minimize injuries. This means that individuals must be properly trained in handling procedures, operation of loading equipment, and methods for transport of birds (Nijdam et al., 2004). In addition, the transport truck must be capable of providing protection for the birds from extremes in temperature during transport by using side panels or curtains and to ensure adequate air movement in the center of the loads during warm and cold weather.

REFERENCES

Adams A.W., and J.V. Craig. 1985. Effects of crowding and cage shape on productivity and profitability of caged layers: A survey. *Poult. Sci.* 64: 238–242.

Agriculture Canada Publication 1757E. 1989. Recommended code of practice for the care and handling of poultry from hatchery to processing plant. Communications Branch, Agriculture Canada, Ottawa.

AVMA (American Veterinary Medical Association). 2007. AVMA Guidelines on Euthanasia. http://www. avma.org/issues/animal_welfare/euthanasia.pdf

Anderson, K.E. 1991. 28th North Carolina Layer Performance and Management Test: Final Production Report. 28(3). http://www.ces.ncsu.edu/depts/poulsci/tech_manuals/layer_reports/28_final_report.pdf

Anderson, K.E. 1996. Final Report of the Thirty-First North Carolina Layer Performance and Management Test: Production Report. 31(4). Accessed November 16, 2010. http://www.ces.ncsu.edu/depts/poulsci/ tech_manuals/layer_reports/31_final_report.pdf

Anderson, K.E. 1998. Final Report of the Thirty-Second North Carolina Layer Performance and Management Test. North Carolina Cooperative Extension Service. 32(4). http://www.ces.ncsu.edu/depts/poulsci/tech_ manuals/layer_reports/32_final_report.pdf

Anderson, K.E. 2000. Final Report of the Thirty-Third North Carolina Layer Performance and Management Test: Production Report. 33(4). http://www.ces.ncsu.edu/depts/poulsci/tech_manuals/layer_reports/33_ final_report.pdf

Anderson, K.E. 2002. Final Report of the Thirty-Fourth North Carolina Layer Performance and Management Test. 34(4). North Carolina Cooperative Extension Service. http://www.ces.ncsu.edu/depts/ poulsci/tech_manuals/layer_performance_tests.html#layer_34

Anderson, K.E. 2007. Final Cycle Report of the Thirty-Sixth North Carolina Layer Performance and Management Test. 36(5). http://www.ces.ncsu.edu/depts/poulsci/tech_manuals/layer_reports/36_final_report.pdf

Anderson, K.E. 2009a. Single Production Cycle Report of the Thirty Seventh North Carolina Layer Performance and Management Test. 37(4). http://www.ces.ncsu.edu/depts/poulsci/tech_manuals/layer_reports/37_ single_cycle_report.pdf

Anderson, K.E. 2009b. Final Report of the Thirty-Seventh North Carolina Layer Performance and Management Test. 37(5). Accessed November 16, 2010. http://www.ces.ncsu.edu/depts/poulsci/tech_manuals/layer_ reports/37_final_report.pdf

Anderson, K.E. 2009c. Overview of natural and organic egg production: Looking back to the future. *J. Appl. Poult. Res.* 18: 348–354.

Anderson, K.E., and G.B. Havenstein. 2007. Effects of alternative molting programs and population on layer performance: Results of the Thirty-Fifth North Carolina Layer Performance and Management Test. *J. Appl. Poult. Res.* 16: 365–380.

Appleby, M.C., and B.O. Hughes. 1991. Welfare of laying hens in cages and alternative systems: Environmental, physical and behavioural aspects. *World's Poult. Sci. J.* 47: 109–128.

Arrington, L.R., R.A. Santa Cruz, R.H. Harms, and H.R. Wilson. 1967. Effects of excess dietary iodine upon pullets and laying hens. *J. Nutr.* 92: 325–330.

Beckman, B. 2010. Chick well-being at the hatchery. Hy-Line North America, Hy-Line Technical School, Des Moines, IA, May 2010. http://www.hyline.com/userdocs/library/Beckman.pdf

Bell, D. 1970. Further investigations of molting methods. *Proceedings of the Poultry Institute*, University of California.

Bell, D. 1984. Fast/slow molting techniques. *Proceedings of the Poultry Institute*, University of California.

Bell, D. 1987. Is molting still a viable replacement alternative? *Poultry Tribune* 93: 32–35.

Bell, D., D. McMartin, F. Bradley, and R. Ernst. 1995. Animal care series: Egg-type layer flock care practices. California Poultry Workgroup, Cooperative Extension, University of California.

Bell, D., M.H. Swanson, and D.R. Kuney. 1980. A comparison of force molting methods II. *Progress in Poultry* No. 21 (May), University of California.

Bell, D.D., and W.D. Weaver, Jr. 2002. *Commercial Chicken Meat and Egg Production*, 5th ed. Norwell, MA: Kluwer Academic Publishers.

Benson, E.R., R.L. Alphin, M.D. Dawson, and G.W. Malone. 2009. Use of water-based foam to depopulate ducks and other species. *Poult. Sci.* 88: 904–910.

Berry, W.D., and J. Brake. 1985. Comparison of parameters associated with molt induced by fasting, zinc, and low dietary sodium in caged layers. *Poult. Sci.* 64: 2027–2036.

Berry, W.D., R.P. Gildersleeve, and J. Brake. 1987. Characterization of different hematological responses during molts induced by zinc or fasting. *Poult. Sci.* 66: 1841–1845.

Bierer, B.W., and T.H. Eleazer. 1966. The relationship of feed and water deprivation on greenish gizzard mucous membrane and contents in chickens. *Poult. Sci.* 45: 379–380.

Biggs, P.E., M.W. Douglas, K.W. Koelkebeck, and C.M. Parsons. 2003. Evaluation of nonfeed removal methods for molting programs. *Poult. Sci.* 82: 749–753.

Biggs, P.E., M.E. Persia, K.W. Koelkebeck, and C.M. Parsons. 2004. Further evaluation of nonfeed removal methods for molting programs. *Poult. Sci.* 83: 745–752.

Bogner, V.H., W. Peschke, V. Seda, and K. Popp. 1979. Studie zum flächenbedarf von legehennen in käfigen bei besttimmten aktivitäten. *Berl. Münch. Tierärztl. Wochenschr.* 92: 340–343.

Brake, J.T. 1994. Progress in induced moulting. *Poultry Intl.* 10(6): 44–46.

Brake, J., and J.B. Carey. 1983. Induced molting of commercial layers. North Carolina Agricultural Extension Service. Poult. Sci.ence and Technology Guide No. 10. Extension Poult. Sci.ence, Raleigh, NC.

Brake, J., and P. Thaxton. 1979a. Physiological changes in caged layers during a forced molt. 1. Body temperature and selected blood constituents. *Poult. Sci.* 58: 699–706.

Brake, J., and P. Thaxton. 1979b. Physiological changes in caged layers during a forced molt. 2. Gross changes in organs. *Poult. Sci.* 58: 707–716.

Brake, J., P. Thaxton, and E.H. Benton. 1979. Physiological changes in caged layers during a forced molt. 3. Plasma thyroxine, plasma triiodothyronine, adrenal cholesterol, and total adrenal steroids. *Poult. Sci.* 58: 1345–1350.

Breeding, S.W. 1991. Molt induced by dietary zinc in a low calcium diet. Masters of Science Thesis, Physiology Program. North Carolina State University, Raleigh, NC.

Buchholz, R. 1996. Thermoregulatory role of the unfeathered head and neck in male wild turkeys. *The Auk* 113(2): 310–318.

CCAC (Canadian Council on Animal Care). 2010. CCAC guidelines on:euthanasia of animals used in science. htta://www.ccac.ca/Documents/Standards/Guidelines/Euthanasia.pdf (Accessed December 2010).

Carey, J.B. 1990. Influence of age at final beak trimming on pullet and layer performance. *Poult. Sci.* 69: 1461–1466.

Carey, J.B., and Brake, J. 1989. Induced molting of commercial layers. North Carolina Agricultural Extension Service. Poult. Sci.ence and Technology Guide No. 10. Extension Poult. Sci.ence, Raleigh, NC.

Chamberlain, F.W. 1943. *Atlas of Avian Anatomy, Osteology, Arthrology, and Myology*. Michigan State College, Ag. Exp. Station.

Chapman, C.R., K.L. Casey, R. Dubner, K.M. Foley, R.H. Gracely, and A.M. Reading. 1985. Pain measurement: An overview. *Pain* 22: 1–31.

Clarenburg, R. 1986. Syllabus for the Course of Veterinary Physiology I. Eleventh Ed. Department of Anatomy and Physiology, College of Veterinary Medicine, Kansas State University, Manhattan.

Cole, R.K., and F.B. Hutt. 1954. The effect of dubbing on egg production and viability. *Poult. Sci.* 33: 966–972.

Craig, J.V., and H.-Y. Lee. 1989. Research note: Genetic stocks of white leghorn type differ in relative productivity when beaks are intact versus trimmed. *Poult. Sci.* 68: 1720–1723.

Craig, J.V., and H.-Y. Lee. 1990. Beak trimming and genetic stock effects on behavior and mortality from cannibalism in white leghorn type pullets. *Appl. Anim. Behav. Sci.* 25: 107–123.

Craig, J.V., and W. Muir. 1996. Group selection for adaptation to multiple-hen cages: Beak related mortality, feathering, and body weight responses. *Poult. Sci.* 75: 294–302.

Davis, G.S., K.E. Anderson, and A.S. Carrol. 2000. The effects of long term caging and molt of single comb white leghorn hens on heterophil to lymphocyte ratios, corticosterone and thyroid hormones. *Poult. Sci.* 79: 514–518.

Davis, G.S., K.E. Anderson, and D.R. Jones. 2004. The effects of different beak trimming techniques on plasma corticosterone and performance criteria in single comb white leghorn hens. *Poult. Sci.* 83: 1624–1628.

Douglas, C.R., R.H. Harms, and H.R. Wilson. 1972. The use of extremely low dietary calcium to alter the production pattern of laying hens. *Poult. Sci.* 51: 2015–2020.

Dryden, J. 1916. *Poultry Breeding and Management.* New York: Orange Judd Company.

Frasier, F. 1948. How force molting works. *Pacific Poultryman*, April, pp. 7, 18–20.

Freeman, B.M., 1987. The stress syndrome. *Worlds Poult. Sci.* 43(1): 15–19.

Garlich, J.D., J. Brake, C.R. Parkhurst, J.P. Thaxton, and G.W. Morgan. 1984. Physiological profile of caged layers during one production year, molt, and postmolt: Egg production, egg shell quality, liver, femur, and blood parameters. *Poult. Sci.* 63: 339–343.

Gentle, M.J. 1986a. Beak trimming in poultry. *Worlds Poult. Sci. J.* 42: 268–275.

Gentle, M.J. 1986b. Neuroma formation following partial beak amputation (beak trimming) in the chicken. *Res. Vet. Sci.* 41: 383–385.

Gentle, M.J. 1991. The acute effects of amputation on peripheral trigeminal afferents in *Gallus gallus domesticus. Pain* 46: 97–103.

Gentle, M.J., and J. Breward. 1985. The bill tip organ of the chicken. *J. Anatomy* 145: 79.

Gentle, M.J., and L.H. Hunter. 1988. Neural consequences of partial toe amputation in chickens. *Res. Vet. Sci.* 45: 374–376.

Gibson, S.W., P. Dunn, and B.O. Hughes. 1989. The performance and behavior of laying fowls in a covered strawyard system. *Res. Dev. Agric.* 5: 153–163.

Goodman, B.L., R.A. Norton, Jr., and O.H. Diambra. 1986. Zinc oxide to induce molt in layers. *Poult. Sci.* 65: 2008–2014.

Govrin-Lippmann, R., and M. Devor. 1978. Ongoing activity in severed nerves: Source and variation with time. *Brain Res.* 159: 406–410.

Gottschaldt, K-M., H. Fruhstorfer, W. Schmidt, and I. Kruft. 1982. Thermosensitivity and its possible fine structure basis in mechanoreceptors in the beak and skin of geese. *J. Comp. Neurol.* 205: 219–245.

Havenstein, G.B., P.R. Ferket, and M.A. Qureshi. 2003. Growth, livability, and feed conversion of 1957 versus 2001 broilers when fed representative 1957 and 2001 broiler diets. *Poult. Sci.* 82: 1500–1508.

Hester, P.Y. 2005. Impact of science and management on the welfare of egg laying strains of hens. *Poult. Sci.* 84: 687–696.

Honaker, C.F., and P.L. Ruszler. 2004. The effect of claw and beak reduction on growth parameters and fearfulness of two Leghorn strains. *Poult. Sci.* 83: 873–881.

Humane Slaughter Association. 2002. *Codes of Practice for the Disposal of Chicks in Hatcheries*, 2nd ed. Wheathampstead, UK: Humane Slaughter Association.

Hughes, B.O., and P. Dun. 1986. A comparison of hens housed intensively in cages or outside on range. *Zootech. Int.* Feb: 44–46.

Hughes, B.O., and M.J. Gentle. 1995. Beak trimming of poultry: Its implications for welfare. *Worlds Poult. Sci. J.* 51: 51–61.

Hughes, B.O., and C.C. Whitehead. 1974. Sodium deprivation, feather pecking and activity in laying hens. *Br. Poult. Sci.* 15: 435–439.

Hurwitz, S., E. Wax, Y. Nisenbaum, M. Ben-Moshe, and I. Plavnik. 1998. The response of laying hens to induced molt as affected by strain and age. *Poult. Sci.* 77: 22–31.

Hurwitz, S., E. Wax, Y. Nisenbaum, and I. Plavnik. 1995. Responses of laying hens to forced molt procedures of variable length with or without light restriction. *Poult. Sci.* 74: 1745–1753.

Jensen, T.S., B. Krebs, J. Nielsen, and P. Rasmussen. 1985. Immediate and long-term phantom limb pain in amputees: Incidence, clinical characteristics and relationship to pre-amputation limb pain. *Pain* 21: 267–278.

Jull, M.A. 1951. *Poultry Husbandry,* 3rd ed. New York: McGraw-Hill Book Company.

Kannan, G., J.L. Heath, C.J. Wabeck, M.C.P. Souza, C.J. Howe, and J.A. Mench. 1997. Effects of crating and transport on stress and meat quality characteristics in broilers. *Poult. Sci.* 76: 523–529.

King, D.F., and G.A. Trollope. 1934. Force molting of hens and all night lighting as factors in egg production. Circ. No. 64, Alabama Polytechnic Inst., Auburn.

Koelkebeck, K.W., C.M. Parsons, and R.W. Leeper. 1993. Effect of early feed withdrawal on subsequent laying hen performance. *Poult. Sci.* 72: 2229–2235.

Krueger, K.K. 2008. High feed prices! Reevaluating turkey slaughter ages and weights. *Proceedings of the NC Turkey Industry Days*, J. Grimes, Ed. Wilmington, NC, Sept. 24–25, pp. 19–28.

Kuney, D.R., and D.D. Bell. 1987. Effect of molt duration on performance. University of California, Extension Bulletin, pp. 1–10.

Kuo, F.-L., J.V. Craig, and W.M. Muir. 1991. Selection and beak-trimming effects on behavior cannibalism and short term production traits in white leghorn pullets. *Poult. Sci.* 70: 1057–1068.

Lacy, M.P., and M. Czarick. 1998. Mechanical harvesting of broilers. *Poult. Sci.* 77: 1794–1797.

Lagadic, H., and J.M. Faure. 1987. Preferences of domestic hens for cage size and floor types as measured by operant conditioning. *Appl. Anim. Behav. Sci.* 19: 147–155.

Lee, K. 1980. Long term effects of Marek's disease vaccination with cell-free herpesvirus of turkey and age at debeaking on performance and mortality of white leghorns. *Poult. Sci.* 69: 2002–2007.

Lee, K. 1982. Effects of forced molt period on postmolt performance of Leghorn hens. *Poult. Sci.* 61: 1594–1598.

Lee, K. 1984. Feed restriction during the growing period, forced molt, and egg production. *Poult. Sci.* 63: 1895–1897.

Lilburn, M.S., K.E. Nestor, R. Stonerock, and T. Wehrkamp. 1993. The effect of feed and water withdrawal on body weight and carcass characteristics of molted turkey breeders. *Poult. Sci.* 77: 30–36.

Marble, D.R. 1963. Comparison of pullet and hen flocks at the New York Random Sample Tests. Cornell University Agricultural Experiment Station, New York State College of Agriculture, Ithaca, NY. Bulletin 979.

McEwen, S. A., and S. Barbut. 1992. Survey of turkey down grading at slaughter: Carcass defects and associations with transport, toenail trimming, and type of bird. *Poult. Sci.* 71: 1107–1115.

Mitchell, M.A., and P.J. Kettlewell. 2004. Transport of chicks, pullets and spent hens. In: *Welfare of the Laying Hen*, G.C. Perry, Ed. Abingdon, Oxfordshire, England: Carfax Publishing Co., pp. 361–374.

Najdam, E., P. Arens, E. Lambooij, E. Decuypere, and J.A. Stegeman. 2004. Factors influencing bruises and mortality of broilers during catching, transport, and lairage. *Poult. Sci.* 83: 1610–1615.

Nesbeth, W.G., C.R. Douglas, and R.H. Harms. 1976a. Response of laying hens to a low salt diet. *Poult. Sci.* 55: 2128–2132.

Nesbeth, W.G., C.R. Douglas, and R.H. Harms. 1976b. The potential use of dietary salt deficiency for the force resting of laying hens. *Poult. Sci.* 55: 2375–2380.

Noles, R.K. 1966. Subsequent production and egg quality of forced molted hens. *Poult. Sci.* 45: 50–57.

Ouart, M.D., G.B. Russell, and H.R. Wilson. 1989. Mating behavior in response to toe nail removal in broiler breeders. *Zootechnica International* 7: 35–37.

Palafox, A.L. 1976. Comparative performance of non-force molted and force-molted White Leghorn Hens. *Poult. Sci.* 55: 2076.

Poole, G.H., and D.L. Fletcher. 1998. Comparison of a modified atmosphere stunning-killing system to conventional electrical stunning and killing on selected broiler breast muscle rigor development and meat quality attributes. *Poult. Sci.* 77: 342–347.

Prinz, S., G. Van Oijen, F. Ehinger, A. Coenen, and W. Bessei. 2010 Electroencephalograms and physical reflexes of broilers after electrical waterbath stunning using and alternating current. *Poult. Sci.* 89: 1265–1274.

Raj, A.B.M., and P.E. Whittington. 1995. Euthanasia of day-old chicks with carbon dioxide and argon. *Veterinary Record* 136: 292–294.

Raj, M. 1998. Welfare during stunning and slaughter of poultry. *Poult. Sci.* 77: 1815–1819.

Rice, J.E. 1905. In: *The Feeding of Poultry, The Poultry Book,* W.G. Johnson and G.O. Brown, Eds. New York: Doubleday, Page and Co.

Rice, J.E., C. Nixon, and C.A. Rogers. 1908. The molting of fowls. Bulletin 238, Cornell University, Agricultural Experiment Station of the College of Agriculture, Department of Poult. Sci.ence, Ithaca, NY.

Roland, D.A., Sr., and J. Brake. 1982. Influence of premolt production on postmolt performance with explanation for improvement in egg production due to force molting. *Poult. Sci.* 61: 2473–2481.

Rolon, A., R.J. Buhr, and D.L. Cunningham. 1993. Twenty-four-hour feed withdrawal and limited feeding as alternative methods for induction of molt in laying hens. *Poult. Sci.* 72: 776–785.

Said, N.W., T.W. Sullivan, H.R. Bird, and M.L. Sunde. 1984. A comparison of the effect of two force molting methods on performance of two commercial strains of laying hens. *Poult. Sci.* 63: 2399–2403.

Savage, S. 1992. Molting programs: Get more uniform hens with better bones. *Poultry Digest* 51: 16.

Scadding, J.W. 1981. Development of ongoing activity, mechanosensitivity, and adrenaline sensitivity in severed peripheral nerve axons. *Exp. Neurol.* 73: 345–364.

Scott, G.B., B.J. Connell, and N.R. Lambe. 1998. The fear levels after transport of hens from cages and a free-range system. *Poult. Sci.* 77: 62–66.

Selye, H. 1973. The evolution of the stress concept. *Am. Sci.* 61: 692–699.

Shippee, R.L., P.R. Stake, U. Koehn, J.L. Lambert, and R.W. Simmons. 1979. High dietary zinc or magnesium as forced-resting agents for laying hens. *Poult. Sci.* 58: 949–954.

Spratt, D. 1993. Basic Husbandry for Turkeys: Factsheet. February 1993. Ministry of Agriculture, Food and Rural Affairs, Ontario.

Statutory Instruments. 2007. The Mutilations (Permitted Procedures) (Wales) Regulations 2007. No. 1029 (W.96) http://www.opsi.gov.uk/legislation/wales/wsi2007/wsi_20071029_en_1

Stevens, L.. 1996. *Avian Biochemistry and Molecular Biology.* Cambridge, England: Cambridge University Press.

Summers, J.D., and S. Leeson. 1977. Sequential effects of restricted feeding and force-molting on laying hen performance. *Poult. Sci.* 56: 600–604.

Swanson, M. H. and D.D. Bell. 1970. Field tests of forced molting practices and performance in commercial egg production flocks. Proceedings of the World's Poult. Sci.ence Meeting, Spain.

Swanson, M.H., and D.D. Bell. 1974a. Force molting of chickens. 1. Introduction. *Univ. of Calif. Coop. Ext. Bull.* AXT-410.

Swanson, M.H., and D.D. Bell. 1974b. Force molting of chickens. 2. Methods. *Univ. of Calif. Coop. Ext. Bull.* AXT-411.

Swanson, M.H., and D.D. Bell. 1974c. Force molting of chickens. 3. Performance characteristics. *Univ. of Calif. Coop. Ext. Bull.* AXT-412.

Swanson, M.H., and D.D. Bell. 1974d. Force molting of chickens. 4. Egg quality. *Univ. of Calif. Coop. Ext. Bull.* AXT-413.

Swanson, M.K., D.D. Bell, and D.R. Kuney. 1978. Effects of water restriction and feed withdrawal time on forced molt performance. *Poult. Sci.* 57: 1116.

Tauson, R. 2000. Producción, salud y manejo en jaulas equipadas. Congreso Internacional de Produccion y Sanidad Animal. XXXVI Symposium of the World's Poult. Sci.ence Association, Barcelona, Spain, pp. 32–46.

TNAU Agritech Portal. 2010. Turkey Management. http://agritech.tnau.ac.in/animal_husbandry/animhus_tur_management.html (Accessed 10/15/2010).

United Egg Producers. 2010. Animal Husbandry Guidelines for U.S. Egg Laying Flocks, 2010 Edition. United Egg Producers, Alpharetta, GA. http://www.unitedegg.org/information/pdf/UEP_2010_Animal_Welfare_Guidelines.pdf

van Kampen, M. 1983. Heat stress, feed restriction, and the lipid composition of egg yolk. *Poult. Sci.* 62: 819–823.

Wakeling, D.E. 1978. The use of low calcium and low sodium diets to induce an egg production pause in commercial layers. Agricultural Development and Advisory Service Bull. No. 8L-51-8-76, Starcross, Exeter, Devon.

Wang, B., B.M. Rathgeber, T. Astatkie, and J.L. Macisaac. 2008. The stress and fear levels of microwave toe-treated broiler chickens grown with two photoperiod programs. *Poult. Sci.* 87: 1248–1252

Washburn, K.W., R. Peavey, and G.M. Renwick. 1980. Relationship of strain variation and feed restriction to variation in blood pressure and response to heat stress. *Poult. Sci.* 59: 2586–2588.

Whitehead, C.C., and D.W.F. Shannon. 1974. The control of egg production using a sodium-deficient diet. *Br. Poult. Sci.* 15: 429–434.

Whitehead, C.C., and P.J. Sharp. 1976. An assessment of the optimal range of dietary sodium for inducing a pause in laying. *Br. Poult. Sci.* 17: 601–611.

Zimmerman, N.G., D.K. Andrews, and J. McGinnis.1987. Comparison of several induced molting methods on subsequent performance of single comb white Leghorn hens. *Poult. Sci.* 66: 408–417.

BEEF CATTLE

Terry Engle

Animal agriculture is one of the fundamental cornerstones that have helped shape the development of the United States. Over the last 100 years, animal agriculture has changed in dramatic ways. Consolidation of livestock production facilities has increased production efficiency while maintaining low costs of meat, milk, and eggs to the consumer. However, consolidation has yielded fewer people working directly in animal agriculture and has shifted the focus of animal care from animal husbandry to animal productivity. This disconnect has caused societal concerns for animal well-being and lack of citizen understanding of, and support for, animal agriculture. This section will discuss ways in which animal comfort can be practically vectored into beef cattle production.

Beef cattle production has drastically changed over the past 50 years. The implementation of new technologies and production techniques has enhanced the efficiency of production of meat products. The increase in production efficiency has enabled producers to produce more products with fewer animals, while maintaining a high-quality product at a low cost for the consumer. Enhanced beef cattle production efficiency is primarily a result of improvements in feed technologies, genetic selection, animal health, and management.

With the increased focus on enhancing production efficiency, the individual animal itself cannot be forgotten. The basic beef cattle husbandry principles still apply to modern beef cattle production today: Provide the basic needs for cattle (feed, protection, medical assistance, etc.) and the animal will provide product for human consumption. Thus, it is in the producers' best interest to maintain an environment wherein beef cattle can thrive—where disease is kept to a minimum, moribund animals are expeditiously treated or humanely euthanized, and feed, water, and shelter are in adequate supply.

Several food animal production systems have evolved into systems where environmental conditions, feeding regimes, and animal activities are tightly controlled in order to increase production efficiency. Beef cattle production has taken a different approach to increase production efficiency. Typically, a cow-calf operator confines cattle in open pastures and allows the animals to harvest native forage. When indigenous feedstuffs become incapable of supporting proper cattle nutrition, the rancher supplies stockpiled feedstuffs to compensate for the nutrient void until the indigenous forages are replenished. Stockpiled feedstuffs can be items such as hay, by-products from other industries such as cull vegetables, fermentation by-products, bakery waste, etc. The ability of these animals to harvest their own feed as well as their ability to utilize by-products from other industries has been instrumental in enhancing cow-calf production efficiency.

In a commercial cow-calf operation, a certain percentage of the female calves born each year are retained in the cow herd as replacement females. At weaning, females not retained as replacement animals, cows being removed from the production herd, and the majority of male calves (typically castrated at or shortly after birth), enter the cattle-feeding sector of beef production. In general, these animals can be marketed through an auction system, transported directly to a feedlot setting, or allowed to graze crop residues throughout the winter to increase body weight and, therefore, enter the feedlot at a heavier weight at some time in the future. Nevertheless, calves entering the feedlot sector are transported from pasture-based production settings to feedlot settings where cattle are housed in group pens, cared for daily, sometimes comingled with cattle from other geographic locations, and a total mixed ration containing all the appropriate nutrients is delivered daily, thus eliminating the need for the animal to harvest feed on its own via grazing. Cattle typically spend approximately 140 to 200 days (depending on the weight at which they enter the feedlot) in a feedlot setting until slaughtered at approximately 14 to 16 months of age (heifers and steers).

Due to the length of time that it takes to produce beef for human consumption (from breeding to slaughter), proper nutrition and abatement of animal stressors are fundamental animal husbandry components essential for optimizing animal health and productivity. Environmental and management stressors can increase disease outbreaks and decrease efficiency of food producing animals, thus increasing the cost of production and ultimately affecting animal welfare. Adverse weather conditions, including both the effects of hot and cold climatic conditions, are particularly difficult for grazing animals as well as confinement-fed animals housed in outdoor facilities. Prolonged hot or cold environmental conditions can decrease nutrient quality of feedstuffs as well as alter the nutrient utilization of feed by the animal. Decreased nutrient quality and the need to metabolically repartition nutrients to cope with extreme climatic conditions diminish the ability of the animal to immunologically protect itself from environmental pathogens, ultimately compromising animal health and overall productivity. Therefore, the subsequent sections in this chapter are devoted to discussing practical ways to enhance animal comfort in beef cattle production systems by minimizing animal stress.

Stress and its relationship to the occurrence of disease have long been recognized. Stress is the nonspecific response of the body to any demand made upon it (Selye, 1973). Stressors relative to animal production include infection, environmental factors, parturition, lactation, weaning, transport, and handling. Stress has been reported to decrease animal production (growth, reproduction, efficiency, etc.) and overall animal welfare.

SOCIAL BEHAVIOR

Beef cattle are social, gregarious animals that can thrive in various environmental conditions. Since cattle are social animals that develop hierarchies within the herd, introducing new animals to an established herd or pen of cattle can be stressful to both resident animals and new arrivals. Numerous dominance-subordination experiments from the late 1950s and 1970s (Wieckert, 1970) indicate that a hierarchy is established within a few days of animals being comingled and that dominant animals do stake out a "territory." New animals introduced into an established group will spend time and energy learning the established hierarchy. This can be accomplished within a few days, but noticeable agitation across the group will be observed until the new animal learns the hierarchy and is accepted into the group. Therefore, introducing new animals to established groups of animals as infrequently as possible can help minimize stress.

ENVIRONMENTAL STRESSORS

As indicated earlier, beef cattle production takes place outdoors in pastures or large feedlot pens. Therefore, beef cattle are exposed to various environmental conditions throughout the course of a year. Depending on the geographical location, cattle can be exposed to ambient temperatures below freezing or in excess of 38°C for prolonged periods of time. When climatic conditions exceed upper and lower critical temperatures for cattle, the animal needs to compensate metabolically for such a deviation. Any time an animal has to expend energy to heat or cool itself, the overall production efficiency of that animal is decreased.

COLD STRESS

Cattle are typically cold-hardy animals (Young, 1981). However, the ability of cattle to tolerate cold temperatures requires that they remain well insulated from the environment. Maintaining effective insulation requires protection from the wind, maintenance of a dry hair coat, and protection from cold and frozen or wet and muddy conditions (Wagner, Grubb, and Engle, 2008). Providing shelter during times of inclement weather will improve animal efficiency (Young, 1981) and well-being. However, building extensive structures for beef cattle in cow-calf operations is not economically feasible. Allowing range cows and calves access to natural structures such as trees, rocks, etc., and utilizing existing structures such as stockpiled hay and buildings as windbreaks can be very effective at minimizing the impact of cold weather. Furthermore, providing bedding, such as straw, can help keep cattle dry during times of wet, muddy conditions.

Feedlot operators may be reluctant to provide bedding and windbreaks for cattle during the winter months because, although windbreaks can effectively alleviate the negative impact of wind on winter performance, airflow in the summer months can be compromised and performance reduced (Mader et al., 1999). Therefore, unless portable, windbreaks will not likely become common in areas that experience cold climates in the winter months and hot climates in the summer months. Providing bedding to cattle can effectively combat cold stress in northern climates (Birkelo and Lounsbery, 1992). However, feedlot operations may be reluctant to use bedding due to the cost of removing bedding plus manure from the pens. Furthermore, bedding may retain moisture in pens and delay drying of the pen surface. Providing bedding as a routine management strategy will likely not become common during times of typical inclement weather. However, the economics of providing bedding in the aftermath of a catastrophic winter storm should be evaluated. Wagner et al. (2008) reported net energy requirements for maintenance of feedlot cattle exposed to a storm in southeast Colorado in December 2006 and January 2007. Average high and low temperatures from December 26, 2006, through February 22, 2007, were −2.16°C and −14.69°C, respectively. Furthermore, snowfalls of 25.4 and 5.08 cm were recorded on December 20 and 21, 2006. An additional 25.4, 30.48, and 30.48 cm of snow fell on December 29, 30, and 31, 2006, respectively. Additional snow events occurred on January 13 and 14, January 21, and February 14 and 15, 2007. The snow pack peaked at 91.44 cm on December 31, 2006, and averaged 32.33 cm ± 0.26 from December 26, 2006, through February 22, 2007. Net energy required for maintenance (NEm) was approximately 21.92 Mcal/hd/d or 0.21 Mcal per kg $EBW^{0.75}$. These data indicate that NEm required during and in the aftermath of a major winter weather event may be 2.7-fold higher than NEm required ($0.077 \times EBW^{0.75}$) under thermal neutral conditions. Calculations of lower critical temperature and external insulation indicate that the insulation value of the hair coat of these cattle may have been inhibited by the moisture, mud, and snow following the storm. Table 8.1 describes the effect of corn and feeder cattle prices on economic losses ($ per head) associated with a catastrophic winter storm. These data indicate that applying bedding to feedlot pens after an extensive cold/snowfall event needs to be considered.

Heat Stress

Cattle raised in most portions of the United States can be exposed to heat stress during certain times throughout the year. Typically cattle in cow-calf operations have access to shade provided by natural (trees, berms, etc.) or constructed (buildings, stockpiled feed, etc.) structures and during the summer months are exposed to moderate wind speeds that help with cooling. Furthermore, genetic selection has helped to reduce the impact of heat stress on beef cattle. In general, *Bos indicus* cattle are more heat tolerant and parasite resistant than are *Bos taurus* cattle. Typically, cattle raised in hot and dry desert climates or hot and humid semi-tropical climates have a certain percentage of *Bos indicus* genetics to assist with minimizing heat stress.

Feedlot cattle are typically finished in the high plains of the western United States due to the dry climate (low precipitation—rain and snow and low humidity). However, periodically cattle finished in the high plains are exposed to ambient temperatures at or above the thermal neutral zone for cattle for prolonged periods of time. Feedlot cattle performance can be adversely affected during prolonged periods of elevated ambient temperatures, especially if the elevated ambient temperature is coupled with low wind speeds and high humidity (Hahn and Mader, 1997; Mader et al., 1999). Enhancing an animal's ability to dissipate heat or reduce solar radiation load can help to diminish the impact of heat stress on overall animal performance and well-being. Several management strategies have been implemented by feedlot producers to reduce the effect of heat stress on feedlot cattle. Providing shade to decrease solar load, but not airflow (i.e., overhead structures), sprinkling pen surfaces and cattle with water, and restricted or managed feeding programs (Mader et al., 2002; Davis et al., 2003) are common techniques used to help mitigate heat stress in feedlot cattle. For an in-depth review of the aforementioned strategies to mitigate heat stress in cattle, see Mader (2003).

TABLE 8.1
The Effect of Corn and Feeder Cattle Prices on Economic Losses ($ Per Head) Associated with a Catastrophic Winter Storm

Item	$ per 45.45 kg	Cattle Price[a] 2.50	3.50	Corn[b] Price ($ per 25.41 kg) 4.50	5.50	6.50
Feed costs[c]	—	91.08	111.79	132.51	153.22	173.94
Yardage[d]	—	20.30	20.30	20.30	20.30	20.30
Interest[e]	80.00	9.05	9.05	9.05	9.05	9.05
	100.00	11.31	11.31	11.31	11.31	11.31
	120.00	13.58	13.58	13.58	13.58	13.58
	140.00	15.84	15.84	15.84	15.84	15.84
Death loss[f]	80.00	65.27	67.69	70.12	72.54	74.97
	100.00	77.99	80.41	82.84	85.26	87.69
	120.00	90.71	93.13	95.56	97.98	100.40
	140.00	103.42	105.85	108.27	110.70	113.12
Total costs[g]	80.00	185.69	208.83	231.98	255.12	278.26
	100.00	200.68	223.82	246.96	270.10	293.24
	120.00	215.66	238.80	261.94	285.08	308.22
	140.00	230.64	253.78	276.92	300.06	323.20

Source: Adapted from Wagner et al., 2008. *Professional Animal Scientist.* 24: 494–499.

[a] 403.8 kg pay weight.
[b] 15% moisture.
[c] 9.67 kg per day dry matter intake for the 58-day study period and diet dry matter concentration was 70%.
[d] $0.35 per head daily for the 58-day study period.
[e] 8% on initial calf value.
[f] 7% of the steer value at the start of the study period calculated from initial calf value and production costs up to the start of the study.
[g] Feed plus yardage, interest, and death loss costs.

PEN DESIGN

Three very effective methods commonly utilized by feedlot operators to help keep cattle dry during times of wet, muddy conditions are mounding within pens, pen slope, and concrete pads adjacent to the feed bunk. Constructing mounds of dirt and dried manure in pens coupled with the appropriate slope of a feedlot pen surface where water can be diverted out of the pen, minimizing standing water and maximizing pen surface drying, allows cattle to avoid muddy pen surfaces. Furthermore, it is common practice to have a concrete apron adjacent to the feed bunk, which allows cattle a solid foundation to stand on while consuming feed.

MANAGEMENT STRESSORS

Castration, dehorning, branding, handling, and transportation are common management practices used in the beef cattle industry. Pain and distress associated with these management techniques are difficult to quantify and have been the center of much debate regarding animal welfare. Castration induces physiological stress and alters several physiological and behavioral responses indicative of pain (Melony, Kent, and Robertson, 1995; Fisher et al., 1996, 1997a,b). However, attempting to alleviate the stress of castration with local anesthesia or analgesics pre- and post-castration has been challenging and results have been variable. Ting et al. (2003a,b) reported that systemic analgesia with ketoprofen, a nonsteroidal anti-inflammatory drug, was an effective method for alleviating acute inflammatory stress associated with castration. Earlier research by Earley and Crowe (2002) indicated that ketoprofen was superior to local anesthesia with lidocaine in suppressing increases in plasma cortisol (an acute stress indicator) and decreasing abnormal standing post-castration. Other researchers have reported similar results (Gonzalez et al., 2010; Stafford et al., 2002). Furthermore, plasma cortisol response to castration increases as the age of the animal at castration increases (King et al., 1991). This is most likely due to an increase in soft tissue damage (greater tissue innervation and blood flow) at the time of castration in older compared to younger animals (Ting et al., 2003a,b; Weissman, 1990; Fisher et al., 1996). It is evident that castration is painful to cattle based on physiological and behavioral observations reported in the literature. Utilization of analgesics should be implemented to minimize the pain experienced by castration. Furthermore, if castration is going to be used as a management tool, it should be performed at the earliest age possible. Future research should focus on determining the method and duration of analgesics in order to minimize pain in castrated animals. Possible means of chemical or immunological castration should also be investigated.

Removing horns from cattle (dehorning) is a management practice to help prevent bruising of cattle when they are transported together in close quarters, as well as to reduce the risk of injury to other animals and employees. In general, horns can be removed by disbudding (destroying the horn-producing cells) at 6 to 8 weeks of age, or by removing established horns. Hot iron and chemical forms of disbudding are common methods of preventing horns from growing. Once horns are mature, horn removal is more challenging. Horn buds and the base of mature horns are highly vascularized and innervated and mature horns are linked to the frontal sinuses. Due to the innervation, vascularization, and relationship to the sinus, dehorning can be painful and increase the risk of infection and excessive bleeding. Results of numerous experiments indicate that dehorning causes an increase in plasma cortisol (Wohlt et al., 1994; McMeekan et al., 1997; McMeekan et al., 1998; Mellor et al., 2002; Sylvester, et al., 1998; AVMA, 2011). Local anesthesia, analgesics, cauterization, and a combination thereof, have been reported to assist with pain management in cattle that have been disbudded or dehorned. Due to the labor costs and reduced production efficiency, genetic selection for cattle with no horns (polled) is becoming popular.

Hot iron and freeze branding are common management practices for permanently identifying cattle. However, as discussed with castration and dehorning, both forms of branding can be painful as indicated by increased heart rates and plasma epinephrine and cortisol concentrations, which are indicative of pain (Lay et al., 1992 a,b). Therefore, similar pain abatement strategies as describe previously should be utilized when branding cattle. Alternatively, other less painful permanent identification systems could be utilized such as genetic or digital technologies.

Animal handling and transportation can also induce stress in beef cattle. For an extensive review of this topic, see Grandin (1997). If possible, habituating animals to handling equipment, people, and routine handling events can help decrease animal fear, which in turn helps to decrease animal stress. Regardless of acclimatization status to handling, it is imperative that all equipment be functioning appropriately when animals are being handled. Slipping or falling in a squeeze chute or on a cattle trailer can be extremely stressful to cattle (Grandin, 1993, 1997, 2001). Removing or minimizing objects that cattle may find frightening (swinging ropes, shadows, etc.) will also help decrease animal stress during handling. Furthermore, people handling animals need to be appropriately trained in cattle handling techniques, and remain calm and quiet. This will decrease the likelihood of animals having a negative experience during the handling or transportation event. Cattle that have a negative experience during handling and transportation (i.e., falling, slipping, rough handling, etc.) will remember the event and become more stressed during subsequent handling events. If cattle are extensively managed and not handled as frequently as intensively managed cattle are, it is important that the above-mentioned strategies for minimizing stress be implemented in conjunction with understanding the fear response described by Grandin (1997).

The Challenge

It is apparent that beef producers understand the importance of minimizing stress on beef cattle. By doing so, production efficiency is enhanced. However, over the last 10 years societal/consumer concerns for animal well-being and lack of understanding of animal agriculture have increased exponentially (Rollin, 1990, 2004). Society as a whole has begun to question how food animals are raised. In doing so, animal welfare has been moved to the forefront of topics that the beef industry must address. It is no longer satisfactory to consumers to justify beef production practices based on animal performance—the welfare of each individual animal needs to be vectored into production practices. Humane treatment of animals has always been an ingrained social ethic among beef producers. However, more attention needs to be given to pain management and abatement of environmental stressors as they relate to beef cattle production. By implementing these strategies into production practices and communicating them to the consumer, animal welfare will be improved and consumer confidence will be enhanced.

References

AVMA. 2011. Backgrounder: Welfare implications of the dehorning and disbudding of cattle. http://www.avma.org/reference/backgrounders/dehorning-cattle-bgnd.asp

Birkelo, C.P., D.E. Johnson, and H.P. Phetteplace. 1991. Maintenance requirements of beef cattle as affected by season on different planes of nutrition. *J. Anim. Sci.* 69: 1214–1222.

Birkelo, C.P., and J. Lounsbery. 1992. Effect of straw and newspaper bedding on cold season feedlot performance in two housing systems. *South Dakota State Univ. Beef Rep. Cattle.* 92–11.

Davis, M.S., T.L. Mader, S.M. Holt, and A.M. Parkhurst. 2003. Strategies to reduce feedlot cattle heat stress: Effects on tympanic temperature. *J. Anim. Sci.* 81: 649–661.

Earley, B., and M.A. Crow. 2002. Effects of ketoprofen alone or in combination with local anesthesia during the castration of bull calves on plasma cortisol, immunological, and inflammatory responses. *J. Anim. Sci.* 80: 1044–1052.

Fisher, A.D., M.A. Crowe, M.E. Alonso de la Varga, and W.J. Enright. 1996. Effect of castration method and the provision of local anesthesia on plasma cortisol, scrotal circumference, growth, and feed intake of bull calves. *J. Anim. Sci.* 74: 2336–2343.

Fisher, A.D., M.A. Crowe, E.M. O'Nuallain, M.L. Monaghan, J.A. Larkin, P. O'Kiely, and W.J. Enright. 1997a. Effects of cortisol on in vitro interferon-gamma production, acute-phase proteins, growth, and feed intake in a calf castration model. *J. Anim. Sci.* 75: 1041–1047.

Fisher, A.D., M.A. Crowe, E.M. O'Nuallain, M.L. Monaghan, D.J. Prendiville, P. O'Kiely, and W.J. Enright. 1997b. Effects of suppressing cortisol following castration of bull calves on adrenocorticotropic hormone, in vitro interferon-gamma production, leukocytes, acute-phase proteins, growth, and feed intake. *J. Anim. Sci.* 75: 1899–1908.

González, L.A., K.S. Schwartzkopf-Genswein, N.A. Caulkett, E. Janzen, T.A. McAllister, E. Fierheller, A.L. Schaefer, D.B. Haley, J.M. Stookey, and S. Hendrick. 2010. Pain mitigation after band castration of beef calves and its effects on performance, behavior, *Escherichia coli*, and salivary cortisol. *J. Anim. Sci.* 88: 802–810.

Grandin, T. 1993. Teaching principles of behavior and equipment design for handling livestock. *J. Anim. Sci.* 71: 1065–1070.

Grandin, T. 1997. Assessment of stress during handling and transport. *J. Anim. Sci.* 75: 249–257.

Grandin, T. 2001. Livestock-handling quality assurance. *J. Anim. Sci.* 79: E239–E248.

Hahn, G.L., and T.L. Mader. 1997. Heat waves in relation to thermoregulation, feeding behavior, and mortality of feedlot cattle. *Proc. 5th Int. Livest. Environ. Symp., Am. Soc. Agric. Eng.*, St Joseph, MI. pp. 563–571.

King, B.D., R.D.H. Cohen, C.L. Guenther, and E.D. Janzen. 1991. The effect of age and method of castration on plasma cortisol in beef calves. *Can. J. Anim. Sci.* 71: 257–263.

Lay, D.C., Jr., T.H. Friend, C.L. Bowers, K.K. Grissom, and O.C. Jenkins. 1992a. A comparative physiological and behavioral study of freeze and hot-iron branding using dairy cows. *J. Anim. Sci.* 70: 1121–1125.

Lay, D.C., Jr., T.H. Friend, R.D. Randel, C.L. Bowers, K.K. Grissom, and O.C. Jenkins. 1992b. Behavioral and physiological effects of freeze or hot-iron branding on crossbred cattle. *J. Anim. Sci.* 70: 330–336.

Mader, T. L., J.M. Dahlquist, G.L. Hahn, and J.B. Gaughan. 1999. Shade and wind barrier effects on summertime feedlot cattle performance. *J. Anim. Sci.* 77: 2065–2072.

Mader, T. L., S.M. Holt, G.L. Hahn, M.S. Davis, and D.E. Spiers. 2002. Feeding strategies for managing heat load in feedlot cattle. *J. Anim. Sci.* 80: 2373–2382.

Mader, T.L. 2003. Environmental stress in confined beef cattle. *J. Anim. Sci.* 81: E110–119E.

McMeekan, C.M., D.J. Mellor, K.J. Stafford, R.A. Bruce, R.N. Ward, and N.G. Gregory. 1997. Effects of shallow scoop and deep scoop dehorning on plasma cortisol concentrations in calves. *NZ Vet. J.* 45: 72–74.

McMeekan, C.M., K.J. Stafford, D.J. Mellor, R.A. Bruce, R.N. Ward, and N.G. Gregory. 1998. Effects of regional analgesia and/or non-steroidal anti-inflammatory analgesic on the acute cortisol response to dehorning calves. *Res Vet Sci.* 64: 147–150.

Mellor, D.J., K.J. Stafford, S.E. Lowe, N.G., Gregory, R.A. Bruce, and R.N. Ward. 2002. A comparison of catecholamine and cortisol responses of young lambs and calves to painful husbandry procedures. *Aust. Vet. J.* 80: 228–233.

Melony, V., J.E. Kent, and I.S. Robertson. 1995. Assessment of acute and chronic pain after different methods of castration of calves. *Appl. Anim. Behav. Sci.* 46: 33–48.

Rollin, B.E. 1990. Animal welfare, animal rights and agriculture. *J. Anim. Sci.* 68: 3456–3461.

Rollin, B.E. 2004. Annual meeting keynote address: Animal agriculture and emerging social ethics for animals. *J. Anim. Sci.* 82: 955–964.

Selye, H. 1973. The evolution of the stress concept. *Amer. Sci.* 61: 692–699.

Stafford, K., D. Mellor, S. Todd, R. Bruce, and R. Ward. 2002. Effects of local anesthesia or local anesthesia plus a non-steroidal anti-inflammatory drug on the acute cortisol response of calves to five different methods of castration. *Res. Vet. Sci.* 73: 61–70.

Sylvester, S.P., K.J. Stafford, D.J. Mellor, R.A. Bruce, and R.N. Ward. 1998. Acute cortisol responses of calves to four methods of dehorning by amputation. *Australian Vet. J.* 76: 123–126.

Ting, T.L., B. Earley, and M.A. Crowe. 2003a. Effect of repeated ketoprofen administration during surgical castration of bulls on cortisol, immunological function, feed intake, growth, and behavior. *J. Anim. Sci.* 81: 1253–1264.

Ting, T.L., B. Earley, J.M.L. Hughes, and M.A. Crowe. 2003b. Effect of ketoprofen, lidocaine, local anesthesia, and combined xylazine and lidocaine caudal epidural anesthesia during castration of beef cattle on stress responses, immunity, growth, and behavior. *J. Anim. Sci.* 81: 1281–1293.

Wagner, J.J., P.T. Grubb, and T.E. Engle. 2008. Case study: The effects of severe winter weather on net energy for maintenance required by yearling steers. *Prof. Anim. Sci.* 24: 1–6.

Weissman, C. 1990. The metabolic response to stress. An overview and update. *Anaesthesiology.* 73: 308–327.

Wieckert, D.A. 1970. Social behavior in farm animals. *J. Anim. Sci.* 32: 1274–1277.

Wohlt, J.E., M.E. Allyn, P.K. Zajac, and L.S. Katz. 1994. Cortisol increases in plasma of Holstein heifer calves from handling and method of electrical dehorning. *J. Dairy Sci.* 77: 3725–3729.

Young, B.A. 1981. Cold stress as it affects animal production. *J. Anim. Sci.* 52: 154–163.

Section Three

Sustainable Plant and Animal
Agriculture for Animal Welfare

9 Symbiosis of Plants, Animals, and Microbes

James Wells and Vincent Varel

CONTENTS

INTRODUCTION

A diversity of plants, animals, and microbes on Earth abounds due to evolution, climate, competition, and symbiosis. Single-cell species such as microorganisms are assumed to have evolved initially. Over time, plants and animals established and flourished. As each new kingdom of life came about, the ecosystem on Earth became more complex and the bionic components became more interactive. Symbiosis, in a broad definition, is "the living together in an intimate association of two or more dissimilar organisms." Symbiosis can result in a relationship in which both organisms benefit. Nitrogen fixation by legumes is a consequence of microbes that fix nitrogen and plants that supply simple carbons. Plants and fungi have established a cooperation in which the plant provides nutrients and the fungi provide alkaloids to deter predation and allow for greater drought tolerance. More generally, plants and herbivores have essentially co-evolved such that the action of herbivores on plants can lead to greater diversity and dispersion of seed. Complex cellulose degradation of plant material

185

by herbivores is accomplished by specialized bacteria in gastrointestinal compartments that are optimally maintained by each host animal for bacterial growth. Within the mammalian digestive tract, commensal microorganisms can provide energy, amino acids, and vitamins for the host, and provide protection against parasitic microorganisms. This chapter focuses on environmental sustainability of the many symbiotic relationships among plants, animals, and microbes that enhance our global food production.

LIFE ON EARTH

Life on Earth is complex and interactive, with organisms forming populations, which in turn form communities, or ecosystems, both locally and globally. The ecology is defined by the interactions between species and their composition within that system that drives natural selection, evolution, and genetic composition. The fitness for survival of an organism in any ecosystem is not dependent solely upon the species, but includes the interactions of other organisms with that species. Interactions between and among species within an ecosystem can be simple or complex, competitive or beneficial, predatory or symbiotic. Within a similar order, such as plants or animals, competition for resources can select for the better-fit species under one set of conditions, whereas predation results in one species consuming another.

Humans, through the development of agriculture, have identified and exploited different species for food production. Consequently, our desire for better production has often required control of the ecosystem. More importantly, the usefulness of a particular plant or animal species is often dependent on interactions with other species, including plant, animal, or microbial organisms. In agriculture, humans control competitive and predatory interactions to minimize the impact of competitive or predatory species on the agricultural species of interest. In contrast, symbiotic relationships are often encouraged and many production traits of interest are the result of symbiotic interactions.

Symbiosis is defined as two different species "living together." These close interactions between two species are often long-term and, for the most part, beneficial to one or more of the symbionts. There are numerous examples of symbiosis in agriculture. Agriculture in a broad sense involves a symbiotic relationship between humans and plants or animals. Humans plant, fertilize, control weeds and pests, and protect crops. Humans also nurture, feed, and protect livestock. The crops and livestock benefit from human interaction by being more productive and, in turn, they are utilized for food, clothing, shelter, and other human needs. Of more importance are symbioses, particularly interactions of lower order organisms, for example, microorganisms, which can impart health or disease in higher organisms.

PARASITISM AND PATHOGENICITY

Symbiotic relationships can be further defined or characterized by the type and level of interaction (see Figure 9.1). *Parasitism* describes a system in which one species benefits at the expense of another over time. *Pathogenic* relationships are often acute interactions in which one species specifically infects and benefits at the expense, and even death, of another. *Commensalism* describes a system in which one species benefits, but not at the expense of the other. *Mutualism* describes a system in which both species benefit. Within these types of symbiotic interactions, the level of interaction can be close contact between the symbionts (ectosymbiosis or exosymbiosis) or it can include one symbiont living inside the other (endosymbiosis).

Exploitation of a host can result in symbiosis that is parasitic in nature, and in production agriculture, these relationships can be costly. In *endosymbiotic* interactions, immature insects and parasitic microbes, such as protozoa or bacteria, can reside in a host for periods of time and compete for nutrients. In *exosymbiotic* instances, parasites, such as pests or insects, can persistently remove nutrients from the host. Regardless of the level of interaction, the loss of nutrients often results in lower yields of crops or reduced performance by the animal. When the parasitic relationship results

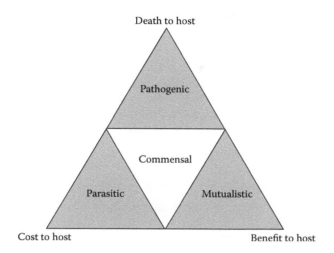

FIGURE 9.1 Trophic relationship between host and symbiont. Each corner of the triangle represents the key interactions, with the peak denoting a necrotrophic relationship that results in host death and the base denoting biotrophic relationships that require survival of the host.

in death of the host, the interaction is *necrotrophic*, whereas in a relationship that requires survival of the host, the interaction is *biotrophic*. The parasitic and pathogenic relationships are costly to agriculture and future efforts will be needed to control these relationships while not affecting the environment negatively.

INSECTS AND PARASITISM WITH PLANTS AND ANIMALS

Plants account for most of the living biomass on Earth, and parasites, insects in particular, have evolved closely with plant hosts (Rosomer and Stoffolano, 1997). *Phytophagous* ("plant eating" or herbivores) insects (anthropods) are the most abundant insects and include locusts, which are not selective and will consume most green plants (*polyphagous*), whereas the western corn rootworm is selective in consuming one species or genus (*monophagous*). Most plant parasites target a variety of plants (*oligotrophic*). Phytophagous insects may be specific in the anatomy of plant tissue that is targeted, and every plant tissue or anatomical part is susceptible to one or more phytophagous insects. These insect herbivores play a key role in the evolution of plant structure and response to herbivory, or being eaten (Stout, Thaler, and Thomma, 2006). Plants signal injury via two main pathways. One pathway is a systemically acquired response utilizing salicylic acid in responses to acute structural damage to elicit antimicrobial release and control microbial entry, and to deter insect attack. The second pathway is an induced resistance response utilizing jasmonic acid in response to chronic herbivory to deter the insect or hinder its development or reproduction. In contrast to the phytophagous insects, predatory insects such as spiders prey on other insects, particularly the phytophagous insects, and they can be beneficial to agriculture. Even the lowly household pest, the Asian cockroach, can be beneficial because it preys on bollworm and armyworm pests of agricultural crops.

Zoophagous insects extract nutrients from a living animal host and represent a broad group of parasitic insects, including the predatory insects that feed on other insects (Rosomer and Stoffolano, 1997). Most zoophagous insects that feed on vertebrate animals are biotrophic and live on the external surface (*exoparasitic*). Among animals common to production agriculture, suckling lice (Anoplura), chewing lice (Mallophaga), flies and mosquitoes (Diptera), ticks (Ixodidea), and fleas (Siphonaptera) are common parasitic insects and most are blood feeders *(hematophagous)*. Zoophagous insects have adapted a variety of host interactions. Lice can spend their entire lifecycle on a single host and are typically species specific. Host-specific zoophagous insects have often co-evolved with the host species and in some cases have evolved special structures for feeding on the host. Other zoophagous insects, such as fleas, may only live on the host for a short period of their lifecycle. Free living zoophagous insects, such as mosquitoes and flies, may only utilize the mammalian host for meals and not live on the host per se. Individually, the zoophagous insect may be insignificant relative to the much larger mass of the host, but collectively, these parasites can carry disease and in large numbers over extended periods can be a nutrient drain to the host. Identifying and exploiting predatory insects that target blood-feeding zoophagous insects would reduce the therapeutic use of pesticides for pest control in animal agriculture, but no significant predatory insect has been identified.

MICROBES AND PARASITISM/PATHOGENICITY WITH PLANTS AND ANIMALS

Parasitism, in its strictest definition, describes an interaction of one organism surviving at the expense of the other. However, when microbes are involved, the level of interaction is less clear and the definition less strict. To some extent, the definition of a parasite may depend on the point of view and the depth of understanding of the symbiosis. In *biotrophic* parasitism, what one may perceive as one organism surviving at the expense of the other may, in reality, be an interaction in which both organisms benefit, but the full symbiosis is not known. In addition, some interactions may be mutualistic (both organisms benefit) at certain times, but parasitic at other times. In contrast, pathogenic organisms represent examples of necrotrophic parasitism, but not all pathogenic microbes cause death of the host. Regardless, pathogen interaction represents greater loss to the host and needs to be considered differently than a parasitic interaction. Among plants and animals, particularly those of agricultural importance, a number of parasitic and pathogenic microbes are important relative to disease and host health, but a thorough discussion is beyond the scope of this chapter. A few interactions of economic importance will be discussed. A further note regarding animals: A number of pathogens may survive and multiply in one host as a commensal and not cause disease, but can cause disease or death in another host. Understanding these zoonotic pathogens is important not only to animal health, but also to providing a safe food and water supply to humans.

Plant Microbial Parasites and Pathogens

Plants are the most abundant form of terrestrial life and are plagued by numerous opportunistic organisms colonizing the leaves, stems, and roots. Fungi are associated with spotting, rusting, wilting, and rotting of plants. Fungi in the phylum of Ascomycota (commonly called Ascomycetes) are a diverse group known for a sac structure and include important decomposers in nature and sources for important medicinal uses. Species associated with plant disease include *Aspergillus*, *Fusarium*, *Thielaviopsis*, *Uncinula*, and *Verticillium*. Fungi in the phylum of Basidiomycota (commonly called Basidiomycetes) are a diverse group known for a "club" or "fruiting" structure, and include plant disease-causing species of *Rhizoctonia*, *Phakospora*, and *Puccinia*. Oomycetes are small eukaryotic organisms, or protists, that are fungal-like, include species of *Pythium* and *Phytophthora*, and are associated with rusts, rots, and blights. Numerous bacteria cause diseases in plants, but species belonging to *Agrobacterium*, *Burkholderia*, *Clavibacter*, *Erwinia*, *Phytoplasma*, *Pseudomonas*, *Spiroplasma*, and *Xanthomonas* can cause significant damage or death to plants. Disruption of plant colonization is important for control of many diseases caused by microbes.

Animal Microbial Parasites and Pathogens

Animals are susceptible to a number of microbes that colonize and alter the health of the host. Animals represent nutrient-dense targets for opportunistic organisms, and bacteria, protozoa, fungi, protists (protozoa), and helminths (parasitic worms) have adapted opportunistic lifestyles that target animals. Primary targets include the pulmonary and digestive systems, but invasive microbes can penetrate the skin where lesions or abrasions have occurred. Among animals of agricultural importance, vaccinations and antibiotics have reduced the incidence of parasites and pathogens. However, concern regarding the use of antibiotics in animals and the potential consequences of antibiotic resistance reducing antibiotic effectiveness in humans has led to mandated reductions in antibiotic use in animal agriculture. Alternatives to antibiotics include prebiotics (dietary component that alters microbial composition) and probiotics (microbial additive to alter microbial composition) in animal diets. Microbial interactions, specifically in the intestinal tract, that may reduce parasites and pathogens are discussed next.

Animals and Microsporidia

Microsporidia are unicellular organisms and intracellular parasites found in all major animal groups (Williams, 2009). Microsporidia are a common parasite in insects and fish, and a particular problem for farm-raised fish. Infection is associated with chronic, persistent illness and the parasite, although not directly lethal, has been shown to result in 30% mortality in farmed salmon. In most cases, the host exhibits reduced weight, vigor, and fertility. In addition to fertility issues, transmission can be vertical to the offspring, particularly in insect and crustacean hosts, and this parasite can change the sex of hosts via suppression of androgenic gland development.

These organisms represent a large group of microbes that are related to fungi phylogenetically, but they are atypical fungi in cell structure. Formerly thought of as protists and called microspora, these eukaryotic organisms lack mitochondria (they have mitosomes), are non-motile, and form spores with thick cell walls that can survive outside the hosts for years. Shifts in pH can prime the spores to germinate and inject the microsporidia into the host cell, which is typically a mucosal epithelial cell. Once colonized in the host, the parasite can exploit the host cell for nutrients and energy. Exploitation unique to microsporidia is the gathering of host cell mitochondria and accumulating the mitochondrial ATP, which may have allowed this eukaryotic organism to shed its endogenous mitochondria long ago.

Animals and Zoonotic Bacterial Pathogens

Zoonotic pathogens are transmissible between animals and humans (Wells and Varel, 2005). Plants may harbor pathogens harmful to humans, but most originate from animal sources. In animals, a variety of zoonotic pathogens has been observed, and in many cases, the pathogen may not be harmful to the animal, but may cause much harm to infected humans. Species of *Salmonella*, *Campylobacter*, *Enterococcus*, *Escherichia*, and *Yersinia* are excreted by animals and are potential pathogens to humans.

Many of these human-pathogenic bacteria can reside in production animals with little or no obvious signs of disease. In particular, *Escherichia coli* O157:H7 can reside in the bovine gastrointestinal tract of some animals for weeks at numbers greater than 100,000 organisms per gram feces, but less than 10 organisms can cause severe gastroenteritis, and even death, in humans. In the case of *E. coli* O157:H7, the zoonotic pathogen does not cause disease and provides little benefit to the animal carrier, but does compete for nutrients. In contrast, *Salmonella* are disease-causing organisms in humans and production animals, and some *Salmonella* strains have evolved adaptations for different hosts. In recent years, *Salmonella* serotypes Typhimurium, Enteritidis, Newport, and Heidelberg account for nearly 50% of the serotypes found in humans, and most likely originate from animal sources. *Salmonella* serotype Typhimurium is most often found in cattle and swine; Heidelberg and Enteritidis are found in chickens; and Newport is most often found in cattle. Controlling pathogen incidence and load in animal reservoirs is important for the safety of the environment, water, and human food, and understanding how the host diet or its gastrointestinal ecology may deter colonization of these pathogens is important for sustainable agriculture.

COMMENSALISM AND MUTUALISM

Typically, symbiosis is thought of as interactions that impart no negative effect to either symbiont, and these interactions can be described as commensal or mutual. These interactions are beneficial to one (commensalism) or both (mutualism) species. In agriculture, we promote or select for these types of interactions that benefit the crop or animal. Although strictly defined as commensal or mutual, these symbiotic interactions are not clearly distinct, and the symbiotic relationship can have shades of both types of symbiotic interactions. In some interactions, the benefit to one species is obvious whereas the benefit to both species may not be as clear. The sustainability of agriculture for future generations is highly dependent on identifying and maximizing commensal and mutual relationships that improve agricultural production while minimizing the environmental footprint.

COMMENSALISM AND MUTUALISM AMONG PLANTS AND ANIMALS

Plants represent a significant amount of biomass on Earth, and are subject to a variety of interactions. As mentioned previously, phytophagous insects target plants and represent a parasitic type of interaction. However, plants attract and utilize insects, such as bees and butterflies, for pollination and these interactions distribute plant genetic material in the form of pollen. Ants also play an important mutualistic role with dispersion of plant seeds (*myrmecochory*) in the terrestrial ecosystem, particularly for flowering plants.

Animals have evolved to exploit plants as well, and in a balanced ecosystem, this interaction can be viewed as commensal to animals, if not possibly mutual to both plant and animal. Animals differ in their dietary adaptations. Carnivores are predominantly meat eaters and have little capacity to utilize plant material, but omnivores, which also cannot digest plant material, may feed on fruiting bodies of plants and, consequently, distribute seeds in the stool. Herbivores in contrast, and ruminants in particular, have developed the capacity to digest and utilize plant biomass for their own nutrient needs. These latter adaptations involve microbial symbioses, to be discussed later, but in relation to the plant, they can be commensal in nature.

Mammalian herbivores differ in their grazing strategies, and thus, can differ in how they affect or benefit the plant or forage consumed (Asner et al., 2009; Augustine and McNaughton, 2004; Bailey, 2005; Kant and Baldwin, 2007; Parker, Burkepile, and Hay, 2006; Rinella and Hileman, 2009). In feeding studies, herbivores tend to selectively consume exotic (non-native) plants over the native plants from their environment. In a larger ecosystem, native herbivores can suppress invasive plants by selectively consuming more of the exotic plants. This action will result in retardation of invasive exotic plants. In contrast, exotic (non-native) herbivores are more likely to selectively graze the native (but exotic to them) plants. This selection can result in greater abundance and overpopulation of exotic plants. In ecosystems that are more complex, such as the African savannas, the combination of native browsers and bulk-feeders retards shrub and woody plant encroachment onto rangelands used for agricultural grazing. On rangelands in the western United States, goats and sheep can improve grasslands production by selectively consuming pines, junipers, and forbs (herbaceous flowering plants) that invade these grazing areas.

The negative implications of animals on plants are widely recognized. Animal movements can trample plants and grazing by herbivores can result in plant injury and loss of the plants' reproductive organs. However, omnivores and mammalian herbivores can benefit plants by dispersing seeds (Pakeman, Digneffe, and Small, 2002). Some plants have evolved structures such as hooks to facilitate attachment to animals for dispersal by *exozoochory*, or transport outside the animal, whereas other plants have evolved to utilize *endozoochory*, or transport by animal ingestion. Fruit-bearing plants often have fleshy fruit to attract a variety of birds and mammals that consume (*frugivory*), transport, and defecate the seeds, a process known as *direct endozoochory*. Likewise, mammalian herbivores graze and consume plants (*herbivory*) including the seeds of those plants, and transport those seeds until defecated, a process known as *indirect endozoochory*. The movement of the seed through the digestive tract may damage the seed and prevent germination, but the nutrient-rich environment is likely more conducive to germination and growth for those seeds that survive. Grasses and plants, particularly annuals, may be widely distributed in dung from herbivores, but plant species with smaller seeds typical of many pasture weeds are better adapted to surviving the gastrointestinal tract. Whether or not seeds are dispersed by an animal, hoof action by animals disrupts the soil surface and can serve to bury seeds for later germination. In agriculture, properly managing forage lands and foraging animals will minimize the environmental impact of animal agriculture and sustain a productive system.

MICROBES AND MUTUALISM/COMMENSALISM WITH PLANTS

In horticulture, combating plant diseases that arise from microbial infections has been the plant/microbial interaction of greatest interest. In recent years, the diversity and economic importance of mutualistic interactions between plants and microbes have been recognized, but with the exception of nitrogen fixation, these interactions have not been as well studied. In plants, the *rhizosphere* describes the soil around the plant roots and this ecosystem represents a community of bacteria, fungi, protozoa, and nematodes that interact with each other and the plant roots (Barea et al., 2005; Berg and Smalla, 2009; Mocali and Benedetti, 2010). The biotic factors that affect the community structure include plant species and cultivar, stage of plant development and health, and animal activity. Abiotic factors include climate, geography, soil type, and amendments made because of human activities (e.g., pesticides, fungicides, or herbicides). The plant may control the predominant interactions via exudates from the roots, which can serve as signals for beneficial bacteria. Exudates include ions, oxygen, water, mucilage, and carbon compounds. The carbon excreted by roots can be variable, but can account for more than 25% of the carbon fixed by the plant. Plants can benefit from microbial interactions due to the release of phytohormones, availability of nutrients, micronutrients, and minerals, increased tolerance to stress, and biocontrol of pathogens. Soil-specific inhibition of the plant pathogens *Fusarium*, *Gaeumannomyces*, *Rhizoctonia*, *Pythium*, and *Phytophthora* is in part due to indigenous rhizosphere microbes.

Nitrogen Assimilation

Carbon and nitrogen are the building blocks of life on Earth. Plants use photosynthesis to transform light energy and carbon dioxide into carbon building blocks. However, plants, like all other eukaryotes, cannot directly assimilate nitrogen, and require nitrogen in the form of nitrates or ammonia for nitrogen assimilation. The biological ability to fix nitrogen to ammonia is limited to prokaryotes that express nitrogenase enzymes (Barea et al., 2005; Hurek and Reinhold-Hurek, 2003; Lindström et al., 2010). Numerous free-living bacteria in the soil (e.g., *Azobacter*, *Clostridium*, *Klebsiella*, and *Rhodospirillum*) have developed abilities to fix nitrogen, which can diffuse to surrounding plants. A variety of plants has evolved extracellular symbiotic (*epiphytic*) relationships with nitrogen-fixing cyanobacteria, most of which involve heterocysts, or cavities formed in the leaf, to house the bacteria (e.g., *Anabeana azollae* and the waterfern *Azolla*). A few flowering woody shrubs and trees have adapted intracellular symbiotic (*endophytic*) strategies with the filamentous antinomycete *Frankia* in large root nodules, but this interaction is limited to some species of angiosperms in the plant kingdom. More agriculturally important plants in the legume family (*Fabaceae*) have widely evolved symbiotic relationships to exploit microorganisms. In general, these plants have specialized nodules in their roots where atmospheric nitrogen is fixed and assimilated by a variety of rhizobia bacteria. Rhizobia are Gram-negative rod-shaped bacteria and the nitrogen-fixing species are distributed among *Azorhizobium*, *Bradyrhizobium*, *Mesorhizobium*, and *Rhizobium* groups.

The symbiosis with legumes and rhizobia occurs in specialized root tissue called *nodules*. The development of this symbiosis begins with plants secreting an exudate from the root hairs that chemotactically attracts the rhizobia. The rhizobia colonize and multiply on the root hair. Flavonoids produced by the plant induce the nodulation (*nod*) genes in the rhizobia to produce Nod factors, which result in a sequential series of plant host reactions that result in internalization of the bacteria and nodule development. Specificity between legume species and bacterial species is determined by modifications to the Nod factors that are encoded by host-specific nodulation genes. Nitrogen fixation genes (*nif* and *fix*) encoded by the bacteria are triggered and the highly conserved nitrogenase and accessory proteins are produced. The reduction of nitrogen to ammonia is energy intensive (requiring 16 mole ATP per 1 mole NH_3 produced). The rhizobia require readily available oxygen for catabolism, but the nodules have low oxygen content. To compensate, the host plant produces leghemoglobin to deliver oxygen to the rhizobia. This symbiosis can account for more than 50% of all biologically fixed nitrogen in agriculture, and modern cropping systems implementing a legume in rotation can derive significant savings in nitrogen fertilizer applications.

The ability to fix nitrogen in agricultural crops may not be limited solely to legumes and their mutualistic bacteria (Bhattacharjee, Singh, and Mukhopadhyay, 2008; Hurek and Reinhold-Hurek, 2003; Steenhoudt and Vanderleyden, 2000). The most important agricultural crops are grasses (family Poaceae) and recent evidence suggests that bacterial species belonging to *Azospirillum*, *Acetobacter*, *Herbaspirillum*, and *Azoarcus* may form mutualistic relationships with some of these plants. Mutualistic relationships between nitrogen-fixing bacteria and grasses may be concentrated in the tropic regions, and rice is one agricultural crop that may benefit from endophytic bacteria that can fix nitrogen. Some sugar cane varieties in Brazil and Kaller grass common to saline soils in south-central Asia appear to assimilate most of their nitrogen from nitrogen-fixing endophytes, such as *Azoarcus* spp. Nitrogen-fixing bacteria in grasses may be epiphytic or endophytic. Unlike legumes, the bacterial endophytes in the grasses are not housed in specialized structures, but are free-living in the plants' extracellular spaces. In the case of *Azospirillum*, the bacteria first swarm and attach to the root surface and secrete polysaccharides, essentially anchoring the bacteria to the root in a biofilm. Not all *Azospirillum* can internalize, but as noted with *Azoarcus* spp., cellulolytic enzymes appear to aid in their internalization into the plant root.

Arbuscular Mycorrhiza

Fungal-plant mutualistic interactions in the rhizosphere represent a diversity of interactions by a group of fungal taxa and over 90% of plant species (Bonfante and Genre, 2010). Interactions with trees account for the bulk of the variety of interactions with fungi, utilizing ectomycorrhizal mechanisms in the root hair in which the fungal mycelium are extracellular. Most vascular plants have evolved endosymbiotic interactions with arbuscular mycorrhiza, a common fungus in soil. This relationship appears to have occurred early in the development of land plants and represents the most widespread type of symbiosis in nature. Members of the fungi phylum Glomeromycota are part of the soil matrix, and their hyphae can infect the root hair and form arbuscule structures (endomycorrhizobial) in the plant root cells to exchange nutrients. Without the roots of plants, these microorganisms would be unable to complete their lifecycle and they would die. These fungi benefit the host plant by providing additional phosphorus, but can also provide additional micronutrients and water due to the increased surface area of the filamentous mycorrhiza hyphae widely distributed through the soil, whereas the plant can provide the fungi with carbon, often the sole source of carbon, for the arbuscular mycorrhiza. Disruption of the rhizosphere, or the soil surrounding the plant root system, can disrupt the mycorrhiza hyphal network and impede the symbiosis-based development. In particular, tillage, fungicides, and application of phosphorus fertilizers are modern practices that negate the potential benefits of arbuscular mycorrhiza by disrupting fungal growth and minimizing infection of the host plant. In contrast, this symbiosis could be managed and exploited by farming systems where inputs are minimal, such as organic farms, to improve plant growth and crop yields.

Additional Plant/Microbe Interactions

Understanding of the plant/microbial interactions in the rhizosphere is slowly increasing as research in sustainable agricultural systems matures and newer technologies come online (Newton et al., 2010). Varieties of wheat may selectively support the growth of beneficial bacteria, such as *Pseudomonas* species, which in turn may make minerals and nutrients more available to the plant roots and suppress plant pathogens. Consequently, subsequent wheat varieties of different genotypes may not perform as well in the same field if different rhizosphere ecology is needed. In particular, it has been noted that older wheat cultivars appear to be colonized by a variety of rhizobacteria and more recently developed cultivars are associated with members of Proteobacteria, such as *Pseudomonas*. Mutualistic interactions for production crops such as maize, grasses, barley, and oat cultivars may include microbes *Agrobacterium* sp., *Bacillus* sp., *Burkholderia* sp., *Pseudomonas* sp., *Paenibacillus* sp., and *Streptomyces* sp., but additional enrichments for members of the rhizosphere community are likely to exist. Plant root exudates could play a determining role in selecting mutualistic microorganisms, but microbes have to signal back to the plant to initiate colonization. In general, motile soil bacteria, such as *Pseudomonas* strains, appear to be predominant because motility offers a competitive advantage in colonizing the plant and establishing the symbiosis.

Pathogen Suppression

The identification and potential for mutualism between plants and microbes to be exploited in production agriculture has yet to be fully determined. Research to identify beneficial bacteria and fungi will be difficult, but the rewards could be invaluable (Newton et al., 2010). In particular, suppression of plant pathogens by commensal or mutualistic microbes in the rhizosphere is a viable opportunity. Biocontrol by bacteria, such as *Pseudomonas, Agrobacterium, Bacillus, Streptomyces*, and *Burkholderia* strains, or by non-pathogenic fungi, such as *Trichoderma, Pythium*, and *Fusarium*, against plant pathogens may be a useful preventative system to control plant pathogens or reduce the damage inflicted by the pathogen. Numerous mechanisms may explain the antagonisms, and in nature, more than one may be involved. Putative mechanisms may involve competitive exclusion for colonization sites, stimulation of plant defense systems, niche nutrient competition and depletion (in particular, iron sequestering), inhibition via antimicrobials, degradation of virulence factors, and parasitism.

Fungal Endophytes and Plants

Most of the mutalistic microbes described in the previous sections are endophytes (intracellular in the plant) found in the rhizosphere where they may provide nutrients or prevent microbial pathogens. In the stem and leaf, which are primordial (aerial or *phyllosphere*) portions of the grasses (family Poaceae), symbiotic relationships with fungi (family *Clavicipitaceae*) have evolved (Schardl, Leuchtmann, and Spiering, 2004). In numerous examples with endophyte fungi, the specific fungal strain is transmitted with the seed from the host plant. Many of these relationships are commensal or mutualistic, and in the latter case, the fungal endophyte enhances root development, drought tolerance, and resistance to herbivory by insects and animals. Alkaloids, including lolines, peramine, indolediterpenes, and ergotamines, produced by the endophyte fungi can possess antimicrobial activities that reduce pathogen colonization and can be toxic to insect pests that may forage the host plant. The indolediterpenes and ergotamines can also be a problem for foraging livestock and have been implicated in toxicities observed with ryegrass and tall fescue. In these cases, the mutualistic relationship that benefits plant growth and fitness has become a costly problem to grazing livestock.

Modern Agriculture and Symbiosis with Plants

Historically, consideration of mutualistic microbes has played little role in crop production. However, as the understanding of the mutualism grows, the selection of mutualistic relationships with microbes will likely expand. Understanding plant physiology and the function and role of *R*-genes (which encode proteins for disease-resistance), immune peptides and proteins, flavonoids, and other microbial effectors against parasites, pathogens, and mutualistic microbes, as well as the mechanisms by which the symbiotic microbes control plant defenses, will provide the foundation for proper selection of appropriate mutualistic benefactors. However, as noted with endophyte fungi, the implications for foraging animals where the plant host may be used for grazing livestock must also be considered when selecting mutualistic partners.

MICROBES AND MUTUALISM/COMMENSALISM WITH ANIMALS

Microbes are ubiquitous in nature, but the animal has provided the microbe with the most opportunity. Pathogenic or parasitic microbes, as discussed previously, have evolved to take advantage of an animal host, but a greater level of interaction has evolved between microbes and animals in which both organisms benefit to some degree. In mammals, it is widely recognized that microbes play a role in digestion, but mutualistic and commensal relationships are known even in insects. Termites harbor symbiotic protozoa and other cellulolytic microbes to digest wood cellulose; leafcutter ants nurture a fungus to digest freshly cut leaves; and even the plant parasitic aphid has co-evolved with a bacterium (*Buchnera aphidicola*) that provides the sap-sucking aphid host a source of amino acids (Degnan et al., 2010). Of greater interest with the aphid is the propensity to utilize other bacterial strains to manipulate sex ratios and protect against natural enemies, such as wasps and fungal parasites. In these examples, mutualism with microbes has provided a competitive advantage to parasitic organisms that can be damaging to agriculture.

Gastrointestinal Tracts of Production Animals

The gastrointestinal tract (GIT) begins at the mouth and ends at the anus. Animals ingest and digest food for energy and growth, and the gastrointestinal tract provides a system for the consumption, mastication, digestion, and absorption of nutrients to fuel these needs. The gastrointestinal system can vary from species to species of animals, particularly in the upper GIT (oral cavity to stomach). Compartmentalizations of the esophagus and stomach have allowed dietary specialization in some animals (pregastric digestion and fermentation, see Table 9.1). The lower GIT can also vary in size, but common to most animals are the small intestine, large intestine, and colon. These regions of

TABLE 9.1
Relationships between Dietary Strategy for Mammalian Host and Microbiota Types

Herbivores		Omnivore	Carnivore	Microbiota Type
Foregut Fermenters	**Hindgut Fermenters**	**Omnivore**	**Carnivore**	**Microbiota Type**
Cow, sheep, and giraffe				Type 1: Foregut fermenters, such as ruminants, that consume forage materials
	Horse and rhinoceros			Type 2: Hindgut fermenters that consume forage materials
Columbine monkey	Gorilla and orangutan			Type 3: Pseudo-ruminant and hindgut fermenters, includes foliovores and omnivores
		Chimpanzee, human, baboon, spider monkey, and lemur		Type 4: Simple stomached mammals, includes omnivores and frugivores
	Giant panda and red panda	Brown bear	Polar bear, dog, hyena, and lion	Type 5: Simple stomached mammals, mostly carnivores but includes mammals with extensive dietary range

Source: Adapted from Ley, R.E., C.A. Lozupone, M. Hamady et al. 2008. Worlds within worlds: Evolution of the vertebrate gut microbiota. *Nat. Rev. Microbiol.* 6(10):776–788.

Note: Microbiota type is based on the cumulative microbial composition of feces sampled from a variety of mammalian hosts.

the GIT are active in the digestion and absorption processes. Of particular interest for some animal species is the developed cecum, which allows for dietary specialization (postgastric digestion and fermentation).

At birth, the GIT in mammals is sterile, but that quickly changes unless the newborn is delivered by Cesarean section and maintained germ-free. The lumens of the lower gastrointestinal tissues are nutrient-rich and packed with not only digesta, but also bacteria that are degrading and utilizing ingested nutrients. The GIT is an open system and susceptible to microorganisms from outside the host; however, the predominant microflora in the GIT are often permanent residents and, in some cases, unique to certain animal species. The bacterial population in the GIT can outnumber the host cells by as much as 10 to 1, and the populated tract is now commonly recognized as an organ. The microflora can be a source of energy, amino acids, and vitamins; and these bacteria can function to modulate the immune system, regulate the function of the intestinal tissues, and prohibit pathogen colonization.

Pathogens and other opportunistic bacteria can affect animal performance, and prohibition of pathogen colonization by commensal or probiotic strains provides an important benefit to the host animal. The beneficial bacteria can operate by several mechanisms, including competitive exclusion, antimicrobial production (e.g., bacteriocins), and occupation of colonization sites.

Competitive exclusion, or Gause's Law, describes a principle of ecology in which competing species cannot co-exist with the same resources if all other factors are constant, and one organism will out-compete the other for nutrients to the point that the other becomes extinct or evolves. Antimicrobial compounds can be produced by bacteria to inhibit another species, and the most common compounds are proteinaceous bacteriocins. Bacteriocins have been classified as Class I, IIa, IIb, IIc, and III based on synthesis, biochemistry, and mechanism of action. However, categorization of bacteriocins can depend on a number of factors, including mechanism of action and producing species. For example, colicins and microcins are typically produced by *Escherichia coli*; lantibiotics are produced by lactic acid bacteria; and subtilin is produced by *Bacillus subtilis*. Colonization involves attaching or invading the epithelial tissue. Bacteria as a whole express a variety of extracellular proteins for binding different glycoconjugates and epithelial cell components.

Abundant nutrients feed a diverse microflora and recent technologies should allow researchers to understand better the strong relationship between host and gastrointestinal microflora. Phylogenetic analyses utilize sequence information from cell DNA or proteins to study relatedness or classification of different strains, species, genus, or higher orders. In microbiology, the 16s RNA gene is commonly used for classification of related bacteria. The 16s RNA gene sequences are interwoven with conserved and variable regions, and sequencing a specific region allows the study of the diversity in a sample of microflora. In the mammalian lower GIT, the predominant microflora is bacteria, and of the approximately 24 phyla of bacteria, the lower intestine is predominated by the phyla Bacteroidetes and Firmicutes. Overall, most Bacteroidetes in the distal intestine belong to the genera *Bacteroides*, whereas most Firmicutes belong to genera *Clostridium*, *Enterococcus*, *Lactobacillus*, *Peptostreptococcus*, and *Ruminococcus*. Minor phyla of abundance in the intestine, such as Actinobacteria and Fusobacteria, are represented by genera *Bifidobacterium* and *Fusobacterium*, respectively. The abundance of bacteria can vary by animal species and by location from the small intestine to the colon, with species of *Lactobacillus* predominant in the jejunum region of the small intestine and species of *Bacteroides* and *Clostridium* being predominant in other regions.

Historically, studies of microflora have involved isolation and culturing of bacteria. However, these studies are time consuming and not all bacteria are easily cultured. Modern molecular methods for DNA amplification and sequencing have provided a different view of bacterial niches and recent studies of the human intestinal microflora have provided a better understanding of the symbiosis in the intestine (Eckburg et al., 2005; Ley et al., 2008). Obesity in mice has been associated with higher levels of the phyla Firmicutes and lower Bacteroidetes, and when germ-free mice were inoculated with microbes from obese mice, the animals exhibited weight gain and lower food intake. In humans, obese subjects exhibit similar patterns compared to lean subjects, and imparting a dietary regime to obese subjects altered the microflora to higher Bacteroidetes and lower Firmicutes. Changes in microflora composition are believed to be associated with changes in the energy balance in the intestinal tract. Specific changes in bacterial genera or species have not been reported, but based on results of these recent studies, modulation of gastrointestinal microflora may affect weight gain, adipogenesis, and lean accretion.

Establishment of the gastrointestinal microflora is important to the host (March, 1979; Ratcliffe, 1991). Initial inhabitants in mammals are those ingested during passage at birth and from the mother's skin when suckling. Additional bacteria are ingested from the environment, and over time, the gastrointestinal microflora stabilizes. Milk from the mother provides antimicrobial factors to reduce pathogen risk in the neonate, and the newborn is specialized in digesting and absorbing the nutrient-rich milk. Initial colonizers include coliforms (including *E. coli*), clostridia, and streptococci, and are found in stomach and small intestinal contents. Species of *Lactobacillus* and other lactic acid bacteria soon predominate in these tissues and colonize significant portions of the small intestinal mucosa. The small intestine is a major colonization site for pathogenic *E. coli* associated with diarrhea in young mammals (enteropathogenic or enterotoxigenic *E. coli*; EPEC or ETEC, respectively), and the bacteriocins and exclusion by colonized indigenous flora, in particular the *Lactobacillus* spp., are major factors in reducing bacterial disease.

The stomach has several distinct tissue regions, and the acids produced by the secretory regions are lethal to many bacteria. In the monogastric stomach, the bacterial populations are highest after meals (1000 to 1,000,000 colony forming units per gram of luminal contents) when stomach acid is diluted (Katouli and Wallgren, 2005). Bacterial populations are lowest after digestion is complete, with bacteria often undetectable in luminal contents of the stomach. Many of the observed luminal bacteria may originate with the food or feed, or are dislodged from the upper GIT when food is chewed and swallowed. Regardless, bacteria observed in the stomach contents of the piglet are sparse relative to other regions of the GIT. The non-secreting regions harbor a number of bacteria, and the bacterial flora present are mostly attached to the stomach epithelial surface or embedded in these tissue linings. *Lactobacillus* spp. is most often isolated, although *E. coli* and species of *Streptococcus*, *Eubacterium*, *Bifidobacterium*, *Staphylococcus*, *Clostridium*, and *Bacteroides* have been isolated. Although their numbers may be small, these commensal colonizers such as *Lactobacillus* spp. may reduce ulcerations by excluding or preventing colonization by *Helicobacter pyloris* (humans), *H. suis* (swine), and *H. bovis* (ruminants), and similar mucosal irritants. Numerous bacteria have been tested *in vitro*, including *L. johnsonii*, *L. acidophilus*, *L. reuteri*, *L. gasseria*, *Weisella confusa*, and *Bacillus subtilus*. Effective beneficial commensal and probiotic bacteria in the stomach would have to tolerate low pH and rapid luminal turnover, and need to colonize epithelial surface glycolipids targeted by bacterial irritants in stomachs such as *H. pylori*.

Colonization of the stomach is not limited to the monogastric stomach. Numerous animals have evolved specialized stomachs. In particular, compartmentalization of the stomach regions has led to diversity and food specializations in mammals to exploit microbial interactions (Russell and Rychlik, 2001). In particular, mammals that derive some nutrients from pregastric fermentations have evolved to exploit utilization of plant forages and fiber in their diets. The rumen is one such compartmentalization that will be discussed in detail later. Regardless of the animal species, a compartment equivalent to the gastric stomach, or abomasum, serves as a barrier to transient and pathogenic bacteria that would otherwise invade the lower nutrient-rich GIT.

The small intestine is common to most animals and has three physiological regions—the duodenum, jejunum, and ileum—each with distinct roles in digestion and absorption. The duodenum is a primary site for secretions of bile and enzymes that aid digestion. In contrast, the jejunum and ileum are important for absorption of nutrients. Overall, the small intestine has a fast passage rate for digesta, compared to the regions of the lower intestine, and the lumenal contents have fewer bacteria. The commensal bacteria in the small intestine, such as *Lactobacillus* spp., are most often attached to the intestinal epithelial lining and sloughed into the lumen, with the jejunum and ileum being primary sites for bacterial colonization.

The piglet has one of the most frequently studied small intestine systems due to similarities with that of humans (Katouli and Wallgren, 2005). Commensal *Lactobacillus* spp. in the small intestine most often cultured from the piglet include *L. fermentum*, *L. acidophilus*, and *L. delbrueckii*. In comparison, molecular fingerprinting has more recently identified *L. mucosae*, *L. delbrueckii*, *L. salivarius,* and *L. johnsonii* as being most abundant in weaned piglets. Phylogenetically similar species have been observed with young cattle and poultry. These lactobacilli are typically resistant to bile and other intestinal secretions, and bind to the mucosa via mucin and epithelial binding proteins. Many lactobacilli produce antimicrobial compounds, commonly referred to as bacteriocins, and specifically known as lantobiotics for these bacteria. In addition, colonization by *Lactobacillus* spp. may alter host defensive responses, cytokine release, and immune activity. *Bifidobacterium* spp. can also generate similar responses in humans, cattle, and poultry, but these bacteria are rarely abundant in swine.

The large intestine is common to most animals and has three separate regions—the cecum, the colon, and the rectum—each of which aid in absorption of nutrients and water. The cecum is a region of divergent evolution that has allowed for specialization by the host animal. Amphibians lack any cecal structure, and fish have "pyloric ceca," or out-pockets, along the intestine but not a defined cecum. In most animals, with the exception of amphibians and fish, the cecum is a pouch of the large intestine located at the connection between the small intestine and the large intestine. Birds have two ceca, whereas most mammals have only one cecum. The primary function of the cecum is to provide space for post-gastric fermentation and for absorption of volatile fatty acids. Therefore, the cecum varies in size, with specialized herbivores having a large voluminous cecum and carnivores having a small cecum, or in these latter animals, essentially a blind pouch at the proximal end of the colon with a small appendix tube in some cases.

Bacteria in the lower GIT are predominantly strict anaerobes belonging to the Firmicutes and Bacteroidetes phyla at concentrations of 10^{10} to 10^{11} per gram of lumenal content, but can vary between host animal species, with host diet, and from one host GIT region to another (Allison et al., 1979; Katouli and Wallgren, 2005; Robinson, Allison, and Bucklin, 1981). The bulk of microbial diversity is found in the lower GIT, with estimates of 400+ autochthonous, or indigenous, strains in the ecosystem. Colonization of the cecum after birth appears to assist with the development of the immune system, even in carnivores and humans that lack a developed cecum. In the developed cecum, the microfloras for the young pig and the laying hen have been characterized by a number of studies. In classical anaerobic studies with isolated strains, the swine cecal bacterial strains were characterized as predominantly *Prevotella* sp. and *Selenomonas ruminantium*, whereas culture-independent techniques detected not only an abundance of *Prevotella* sp. but also higher levels of low G+C microorganism related to the diverse group of Gram-positive bacteria including *Clostridium* (Leser et al., 2002). In the hen, recent culture-independent techniques recognized *Prevotella/ Bacteroides* members as the predominant genera in the fed hen, and *Bacteroides* as the predominant genera in hens during molting induced by withholding feed (Callaway et al., 2009). The cecum may also harbor certain pathogens, with *Salmonella* sp. and *Clostridium difficile* detectable at high levels in swine and molting hens. In the GIT of swine, *E. coli* and related coliforms (Proteobacteria) tend to be at their highest concentrations in the cecum and decrease in concentration with passage through the colon. Cultured lactobacilli are found at their highest level in the small intestine, and appear to decrease in amount through the cecum and colon. The microflora in the colon, like the cecum, includes variable levels of *Prevotella*, *Bacteroides*, *Clostridium*, and *Lactobacilli* sp., but also includes *Eubacterium* and *Enterococci* sp. not always observed in the cecum.

The microflora in the lower GIT is beneficial to the host in several ways (Wells and Varel, 2005). Autochthonous bacterial strains colonize the mucosal layer and serve as a primary deterrent to pathogen colonization and entry. The volatile fatty acids generated by microflora fermenting fiber in the cecum and colon can contribute 20 to 30% of the total caloric requirement of omnivores and herbivores (Bergman, 1990). In particular, butyrate is a primary energy source for enterocytes, and butyrate-producing bacteria represent an important functional group of diverse genera (*Eubacterium*, *Roseburia*, and *Faecelibacterium* sp.) that promote intestinal growth, development, and health (Louis and Flint, 2009). Microbial activity also leads to vitamin synthesis; however, the impact is limited for some vitamins due to poor absorption from the lower GIT. Animals reared germ-free require vitamin K supplementation, but normally raised animals do not, and germ-free animals require more B vitamins in their diet. The lower intestinal tract has limited ability to absorb amino acids and, in swine, lysine from microbial activity may contribute 10% of a young pig's requirements and most of a grown pig's needs. Coprophagia (consumption of feces) has been observed in a variety of animals; however, the rabbit, like many hares and picas, has adapted a unique version in which cecal contents are passed directly thorough the colon and the "soft feces," or cecotropes, are re-ingested to extract additional protein and vitamins arising from the initial microbial activity in the cecum.

Pregastric Fermentation and the Ruminant

Animals differ in their abilities to digest foods, and some animals have developed specialized regions of the GIT to exploit microorganisms for digestion, fermentation, or production of nutrients. Pregastric compartmentalization allows for microbial activity prior to the digestion and absorption of nutrients by the host animal. Mammals lack enzymes to break down fiber and digest forages, but microorganisms have these enzymes and can perform these activities. In addition, the host animal can digest the microorganisms as they pass into the lower GIT, and these microorganisms are rich in proteins that have amino acid profiles to meet the host animal's requirements.

Ruminants, in particular, have evolved strong symbiotic relationships with microbes for these purposes, and cattle, sheep, goats, and deer are species that provide most of the meat (>50%) and milk (>90%) consumed by humans. Typically described as having four stomachs, the ruminant animal actually has four specialized compartments of the stomach (Hungate, 1966; Russell and Rychlik, 2001). The rumen is the largest compartment, accounting for 15% of the total empty weight of the GIT. This large voluminous compartment is the primary site for microbial activity. In ruminant animals predominantly consuming forage, the products of microbial fermentations are volatile fatty acids (acetate, propionate, and butyrate are the most abundant) and gases (carbon dioxide and methane). The volatile fatty acids from the rumen can account for up to 70% of the host energy requirements (Bergman, 1990), and the host has adopted metabolic pathways to utilize the volatile fatty acids produced. Most of the glucose used by tissues of ruminant animals originates from propionate conversion by the liver. Microbial proteins produced in the rumen can account for 40 to 90% of the animal's protein requirements, and the animal has a protein requirement similar to the amino acid composition of microbial proteins in rumen fluid (Bergen and Wu, 2009; Reynolds and Kristensen, 2008; Wells and Russell, 1996).

To accommodate the microbial activity and fermentations, ruminant animals continuously pass saliva rich in sodium carbonate into the rumen to buffer the acid product. Nitrogen, in the form of urea, continuously flows into the rumen through the saliva and from the blood in the epithelial tissue of the rumen. Urea is rapidly hydrolyzed in the rumen to ammonia and this free ammonia is important in the nitrogen cycle between the rumen microbes and host. The microbes in the rumen have adapted to using sodium gradients across their membranes to drive nutrient uptake systems and to using ammonia as a predominant source of nitrogen for microbial protein synthesis.

The microbial flora in the rumen is a complex milieu of bacteria, protozoa, and some fungi, many of which can be diet specific and unique in nature to the rumen ecosystem (Hungate, 1966; Russell and Rychlik, 2001). Bacteria constitute the bulk of the rumen microbial mass and functionality of the rumen is dependent on the bacterial composition. Forages are predominantly cellulose and hemicellulose in structure and are digested in the rumen by a combination of several bacteria, including *Fibrobacter succinogenes*, *Ruminococcus albus*, *R. flavefaciens*, *Eubacterium ruminantium*, *Prevotella rumincola*, *P. albensis*, *P. brevis*, *P. bryantii*, *Butyrivibrio fibrisolvens*, and *Selenomonas ruminantium*. In ruminant animals being fed concentrate diets, predominant bacteria may include *Ruminobacter amylophilus*, *Succinomonas amylolytica*, *Streptococcus bovis*, *Lactobacillus* sp., *Succinovibrio dextrino-solvens*, *Megasphaera elsdenii*, *Prevotella* sp., *Butyrivibrio fibrisolvens*, and *S. ruminantium*. Other important ruminal bacteria include *Lachnospira multiparus* (pectinolytic), *Anaerovibrio lipolytica* (lipolytic), *Peptostreptococcus anaerobius* (aminophilic), *Clostridium aminophilum* (aminophilic), *C. sticklandii* (aminophilic), *Wolinella succinogenes* (organic acid utilizer), *Methanobrevibacter ruminantium* (methanogen), *Methanomicrobium* sp. (methanogen), *Methanobacterium* sp. (methanogen), and *Methanosarcina* sp. (methanogen). Propionate is important for glucose homeostasis in the host, and production in the rumen is directly, via propionate production, or indirectly, via succinate production, associated with the strains of *F. succinogenes*, *R. flavefaciens*, *Ruminobacter amylophilus*, *S. amylolytica*, the numerous *Prevotella* sp., and *S. ruminantium*, which can convert the ruminal succinate to propionate. Butyrate is important for milk fat synthesis in dairy ruminants, and is produced as a primary metabolite by *Butyrivibrio fibrosolvens* and numerous varieties of *Clostridium*.

While generally recognized as mutualistic, some of the ruminal bacteria can be detrimental to animal performance. *Streptococcus bovis* is associated with rapid lactic acid accumulation and rumen acidosis in grain-fed cattle, while lactate-utilizing bacteria such as *Megasphaera elsdenii* and *Selenomonas ruminatium* can reduce lactic acid accumulation. Opportunistic organisms like *Fusobacterium necrophorum* can also utilize lactate and infect rumen ulcers arising from even minor bouts of acidosis. Another group of ruminal bacteria, the methanogens, provides no net energy to the ruminant animal, and their production of methane represents both an energy loss to the animal and the generation of significant greenhouse gases associated with global warming.

Ruminal protozoa and fungi are less studied components of the rumen milieu, but still important in rumen ecology and animal production (Trinci et al., 1994; Veira, 1986). The protozoa observed in ruminal fluid are, with a few exceptions, unique to the rumen. Rumen protozoa can account for up to 40% of the microbial biomass, and defaunation, or the elimination of the protozoan population, can alter rumen fermentation. Protozoa are highly mobile and attach to feed particles and rumen wall surface, which reduces the washout rate and minimizes protozoan contribution to net rumen output. The rumen protozoa ingest and digest a number of ruminal bacteria for a source of protein and nutrients. Methanogens appear to colonize the body surface of protozoa (Figure 9.2), and appear to have established an intra-ruminal mutualistic relationship with protozoa that predominantly produce acetate and hydrogen gas. Defaunation often results in less energy losses to nitrogen recycling and methane production, but no significant reductions in rumen digestion are apparent because rumen bacteria fill the niche or void.

Rumen protozoa are mostly ciliated protozoa, belonging to either holotrichs or entodiniomorphs. Flagellated protozoa are present, but at low numbers in the rumen (Hungate, 1966; Veira, 1986).

FIGURE 9.2 Microbes attached to the surface of protozoa isolated from the rumen of cow-fed forage. This is an example of a symbiosis within a symbiosis, denoting the complexity of research to determine cost-benefit to symbiotic interactions. (Micrograph by Sharon Franklin and Mark Rasmussen, National Animal Disease Center, ARS, USDA.)

The holotrichs are covered nearly entirely with cilia and comprised of *Isotricha* and *Dasytrichia* species, which are the predominant types observed in the rumen of grazing animals. The entodiniomorphs have cilia localized in specialized bands called syncilia to aid in food ingestion and locomotion. The entodiniomorphs are in greater variety and the abundance of specific genera is dependent on the host diet. The entodiniomorph groups consist of morphologically distinct species of *Entodinium*, *Epidinium*, *Ophryoscolex*, *Diplodium*, *Eudiplodium*, and *Polyplastron*, of which *Diplodium*, *Eudiplodium*, and *Polyplastron* have cellulolytic activities and may play a role in fiber digestion. Many of the holotrichs and entodiniopmorphs can ingest and accumulate starch granules. Strains of *Entodium* are more tolerant of rumen acidity and are most abundant in rumens of animals fed high grain diets. When protozoa accumulate starch, rapid digestion and production of lactic acid is reduced, thereby alleviating clinical and subclinical rumen acidosis.

Anaerobic fungi have been isolated from pregastric and postgastric herbivorous animals, but are most often observed in ruminants consuming high-fiber diets (Trinci et al., 1994). Vegetative fungi, or the thallus-forming bodies associated with colonization and degradation, are present in rumen at levels lower than protozoa. However, these unique microorganisms have adapted to foraging animals and many types are adept at digesting fiber with the invasive filamentous rhizoids, particularly for the most recalcitrant types of cellulose that many bacteria have difficulty digesting. The Neocallimastigaceae family of fungi is the sole family of the phylum Neocallimastigomycota, which includes six genera, including *Anaeromyces*, *Caecomyces*, *Cyllamyces*, *Orpinomyces*, *Piromyces*, and *Neocallimastix*. The *Neocallimastix* are the best described and most often reported filamentous fungi in the rumen. Anaerobic rumen fungi lack mitochondria and, like the ciliated protozoa, use specialized hydrogenosomes that produce hydrogen gas, which, in turn, is converted to methane by rumen methanogens. The presence of fibrolytic species of ruminal bacteria and anaerobic fungi are often associated with increased fiber degradation and utilization.

Birds, some fish, and reptiles have developed compartmented stomachs. The two compartments include the proventriculus, or true stomach, which is secretory, and the ventriculus, or gizzard, which is a muscular stomach for grinding food. Many birds have a muscular pouch preceding the proventriculus called a crop for storing food, but this compartment is an adaptation of the esophagus and not a compartment of the stomach. Since food is stored in the crop, fermentation by microbes is likely to occur. Herbivorous birds like the hoatzin specialize in eating leaves of trees (foliovores) and the crop in these birds contains a diverse microbial ecosystem predominated by Firmicutes and Bacteroidetes that digest the leaves and provide the host with fermentation products for energy and microbial cells for protein (Godoy-Vitorino et al., 2010). In contrast, commercial agricultural birds, such as the chicken and turkey, have crops adapted to omnivorous diets and the crop of these birds is predominantly colonized by species of *Lactobacillus*, similar to the ileum and jejunum (Hilmi et al., 2007). The *Lactobacillus* strains appear to be influential in minimizing colonization by pathogenic *E. coli* and *Salmonella* strains.

Symbiosis and Evolution in Animals

The GIT is one of the best-studied and most-described symbiotic ecosystems. Host adaptations, such as foregut and hindgut fermenters, are obvious for the host to exploit the power of microbial enzymes. However, the complexity of the system and diversity of the microbiota have precluded an understanding of the strength of the host-microbial interaction. Pyrosequencing of complex microbiota samples has allowed a quantifiable measure of diversity and abundance for different microbial members, and studies of feces from intercrossed mice lines identified 13 regions in the mouse genome associated with abundance of one or more of the bacterial groups analyzed (Benson et al., 2010). Mouse chromosomes 1, 7, and 10 contain a number of genomic regions with strong associations to genus groups and higher orders. Concerning individual species, no specific species relationship was observed. However, strong association between a *Lactobacillus johnsonii/gasseri* group and chromosomes 7 and 14 were found, but none with *L. reuteri* or *L. animalis/murinus* group, and results suggest a microbiota associated with heritable genetic factors. Many of the genes within the identified genomic regions are associated with mucosal immunity.

In biology, the hologenome theory of evolution has been proposed and recognizes the close relationship between symbiotic partners, or *holobiont*, and that this affects the combined genome, or *hologenome*, of the partners (Zilber-Rosenberg and Rosenberg, 2008). The theory is based on generalizations that animals and plants establish symbiotic relationships with microbes, that symbiotic microbes are transmitted between generations, that the relationship affects the holobiont, and that variations in the hologenome can result from changes in the host or the genome of microbes. Thus, as proposed by this theory, evolutionary pressure on the host may be compensated not only by the host, but also by the symbionts that comprise the holobiont. In periods of rapid environmental change, quick adaptation by a versatile symbiont would be beneficial to the host and allow for survival, if not expansion, of the host into the new ecosystem.

Bacteroides thetaiotaomicron is a host-adapted microorganism that can predominate in the GIT of humans. This Gram-negative organism has a completed genome sequence and analyses have revealed a diverse arsenal of genes adept at digesting complex polysaccharides, acquiring nutrients, and producing surface adherence factors for colonization and complex regulatory mechanisms to control and modulate gene expression (Comstock and Coyne, 2003). In addition, this organism is adept at assimilating mobile DNA elements that transmit from cell-to-cell via transposons and plasmids, and the plasticity of this genome indicates the versatility that the microbe has evolved to remain a strong host-adapted symbiont. Similar relationships exist between *B. vulgatus* and *B. distasonis* and the human distal intestine, and it is likely that additional relationships between microbes and their host will be forthcoming as molecular tools and modern sciences tease apart the relationships and understand the genomic traits driving symbioses (Xu et al., 2007).

SUMMARY AND CONCLUSIONS

Modern agriculture has led to significant improvements in efficiency of food production. The use of fertilizers, herbicides, fungicides, pesticides, antibiotics, and growth promoters has been instrumental in feeding the world. However, natural selection has established interactions that are often overlooked or disregarded. These symbiotic relationships could potentially reduce our use of synthetic agents and promote a more sustainable productive agricultural system. In addition, dedicating research efforts to better describe and understand beneficial symbiosis would allow more opportunities to exploit mutualistic relationships when they arise. Planting crops in particular rotations and with minimal tillage may sustain rhizospere interactions that promote healthiness and growth of plants. Herbivores, such as the ruminant, have historically benefitted from a variety of symbioses

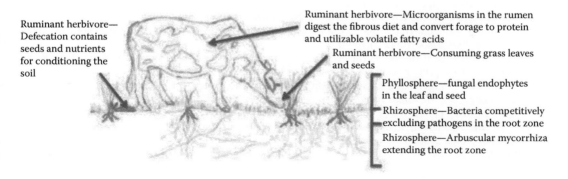

FIGURE 9.3   An ecosystem denoting symbiotic relationships of importance to animal agriculture.

(Figure 9.3), and there are many opportunities to enhance our understanding of these relationships to better utilize non-cultivatable land for animal production. Nature has provided humans with an arsenal of microbes and we need to understand better how to use them with modern practices.

REFERENCES

Allison, M.J., I.M. Robinson, J.A. Bucklin et al. 1979. Comparison of bacterial populations of the pig cecum and colon based upon enumeration with specific energy sources. *Appl. Environ. Microbiol.* 37(6):1142–1151.

Asner, G.P., S.R. Levick, T. Kennedy-Bowdoin et al. 2009. Large-scale impacts of herbivores on the structural diversity of African savannas. *Proc. Natl. Acad. Sci. USA* 106(12):4947–4952.

Augustine, D.J., and S.J. McNaughton. 2004. Regulation of shrub dynamics by native browsing ungulates on East African rangeland. *J. Appl. Ecol.* 41:45–58.

Bailey, D.W. 2005. Management strategies for optimal grazing distribution and use of arid rangelands. *J. Anim. Sci.* 82(E. Supplement):E147–E153.

Barea, J.-M., M.J. Pozo, R. Azcón et al. 2005. Microbial co-operation in the rhizosphere. *J Exp. Bot.* 56(417):1761–1778.

Benson, A.K., S.A. Kelly, R. Legge et al. 2010. Individuality in gut microbiota composition is a complex polygenic trait shaped by multiple environmental and host genetic factors. *Proc. Natl. Acad. Sci. USA* 107(44):18933–18938.

Berg, G., and K. Smalla. 2009. Plant species and soil type cooperatively shape the structure and function of microbial communities in the rhizosphere. *FEMS Microbiol. Ecol.* 68(1):1–13.

Bergen, W.G., and G. Wu. 2009. Intestinal nitrogen recycling and utilization in health and disease. *J. Nutr.* 139(5):821–825.

Bergman, E.N. 1990. Energy contributions of volatile fatty acids from the gastrointestinal tract in various species. *Physiol. Rev.* 70(2):567–590.

Bhattacharjee, R.B., A. Singh, and S.N. Mukhopadhyay. 2008. Use of nitrogen-fixing bacteria as biofertiliser for non-legumes: Prospects and challenges. *Appl. Microbiol. Biotechnol.* 80(2):199–209.

Bonfante, P., and A. Genre. 2010. Mechanisms underlying beneficial plant-fungus interactions in mycorrhizal symbiosis. *Nature Comm.* 1:48 doi: 10.1038/nscomms1046.

Callaway, T.R., S.E. Dowd, R.D. Wolcott et al. 2009. Evaluation of the bacterial diversity in cecal contents of laying hens fed various molting diets by using bacterial tag-encoded FLX amplicon pyrosequencing. *Poultry Sci.* 88(2):298–302.

Comstock, L.E., and M.J. Coyne. 2003. *Bacteroides thetaiotaomicron*: a dynamic, niche-adapted human symbiont. *BioEssays* 25(10):926–929.

Degnan, P.H., T.E. Leonardo, B.N. Cass et al. 2010. Dynamics of genome evolution in facultative symbionts of aphids. *Environ. Microbiol.* 12(8):2060–2069.

Eckburg, P.B., E.M. Bik, C.N. Bernstein et al. 2005. Diversity of the human intestinal microbial flora. *Science* 308(5728):1635–1638.

Godoy-Vitorino, F., K.C. Goldfarb, E.L. Brodie et al. 2010. Developmental microbial ecology of the crop of the folivorous hoatzin. *ISME J.* 4(5):611–620.

Hilmi, H.T.A., A. Surakka, J. Apahalahti et al. 2007. Identification of the most abundant *Lactobacillus* species in the crop of 1- and 5-week-old broiler chickens. *Appl. Environ. Microbiol.* 73(24):7867–7873.

Hungate, R.E. 1966. *The Rumen and Its Microbes*. New York: Academic Press.

Hurek, T., and B. Reinhold-Hurek. 2003. *Azoarcus* sp. Strain BH72 as a model for nitrogen-fixing grass endophytes. *J. Biotechnol.* 106(2-3):169–178.

Kant, M.R., and I.T. Baldwin. 2007. The ecogenetics and ecogenomics of plant-herbivore interactions: Rapid progress on a slippery road. *Curr. Opin. Genet. Dev.* 17(6):519–524.

Katouli, M., and P. Wallgren. 2005. Metabolism and population dynamics of the intestinal microflora in the growing pig. In: *Microbial Ecology in Growing Animals,* W.H. Holzapfel and P.J. Naughton, Eds. New York: Elsevier, pp. 21–53.

Leser, T.D., J.Z. Amenuvor, T.K. Jensen, et al. 2002. Culture-independent analysis of gut bacteria: The pig gastrointestinal tract microbiota revisited. *Appl. Environ. Microbiol.* 68(2):673–690.

Ley, R.E., C.A. Lozupone, M. Hamady et al. 2008. Worlds within worlds: Evolution of the vertebrate gut microbiota. *Nat. Rev. Microbiol.* 6(10):776–788.

Lindström, K., M. Murwira, A. Willems et al. 2010. The biodiversity of beneficial microbe-host mutualism: The case of rhizobia. *Res. Microbiol.* 161(6):453–463.

Louis, P., and H.J. Flint. 2009. Diversity, metabolism and microbial ecology of butyrate-producing bacteria from the human large intestine. *FEMS Microbiol. Lett.* 294(1):1–8.

March, B.E. 1979. The host and its microflora: An ecological unit. *J. Anim. Sci.* 49(3):857–867.

Mocali, S., and A. Benedetti. 2010. Exploring research frontiers in microbiology: The challenge of metagenomics in soil microbiology. *Res. Microbiol.* 161(6):497–505.

Newton, A.C., B.D.L. Fitt, S.D. Atkins et al. 2010. Pathogenesis, parasitism and mutualism I the trophic space of microbe-plant interactions. *Trends in Microbiol.* 18:365–373.

Pakeman, R.J., G. Digneffe, and J.L. Small. 2002. Ecological correlates of endozoochory by herbivores. *Functional Ecol.* 16:296–304.

Parker, J.D., D.E. Burkepile, and M.E. Hay. 2006. Opposing effects of native and exotic herbivores on plant invasions. *Science* 311(5766):1459–1461.

Ratcliffe, B. 1991. The role of the microflora in digestion. In: *In Vitro Digestion for Pigs and Poultry,* M.F. Fuller, Ed. 19-34. Wallingford, Oxfordshire, UK: CAB International.

Reynolds, C.K., and N.B. Kristensen. 2008. Nitrogen recycling through the gut and the nitrogen economy of ruminants: An asynchronous symbiosis. *J. Anim. Sci.* 86(E Supplement):E293–E305.

Rinella, M.J., and B.J. Hileman. 2009. Efficacy of prescribed grazing depends on timing intensity and frequency. *J. Appl. Ecol.* 46:796–803.

Robinson, I.M., M.J. Allison, and J.A. Bucklin. 1981. Characterization of the cecal bacteria of normal pigs. *Appl. Environ. Microbiol.* 41(4):950–955.

Rosomer, W.S., and J.G. Stoffolano. 1997. *The Science of Entolomology.* Dubuque, IA: William C. Brown Pub.

Russell, J.B., and J.L. Rychlik. 2001. Factors that alter rumen microbial ecology. *Science* 292(5519): 1119–1122.

Schardl, C.L., A. Leuchtmann, and M.J. Spiering. 2004. Symbioses of grasses with seedborne fungal endophytes. *Ann. Rev. Plant Biol.* 55:315–340.

Steenhoudt, O., and J. Vanderleyden. 2000. *Azospirillum,* a free-living nitrogen-fixing bacterium closely associated with grasses: Genetic, biochemical and ecological aspects. *FEMS Microbiol. Rev.* 24(4):487–506.

Stout, M.J., J.S. Thaler, and B.P.H.J. Thomma. 2006. Plant-mediated interactions between pathogenic microorganisms and herbivorous antropods. *Ann. Rev. Entomol.* 51:663–689.

Trinci, A.P.J., D.R. Davies, K. Gull et al. 1994. Anaerobic fungi in herbivorous animals. *Mycol. Res.* 98(2):129–152.

Veira, D.M. 1986. The role of ciliate protozoa in nutrition in ruminant. *J. Anim. Sci.* 63(5):1547–1560.

Wells, J.E., and J.B. Russell. 1996. Why do many ruminal bacteria die and lyse so quickly? *J. Dairy Sci.* 79(8):1487–1495.

Wells, J.E., and V.H. Varel. 2005. GI tract: Animal/microbial symbiosis. In: *Encyclopedia of Animal Science,* W.G. Pond and A.W. Bell, Eds. New York: Marcel Dekker, pp. 585–587.

Williams, B.A.P. 2009. Unique physiology of host-parasite interactions in microsporidia infections. *Cell. Microbiol.* 11(11):1551–1560.

Xu, J., M.A. Mahowald, R.E. Ley et al. 2007. Evolution of symbiotic bacteria in the distal human intestine. *PLoS Biol.* 5(7):1574–1586.

Zilber-Rosenberg, I., and E. Rosenberg. 2008. Role of microorganisms in the evolution of animals and plants: The hologenome theory of evolution. *FEMS Microbiol. Rev.* 32(5):723–735.

10 Food Safety Issues in Animal Source Foods Related to Animal Health and Welfare

Jarret D. Stopforth, John N. Sofos, Steve L. Taylor, and Joseph L. Baumert

CONTENTS

MICROBIAL FOOD SAFETY

Jarret D. Stopforth and John N. Sofos

INTRODUCTION

Major current societal issues of concern include animal welfare and food safety (Rostagno, 2009). There is general public consensus that improved animal health and well-being contributes favorably to food safety and, although it is difficult to always draw a direct correlation to this concept, there are examples or evidence supporting this idea (de Passille and Rushen, 2005). Animal welfare is considered an ethical issue; however, scientific evidence linking animal welfare to animal health should create a clearer relationship to its overall effect on food safety aside from the ethical considerations.

Despite the extensive scientific progress and major technological advances of recent years, millions of foodborne disease episodes occur annually in the United States and other countries, causing thousands of deaths and major economic losses (Mead et al., 1999; Scharff, 2010). According to the latest estimates by the United States Centers for Disease Control and Prevention (CDC), foodborne disease agents cause approximately 47 million foodborne illnesses, 127,000 hospitalizations, and 3000 deaths per year in the United States (Morris, 2011; Scallan et al., 2011a, b). A large proportion of the foodborne disease burden may be attributed directly to consumption of animal source food products or indirectly to food animal production through cross-contamination of the environment, other food, or water with pathogens originating from animals (Sofos, 2008, 2009; Sofos and Geornaras, 2010).

As there is a worldwide increase in demand for animal source food products, which is expected to rise even more as disposable income increases, the animal source food product industry is challenged to meet such demands, while concurrently realizing the need for improved welfare for food animals in production animal agriculture (Sofos, 2008, 2009). In brief, the industry is faced with the challenge of increasing production while at the same time considering and addressing issues associated with animal welfare, climate change, environmental pollution, and food safety. Balancing animal welfare, animal health, and food safety is not easily achieved as the assurance of one often may impede another. For example, animal lairage designs for improved food safety (lower external carriage of pathogenic bacteria, which may be transferred to the underlying sterile tissue during carcass dressing) may result in discomfort to the animal. Conversely, the use of antibiotics to control animal diseases (and thus ensure their well-being) may result in development of antibiotic-resistant bacteria pathogens in food from the same animal source. This chapter provides an overview of basic and current food safety issues and explores the relationship between enhancing food safety and implementing proper animal welfare conditions during food production.

FOOD SAFETY: COMPLEX CHALLENGE

According to results from surveys sponsored by the United States Food Marketing Institute (FMI), major food safety concerns of consumers usually include bacterial contamination, pesticide residues, product tampering, and bioterrorism (www.fmi.org). Other concerns or desires expressed by consumers have included hormone residues, "natural" and "organic" products, antibiotic residues, food trace-back, animal welfare, food cost, the environment, and bovine spongiform encephalopathy (BSE). In general, numerous aspects of our food supply have come under intense scrutiny and are being questioned worldwide. Even though food safety is often a fundamental expectation, it remains a complicated challenge for the food industry (Sofos, 2008, 2009). Food safety problems associated with animal source food products can be divided into physical, chemical, and biological hazards (ICMSF, 1996, 2002; NACMCF, 1998).

Physical Hazards

Physical hazards are physical objects introduced into food that may cause injury, but seldom death. Since these hazards can be controlled through good manufacturing practices during production, harvesting, processing, and transportation, and at food service sites, they will not be discussed further.

Chemical Hazards

A wide variety of chemicals may be used in food production and processing. This subject is covered in depth in the section "Chemical Food Safety" of this chapter. Some are acceptable additives while others are strictly forbidden for use in foods. Chemical hazards affect more people than physical hazards, but typically not as many as biological hazards. Chemical hazards should be addressed at each step in the production process: Growth, storage, during use (cleaning agents, sanitizers), prior to receipt (in ingredients and packaging materials), upon receipt of materials, during processing, and prior to shipment of product. Chemicals that should be considered include color additives, direct food additives, indirect food additives, prior-sanctioned substances, allergens, pesticide chemicals, and substances generally recognized as safe. All chemicals used in and around manufactured products should have specifications developed (safe levels of naturally occurring or deliberately added chemicals in food which may be potentially harmful to human health), as well as letters of guarantee (indicating that products are manufactured under sanitary conditions, packaged in approved materials, and comply with government notices/regulations and company standards) from the manufacturer. Foodborne illnesses caused by chemicals or chemical residues are often difficult to link to a particular food because the onset may be gradual and undetected until chronic or permanent damage occurs. On the farm, chemicals of concern are organophosphate pesticides, growth-promoting hormones, antibiotic residues, additives, and naturally occurring toxins (i.e., aflatoxins). These are generally controlled by food safety programs through the enforcement of maximum residue levels based on the assessment of risk that the chemicals pose to human health, yet there is also a need to consider the risk to and health of animals in the production sector due to chemical hazards.

Biological Hazards

Biological hazards associated with foodborne illnesses include pathogenic bacteria, fungi, viruses, parasitic agents, and infectious materials (Bacon and Sofos, 2003). According to United States CDC estimates (Scallan et al., 2011a), based on data for the period 2000–2008, 31 major pathogens acquired in the United States caused 9.4 million episodes of foodborne illness, 55,961 hospitalizations, and 1351 deaths annually. Most (58%) illnesses were caused by noroviruses, followed by non-typhoidal *Salmonella* (11%), *Clostridium perfringens* (10%), and *Campylobacter* spp. (9%). Leading causes of hospitalization were non-typhoidal *Salmonella* (35%), norovirus (26%), *Campylobacter* spp. (15%), and *Toxoplasma gondii* (8%). Leading causes of death were non-typhoidal *Salmonella* spp. (28%), *T. gondii* (24%), *Listeria monocytogenes* (19%), and norovirus (11%). There is a consensus in the food sector that biological hazards should be controlled effectively in order to improve food safety.

FOOD SAFETY ISSUES IN ANIMAL SOURCE FOOD PRODUCTS

The safety of meat products has come under intense scrutiny in recent years because of major bacterial outbreaks that have been associated with their consumption. Food safety concerns associated with animal source food products and other foods are expected to continue. Reasons for the increased importance of food safety concerns include (Samelis and Sofos, 2003; Sofos, 2008, 2009; Sofos and Geornaras, 2010):

- Changes in animal and plant food production and harvesting practices
- Food processing modifications and marketing developments
- Preparation practices and development of new food products to meet consumer demands
- Increased urbanization of society and the associated need for transportation of large amounts of food products from centralized production and processing factories to urban centers
- Increased international trade and associated transportation of foods from exporting to importing countries
- Globalization of the food industry
- Increased travel, which may enhance transfer of pathogens among countries
- Changing consumer demographics, lifestyles, eating habits, and increased life expectancy
- Consumer needs and expectations for foods that have reduced levels of calories, fat, and additives, while being natural, organic, or "healthy"
- Climate changes as well as associated natural environmental stresses, which may induce biological changes and lead to new pathogens
- Decreases in numbers of people directly involved in food production
- Higher numbers of consumers at risk for infection
- Emerging pathogens that may be resistant to control or are more virulent
- Advances in microbial detection methods
- Less food handling education and training of food workers and consumers
- Increased interest, awareness, and scrutiny of food safety issues by consumers, news media, and activist groups

It is widely accepted that, as the world population increases and the standard of living is improved, meat consumption also increases (Rostagno, 2009). Increases in meat consumption are also associated with urbanization, higher disposable income, and the desire for a greater variety in the diet (Sofos, 2008). Although meat consumption may be approaching saturation levels in developed countries, consumers in such countries continue to express a desire for foods: (1) with no additives or chemical residues; (2) exposed to minimal processing; (3) that are convenient and need little preparation; (4) that are safe; and, (5) that are affordable. In general, economically developed societies have undergone major changes in demographics, population numbers, food preferences and expectations, lifestyles, life expectancy, and educational experiences (Sofos, 2008, 2009).

Food safety risks increase as consumers become more sensitive to microbial infection; as the aging population increases in numbers, so does the sensitivity to infection. Our society is composed of more immunosuppressed and chronically ill persons who are more sensitive to foodborne illnesses and their consequences than ever before. Another issue that affects food safety challenges is that the number of people involved in direct food production through agriculture is decreasing dramatically as our total population has become more urban. Furthermore, the composition of households has changed in ways that have led to changes in lifestyles and associated food preferences, food handling practices, and expectations or demands on our food supply. Examples of this include the following:

- Consumers eat more meals outside the home.
- The number of "take home" meals has increased.
- The use of prepared or pre-packaged salads, meals, and other food items, which need minimal preparation and offer convenience, has increased.
- More consumers prefer or follow special diets.
- Present-day consumers are exposed to limited education relative to proper food handling practices.

- An increasing number of consumers prefer minimally processed foods of low fat, reduced salt and other additives, fresh-like properties, convenience, and long shelf life.

However, some of these preferences may be in conflict with food safety. For example, a lower fat content in a food may be associated with higher moisture, which leads to dilution and further reduction of the already lower levels of salt and other additives. This further dilutes the preservative contribution of salt and other additives in a product that may also be minimally processed and needs more attention. Thus, modern consumer preferences may lead to new or increased food safety risks, which become challenges to be addressed by those involved in assuring the safety of our food supply (Samelis and Sofos, 2003; Sofos, 2008, 2009).

In general, approaches to pathogen control may be complicated or fail due to changing needs and expectations from consumers, projected increases in meat consumption worldwide, expanded use of sub-lethal multiple antimicrobial hurdles in food processing and preservation, and the associated potential for stress-adaptation and cross-protection of pathogens exposed to sub-lethal stresses. These issues require optimization of multiple antimicrobial hurdles due to the increasing numbers of consumer groups at risk for severe foodborne illness, the need to follow animal welfare practices during production and processing of foods, and the increasing preference of consumers for organic or natural products (Sofos, 2008, 2009).

A major meat safety issue is the need to control traditional as well as "new" or "emerging" pathogens, which may be of increased virulence, present at low infectious doses, or resistant to antibiotics or food-related stresses. Other microbial pathogen-related concerns include cross-contamination of foods such as produce and water with enteric pathogens of animal origin, and food animal manure disposal and treatment issues. Other issues and challenges include food additives and chemical residues, animal identification and traceability issues, the safety and quality of organic and natural products, products of food biotechnology or genetically modified organisms (GMO), and intentional bioterrorism. As BSE has come under control, efforts should continue for its eradication. Viral agents affecting food animals, such as avian influenza, will always need attention for prevention or containment (Sofos, 2008, 2009). The potential role of improved animal welfare during animal source food production in the safety of derived food products needs to be considered, explored, and improved as necessary.

Additional information on microbial pathogens, including major and emerging pathogens, and their control, as well as other animal source food safety issues, such as chemical additives, GMOs and materials, and transmissible spongiform encephalopathies has been reported by Koutsoumanis and Sofos (2004), Koutsoumanis, Geornaras, and Sofos (2006), Sofos (2002, 2005, 2006, 2008, 2009), Sofos and Geornaras (2010), and Stopforth and Sofos (2006).

MICROBIAL FOOD SAFETY AND ANIMAL WELFARE

General

Food safety is a global issue and, in addition to appearance, sensory, and nutritional characteristics, may be considered as a measure of food quality. Food safety is greatly influenced by the extent of carriage and transmission of foodborne pathogens by animals. Animals are always colonized with microorganisms including pathogens, and carry them in on various external surfaces, including feces from the gastrointestinal tract, hide, hair, fleece, or feathers; therefore, the concept of zero-risk food is nonexistent even when foods are properly and adequately processed. The food industry continually strives to minimize, reduce, or prevent foods from carrying microbial hazards.

Modern animal production practices involve growing or feeding large numbers of animals together, in enclosed or limited environments, and sometimes in nonhomogeneous groups. Such intensive rearing conditions lead to animal welfare concerns due to stressful conditions associated mostly with restricted housing conditions and confined management practices. Important food safety issues associated with animal welfare include increased transmission of foodborne pathogens including bacteria, viruses, parasites, antimicrobial resistant pathogens, and genetic determinants of resistance (Rostagno, 2009).

Intensive livestock production systems (i.e., industrial animal agriculture) are common in developed countries and generate the bulk of the food consumed. However, their implementation has raised various ethical, societal, and practical concerns. A recent review by Davies (2010) concluded that available evidence does not support the hypothesis that intensive pork production practices have increased the risk for foodborne pathogens such as *Salmonella, Campylobacter, Yersinia,* and *Listeria,* which are commensals in pigs, or that alternative systems of animal production reduce the risk of animal colonization with pathogens (Davies, 2010). Concurrently, there is evidence that such production systems have contributed to the improvement of the safety of foods, such as pork, with regard to parasitic and bacterial pathogens (Davies, 2010). The virtual elimination of parasites such as *Trichinella spiralis, Toxoplasma gondii,* and *Taenia solium* from pigs in the United States is attributed to modern intensive rearing systems. Increased animal herd size has also been presented as a risk factor for animal colonization with pathogens but, according to Davies (2010), there is no convincing evidence that it is a risk factor for *Salmonella* prevalence in swine herds.

The potential link between animal welfare and animal health with food safety is well recognized (de Passille and Rushen, 2005; EC, 2000; Passantino, 2009). Improvements in animal welfare have the potential to reduce on-farm risks to food safety, principally through: (1) reduced stress-induced immunosuppression; (2) reduced incidence of infectious disease on farms; (3) reduced shedding of human pathogens by farm animals; and, (4) reduced antibiotic use and antibiotic resistance, although it is not known how reduced use of antibiotics will affect pathogen carriage in animals. The issue of humane treatment of food animals is very important and should receive increased attention worldwide (Grandin, 2006). Evidence suggests that animal stressing may damage meat quality, and lead to more contamination and cross-contamination with pathogens as it may lead to increased pathogen shedding. Irrespective of whether good animal husbandry practices make animal products safer or of better quality, humane treatment of animals is essential and should be practiced by all involved in animal handling.

Animal Stressing

All food animals experience some level of stress, which may lead to reduced performance standards, disease conditions, or death, as well as detrimental effects on animal product quality. Among other consequences, exposure to stress may influence the gastrointestinal tract by disturbing production and action of endogenous hormones, the stomach pH, and the overall immune system, which may affect colonization, infection, and shedding of foodborne microbial pathogens. In general, stressed animals may exhibit reduced performance, health problems or death, and reduced product quality (Rostagno, 2009).

Common stressors include lack of feed or water leading to inadequate nutrition, heat, cold, overcrowding which may also occur during loading, transportation, and unloading, as well as improper handling or contact with humans. Exposure to stressors leads to disturbed homeostasis, and activated adaptive responses to maintain electrolyte balance (Rostagno, 2009). The disturbed homeostatic state may then have adverse effects on animal health, as well as yield, quality, and safety of animal source foods. A reason that food products derived from stressed animals may pose increased safety risks for consumers is associated with the stress of feed deprivation and transportation of animals to the slaughterhouse. In general, it is widely believed that the number of animals carrying and shedding foodborne pathogens, as well as the levels of bacteria in the gastrointestinal system increase as a consequence of stressing, which may also increase animal susceptibility to infection. However, these views are based on limited scientific evidence. Thus, firm documentation is needed, although there is some evidence demonstrating adverse effects on food safety. Results of some studies suggest increased shedding of *Salmonella* during transportation of pigs, while other studies have found conflicting results (Rostagno, 2009). In general, it is difficult to demonstrate a clear relationship between animal stressing and safety of derived products, as the potential mechanisms involved are varied and complex. Knowledge of mechanisms involved in stressing animals in ways that lead to increased food safety risks would allow easier development of interventions against such risks. Similarly, food safety risks may be increased in animal production systems employing improved animal welfare conditions, including organic and natural animal production systems. However, it is a common belief that optimizing animal welfare minimizes losses in yield, quality, and, potentially, food safety. Both issues need attention and further investigation for clarification.

Since there is limited evidence that stressing animals is associated with changes that may have a negative influence on food safety through a variety of mechanisms, it is important to explore this issue through well-designed research. This will allow determination of whether improved animal welfare has a positive impact on food safety and quality (Rostagno, 2009). Research should also explore mechanisms involved in increased pathogen carriage and shedding by farm animals when exposed to stress. As indicated, understanding the underlying physiological mechanisms involved will allow development of effective approaches to enhance the safety and quality of animal source foods.

Animal Manure Issues

In addition to direct exposure of animal food products to enteric pathogens residing in or on animals, such contamination may be introduced to the environment and water and, through these vehicles, to foods of plant origin, either directly during animal shedding and defecation or through animal manure used to fertilize soil (e.g., during organic food production). Microbial pathogen outbreaks associated with contaminated drinking water or consumption of fruits, juices, and vegetables such as lettuce, spinach, and green onions, appear to be more common in recent years. There are estimates that 80% or more of the illnesses are traceable to animals via water contamination, exposure of humans to animals at fairs, and contact with untreated manure. Water may be contaminated with microbial pathogens of concern to humans (Bicudo and Goyal, 2003). Such water may be used to irrigate food crops or wash and otherwise treat plant food products in the field.

Specifically, pathogens such as Enterohemorrhagic *E. coli* O157:H7 and *Salmonella* are associated with illnesses caused through transmission on a variety of food products other than beef, including apple juice/cider, alfalfa/radish sprouts, mayonnaise, watermelon, spinach, lettuce, and onions (www.cdc.gov). Food vehicles for *Salmonella* outbreaks have included cantaloupes, watermelon, alfalfa sprouts, tomatoes, chocolate, and dry breakfast cereal, while *Shigella* has been transmitted by green onions and lettuce (www.cdc.gov). These events demonstrate the role of environmental and cross-contamination on transfer of enteric pathogens from their animal hosts to a variety of food products of non-animal origin (Sofos, 2008).

Animal manure, if not properly composted, processed, and handled leads to environmental pollution. As the world population continues to grow and demand for food increases, the environment will continue being associated with important international health issues, including food safety. The impact of meat animals and wild animals and their manure as sources of environmental, water, and food contamination, as well as direct animal-to-human transmission of pathogens, must be taken under consideration by those involved in the food industry in general, including producers, regulators, public health agencies, and consumers.

Irrespective of whether wild animals and birds or human negligence contribute to the problem, the food animal industry must address this issue considering the recent highly publicized outbreaks of *E. coli* O157:H7 and *Salmonella* in the United States associated with consumption of vegetables (www.cdc.gov).

ANIMAL HEALTH AND FOOD SAFETY IN ORGANIC AND CONVENTIONAL PRODUCTION SYSTEMS

The safety of produce has become a major concern in recent years, as bacterial pathogen outbreaks associated with consumption of plant foods have increased in frequency and size. This development has been associated with increased consumption of uncooked produce and with the increase in organic farming practices. Organic foods are "those grown, raised, and processed without the use of synthetic pesticides and fertilizers, and without the use of growth-promoting hormones and genetic engineering" (Cahill, Morley, and Powell, 2010). Organic produce and livestock production systems are becoming more popular in developed countries because consumers consider them to be healthier or more wholesome than conventional foods (Cahill et al., 2010; Kijlstra and Eijck, 2006).

Organic livestock production is based on numerous rules developed with the objectives of improving animal welfare and the environment, and limiting use of medical drugs and pesticides (Kijlstra and Eijck, 2006). The effects of these rules on animal health have not been well considered or proven. Disease prevention is anticipated based on reduced stress on animals, optimal quality feedstuff, proper feeding, and an increased ability of animals in open-range to deal with health issues such as infections. However, organically managed animals may face important health problems associated with access to the outdoors, which increases exposure to disease-causing agents such as viruses, parasites, and bacteria. Such agents of disease may affect the health of the animals or become food safety risks. Effective controls are needed to address these concerns and may include animal breed selection, optimized environmental conditions, vaccination, and use of pre- and probiotics (Kijlstra and Eijck, 2006). Overall, the implementation of organic versus conventional production systems affects the occurrence of pathogenic organisms (and their development of resistance to antimicrobial compounds) on livestock and thus indirectly food safety as well as the occurrence of zoonotic parasites and animal health.

Impact on Food Safety

As discussed by Sofos (2008), the demand for "organic" and "natural" food products is increasing, but their safety and quality, compared to commercially produced foods, will continue to be controversial among consumers and experts (Winter and Davis, 2006). The reasons that consumers prefer natural beef include absence of added hormones; association with a specific source; no feeding of antibiotics or animal by-products; humane treatment and handling of animals; animal production that considers environmental impact; and sustainability of the production system. It should be considered certain that the popularity of such products will increase; however, the existence and increased production of natural beef will not solve problems of the world's hunger and food safety. According to the Federation of Animal Science Societies (www.fass.org), organic foods offer the consumer a choice, but there is no evidence of nutritional differences between organic and conventionally produced meat, milk, and eggs, or that organic foods are safer than conventional foods. Lund and Algers (2003) reported that, except for parasitic diseases, the health and welfare of organic and conventional food animal herds are similar. In contrast, some have expressed concerns that organically grown products carry heavier microbial populations, as well as pathogens. It should be repeated, however, that a complete conversion to organic or natural agriculture might be impossible considering increases in the human population, urbanization, and needs for efficient food production (Sofos, 2008). Furthermore, organic and natural food products need to be investigated thoroughly for their potential effects on human health in comparison with counterparts from conventional agriculture. The issue needs careful and responsible attention and such products should be researched thoroughly.

One reason that consumption demand for organically produced food is increasing is that many consumers believe that such foods are healthier or safer than conventional products (Young et al., 2009). Hermansen (2003) found that retailers in 7 of 12 European countries used "food safety/health" as their primary argument for promoting and marketing organically grown foods, and 3 of the 12 countries used this as the number two argument. Generally, although the evidence is scarce, consumers consider organic or natural foods healthier, better tasting, more environmentally friendly, and safer than foods produced conventionally (Jacob et al., 2008; Sofos, 2008). In contrast, it has been hypothesized that organically grown produce is more likely to be contaminated with pathogenic bacteria such as *Salmonella, Campylobacter,* and pathogenic *E. coli* compared to conventionally produced counterparts due to the use of animal manure as fertilizer. However, the evidence in support of this is limited (Vaarst et al., 2005). Various researchers have concluded that food safety risks should not differ greatly between conventionally and organically produced foods of plant origin (McMahon and Wilson, 2001; Sagoo, Little, and Mirchell, 2001). Although there were few pathogens in samples of organic and conventionally grown produce examined, *E. coli* was more prevalent in organically grown produce (Mukherjee et al., 2004).

There are aspects of organic animal husbandry, like access to an outdoor run, that can increase risks to food safety. This includes increased carriage of pathogens on animals and thus increased risk of transfer to products derived from the same (Rodenburg, van der Hulst-van Arkel, and Kwakkel, 2004). Organic swine and poultry production, involving increased outdoor access, is associated with issues such as increased parasitic infestation. Greater outdoor access may lead to re-emergence of zoonotic diseases such as toxoplasmosis and campylobacteriosis (Davies, 2010; Kijlstra and Eijck, 2006). Esteban et al. (2008) conducted a survey of the occurrence of *Campylobacter*, *Salmonella*, *Listeria*, and Shiga toxin-producing *E. coli* (STEC) in 60 flocks of free-range chickens from 34 farms in northern Spain. *Campylobacter*, the most prevalent of the four pathogens, was isolated on 70.6% of the farms, followed by *L. monocytogenes* (26.5%), and *Salmonella* (2.9%). No *E. coli* O157 or other STEC was isolated. *Salmonella* was found in the cecal content of only two birds (2.5%) (both from the same conventional farm), whereas 44 (55.7%) birds were infected with *Campylobacter*. The prevalence of *Campylobacter*, the concentration of lactic acid bacteria, the duration of tonic immobility, and the condition of the breasts and footpads did not differ between the production systems (Tuyttens et al., 2008). A study by Rodenburg et al. (2004) in the Netherlands found 13% of 31 organic poultry flocks to be positive for *Salmonella* and 35% positive for *Campylobacter*. Thus, questions exist regarding differences in safety of animal source food products from free-range and organic production systems (Vaarst et al., 2005). Research is needed to examine and improve the safety of organically produced foods. Studies should examine pathogen and disease prevalence, risk factors, and optimization of management practices and control strategies (Kijlstra and Eijck, 2006).

Impact on Antimicrobial Resistance Development

The increasing prevalence and severity of antimicrobial resistant organisms and genetic determinants in the environment need serious consideration in order to maintain the existence of effective therapeutic agents for human and animal clinical intervention. The development of antimicrobial resistance in foodborne and clinically important bacteria is gaining attention and becoming a major challenge. Potential contributors to this concern are the use of antibiotics for therapeutic and nontherapeutic reasons in animal, plant, and aquaculture food production, as well as their abuse in human medicine (IFT, 2006). The issue of antimicrobial resistance and antibiotic use in food animals is complex and heavily debated in the scientific community. The specific concerns with development of antimicrobial resistance in foodborne pathogens are as follows: (1) resistant pathogens contaminating food animals have the potential to be transferred to products derived from the same and consumed by humans; (2) human use of antibiotics increases the risk of acquiring an infection with an antimicrobial-resistant pathogen; (3) human infection by an antimicrobial-resistant pathogen limits treatment options; and (4) antimicrobial-resistant pathogens may develop increased virulence. Admittedly, resistance of pathogens to antibiotics used in animal production or human medicine is of major concern and will continue being important in the future for clinical settings (Doyle and Erickson, 2006). As the issue is complex and there are no simple solutions, it is recommended that one not over-use or abuse antibiotics in food production and human medicine. Prudent use is recommended and actions should consider individual situations based on the principles of risk analysis (Sofos, 2008, 2009).

One major difference between organic and conventional production systems is the use of antimicrobials, which could influence the susceptibility of foodborne pathogens. Organic food production emphasizes animal welfare and ecosystem sustainability while minimizing the use of nonfarm inputs such as use of antimicrobials, which could influence the resistance profile of foodborne pathogens (Young et al., 2009). Antimicrobials are used only for treatment of animal illness when other options fail, and not for prophylactic reasons. The use of growth-promoting hormones is also prohibited and organic animal production requires daily outdoor access and consumption of organic feed. There are further concerns that animal stressing associated with conventional production systems changes the composition of microflora associated with animals, including selection of antimicrobial-resistant pathogenic strains (Rostagno, 2009).

Jacob et al. (2008) reviewed literature that compared the antimicrobial susceptibility of foodborne bacterial pathogens from organic and conventional food animal production systems. They found it to be highly variable in terms of production types and practices and susceptibility associations in only a few studies that compared truly organic and conventional practices. When statistical associations were possible, the isolates from conventionally reared animals and their products were more commonly resistant than were animals reared organically and free of antibiotics. Additional studies are needed to better assess public health consequences of antimicrobial resistance and food animal production systems, specifically organic or natural versus conventional (Jacob et al., 2008). Foodborne pathogens developing the greatest antimicrobial resistance include *Salmonella* and *Campylobacter*. However, one study ranked these two pathogens as number 15 and 18 out of 20 in clinical importance for antimicrobial resistance (Bywater and Casewell, 2000). A meta analysis by Young et al. (2009) concluded that *Campylobacter* isolates from conventional retail chicken meat were more likely to be ciprofloxacin-resistant than those from organically grown broilers. Overall, it was concluded that more bacterial isolates with resistance to antimicrobials were isolated from conventionally grown animals, but some resistant strains were isolated from organically grown animals. These findings should be confirmed with additional research.

It should be noted that antibiotic-resistant foodborne pathogens of concern in food that is cooked or processed also show greater resistance to sanitation and food preservation than do their antibiotic-sensitive counterparts. The need to consider whether antimicrobial-resistant pathogens exhibit increased resistance to subsequent food-processing stresses (such as antimicrobial applications, cooking, etc.) is well accepted in the scientific community (Lou and Yousef, 1997; Skandamis et al., 2008; Stopforth et al., 2003), yet there are limited studies specifically documenting increased resistance of antibiotic-resistant pathogens from animal sources to food-processing related stresses (Arthur et al., 2008; Stopforth et al., 2008). Arthur et al. (2008) reported that the antimicrobial interventions in place at a beef processing facility did not differ in the ability to reduce resistant and susceptible *Salmonella* strains. Stopforth et al. (2008) found no significant differences in overall heat resistance between resistant and susceptible *Salmonella* strains in ground beef although susceptible strains had slightly higher heat resistance at certain temperatures. Thus, increased resistance of antibiotic-resistant foodborne pathogens to subsequent food-processing stresses is not well established and needs further investigation to assess if there is an actual rather than a perceived risk.

Impact on Animal Health

The actual evidence suggesting that organic livestock production is more or less detrimental to the overall animal health and welfare in comparison with conventional systems is limited (Hovi, Sundrum, and Thamsborg, 2003). However, there is strong evidence that parasite control is of greater concern in organically managed animals (Kijlstra, Meerburg, and Mul, 2004). Nematode parasites of domestic food animals pose the greatest worldwide disease problem in grazing livestock systems (Waller and Thamsborg, 2004). The outdoor production of pigs (primarily sows), dairy cattle, and laying hens is associated with higher prevalence of parasites compared to conventional intensive indoor production (Thamsborg, Roepstorff, and Larsen, 1999). Parasites and associated zoonotic diseases are more prevalent in organic production systems compared with conventional systems due to two major factors: (1) withdrawal of preventative drug therapy to control parasites (Thamsborg et al., 1999); and (2) increased exposure of animals to vehicles transferring disease via increased outdoor/open range exposure (Meerburg et al., 2004). Rodents serve as vehicles of transfer of pathogens and parasites to animals and derived products within and between farms. Therefore, effective rodent management programs should be established in food animal growing facilities. This is more difficult to accomplish in open production systems such as those following organic guidelines (Meerburg et al., 2004, 2009). Therefore, there is a need for development and implementation of acceptable control strategies in organic production systems. It is worthwhile to consider that indoor housing of food animals in intensive animal production systems has resulted in effective control of certain pathogens such as *Toxoplasma* and *Trichinella* in swine (Kijlstra et al., 2004). Livestock-production systems operating under better animal welfare guidelines need extra measures to control transfer of parasites. Such measures may include preventing access of rodents and cats to the premises (Kijlstra et al., 2004).

MANAGING ANIMAL HEALTH AND WELFARE AND FOOD SAFETY

Regulations, Standards, and Trade Implications Regarding Food Safety and Animal Health and Welfare

Animal producers and processors are constantly faced with meeting or exceeding regulatory guidelines and market acceptability standards for their products (Farm Foundation, 2006). Food safety and animal health and welfare are being linked in various government agricultural best practices policies and quality management programs, which are also available through animal producer associations such as the National Cattlemen's Beef Association and National Pork Producers Council in the United States. Food safety and animal health and welfare are closely related issues. However, the priorities of each are sufficiently different to warrant separate strategies. Although there are similarities in the approaches that address animal health and food safety, it is important to recognize the conflict in objectives and desired outcomes. Finding approaches to reduce the conflict between regulations and standards governing food safety and animal health and welfare is essential as there is a push for international standards governing animal health and welfare and the quality of products derived from them.

Regulations or standards developed to improve food safety can have a direct impact on animal health and welfare and vice versa and they can sometimes benefit all three simultaneously. However, the impact can also be negative, creating potential conflicts among regulatory objectives (de Passille and Rushen, 2005). Some examples are as follows:

- EU legislation reduces use of fully slatted flooring in pig houses because of their impact on animal welfare (discomfort to animals); however, research revealed that slatted floors reduce the incidence of *Salmonella* infections in growing pigs and help reduce ammonia emissions from pig houses (Ni et al., 1999; Nollet et al., 2004).
- In Canada, some jurisdictions limit the access grazing cattle have to natural water sources in order to prevent water contamination; however, this restricts the cattle's access to fresh, clean water and areas of shade, both negatively impacting animal welfare.
- In many countries, there are bans on the use of antibiotics as growth promotants due to the fear of antibiotic residues in food from these animals; however, this may result in an increased incidence of gastrointestinal illnesses.

There are many more examples of perceived conflicts between food safety, animal health, and animal welfare models that often do not actually occur. The assumption that free-range housing of hens results in increased fecal contamination of eggs is questioned after one study found that the use of free-range eggs was associated with a reduced risk of salmonellosis (Casewell et al., 2003). There is, however, a tendency to consider that food safety and animal health are more important than animal welfare and that animal welfare can be neglected to ensure the primacy of the other two issues (Sorensen, Sandoe, and Halberg, 2001). There is increasing pressure to harmonize animal welfare, animal health, and food safety regulations and standards especially in the context of international trade (Baines and Davies, 2000). Modern consumers require a wide variety of foods throughout the year. In order to meet these demands, foods are sourced globally, which presents challenges for ensuring that the products are produced under best practices for animal health and welfare and are of acceptable quality and safety. Good agricultural practices (GAPs) and sanitary and phytosanitary standards (SPS) affecting animal health and welfare and hazard analysis and critical control point (HACCP) systems affecting product safety and quality have become part of most trade agreements, but the dispute over implementation and standards has and can result in trade restrictions and embargos. These issues will continue to be a cause of concern and become more complicated and difficult to navigate as global trade expands unless there is harmonization of standards and enforcement thereof. It is of paramount importance that animal welfare standards be developed to be consistent with animal health and food safety standards. The complexity of resolving the conflicts in regulations and standards governing animal welfare, animal health, and food safety may be associated with how animal welfare is approached. Currently, focus is given to logistic-based criteria (design and engineering) as opposed to animal-based criteria (de Passille and Rushen, 2005). Rethinking the way we measure animal welfare may allow a resolution for many (although not all) of its conflicts with animal health and food safety.

Improving Animal Health and Welfare to Complement Food Safety

There are many instances in which the regulations and standards to improve animal welfare are in conflict with those to meet animal health and food safety requirements. However, in order to move forward with harmonizing these issues, there is a need to consider aspects of these issues that are complementary rather than conflicting. In general, the relationship between animal welfare, animal health, and food safety has two common scenarios: (1) poor animal welfare results in poor animal health and a higher risk for poor food quality and safety (de Passille and Rushen, 2005); and (2) improved animal welfare results in improved animal health, reduced need for antibiotics, and a lower risk for development of antibiotic resistance in foodborne pathogens (Regula et al., 2003).

Animal welfare is based on the state of biological needs of the animal and aims to provide all livestock with conditions of life that are harmonized with their innate behavior (Bracke et al., 2001). Typically, assessing animal welfare has been oriented toward logistic-based criteria (design and engineering of animal welfare systems) rather than animal-based criteria (animal health). Logistic-based criteria include: (1) the selection of animal breeds; (2) animal housing and transportation; (3) management or stockmanship; (4) diet and nutrition; and (5) veterinary care. These criteria are easier to measure and audit than animal-based criteria and thus are favored as a means of assuring animal welfare. The drawback to such a system is limited evidence establishing a direct correlation between the criteria and overall welfare of the animal. There are standards developed for "organic" livestock production, yet there is no assurance that the standards actually improve animal welfare (Maine, Webster, and Green, 2001). This point is further strengthened if one considers welfare standards from one country that differ vastly in the logistic-based criteria, which makes it impossible to base animal welfare standards on these criteria. The logistic-based criteria are somewhat inflexible as they relate to design or engineering elements and could more often than not be the source of conflict with regulations or standards relating to food safety since animal welfare is about adaptation processes and adaptability. Thus, the emphasis on monitoring animal welfare should be on animals deviating from an acceptable standard of health (von Borell, 2000). Use of animal-based criteria relates more to management practices and techniques and relies on measuring the actual condition of the animal. With this type of system, animal health becomes the indicator of animal welfare and a reliable quantitative measure of problems to allow food safety and animal welfare standards to be balanced. Other criteria for assessing animal welfare include behavioral measures, physiological measures, immune measures, incidence of health problems, and production levels. A good basis for developing a system measuring animal-based criteria is the "Five Freedoms," which state that animals should be free from (Webster, 2001): (1) thirst, hunger, and inappropriate feed;

(2) physical and physiological discomfort; (3) pain, injury, and diseases; (4) fear, distress, and chronic stress; and (5) physical limitations to express normal behavior. De Passille and Rushen (2005) proposed the concept of measuring animal health as an indicator of animal welfare; however, there is a need to establish a critical mass of science-based results for a meaningful understanding of how frequency and type of disease affects animal welfare. Until such a system is developed, it is necessary to improve logistic-based criteria for improving animal health and welfare. Of the logistics-based criteria, the following can be improved to increase overall animal health and welfare, as well as quality and safety of the resulting food products (Kijlstra and Eijck, 2006).

Breeding

Disease prevention in livestock production may be improved by selecting appropriate breeds of animals with resistance characteristics (Magnusson, 2001). Strategies for improved disease resistance in livestock production include: (1) recording disease incidence in progeny and selection of parents that produce progeny with the lowest incidence; (2) using breeds possessing major histocompatibility-complex antigens associated with resistance to disease; and (3) identification and use of highly heritable gene markers for immune parameters for resistance to disease. An effective example of such strategies is genetic control of parasitic diseases in sheep (Windon, 1996) and poultry (Gauly et al., 2002).

Feeding

Almost all nutrients in the diet have a critical role in maintaining "optimal" immune responses in animals and deficiencies or excessive intakes may negatively affect the immune response of an animal to pathogens (Field, Johnson, and Schley, 2002). Intestinal parasites have deleterious effects on an animal's nutrition status and malnutrition of animals predisposes them to intestinal parasitic infections (Koski and Scott, 2001). Animal diets and dietary practices must be optimized for specific livestock production systems to enhance the animal's immune system and thereby improve its overall health and the safety and quality of the resulting food.

Housing and Transportation

A major topic regarding animal welfare is the effect of outdoor versus indoor housing on the quality of the animal's life and its immune status. Housing and transportation of animals should: (1) be comfortable; (2) be clean; (3) minimize stress and fear; (4) minimize pain and injury; and (5) minimize exposure to parasites and pathogens. While outdoor access provides an animal with its natural habitat and ability to express normal behavior, its impact on animal health may be negative since the animal is exposed to parasites and pathogens in the environment (Hoglund, Nordenfors, and Uggla, 1995; Kijlstra et al., 2004; Permin et al., 1999). Housing poses a challenge in strategies to improve animal welfare (comfort). For example, deep litter for pigs is comfortable, but also provides a source of pathogens and air pollution. However, this is one area where management techniques are complementary for animal health and welfare and food safety if risk to animal health is reduced by changing the litter or adding fresh layers on the surface of existing litter (Basset-Mens and van der Werf, 2005).

Medication

There are two major challenges regarding animal welfare in livestock production systems that optimize animal health and food safety through control of zoonotic disease: (1) chemical residues in foods from use of antiparasitic agents and antibiotics; and (2) development of antibiotic resistance from therapeutic and subtherapeutic use of antibiotics for disease control. A strategic focus on animal husbandry and management practices that minimize or replace routine use of medicinal therapies and prophylaxis is needed to control disease in animals while alleviating concerns among consumers regarding chemical residues and antibiotic-resistant organisms.

Harmonizing Animal Health and Welfare and Food Safety through a Hazard Analysis Critical Control Point (HACCP)-Based Approach

To ensure the complementarity of regulations and standards between food safety and animal health and welfare as proposed earlier, it would be beneficial to evaluate the two issues together. While food safety is currently evaluated and monitored in a well-organized manner, animal health and welfare are regulated more haphazardly. Food safety considers product quality and safety (from a microbiological, physical, and chemical standpoint) through approaches based on hazard analysis critical control point (HACCP) principles from farm to table. However, animal welfare, animal health, and food safety are also primary issues in policies, retailer strategies, and consumer concerns. Among producers in the food chain, there is a need to introduce the concept of process quality in the production sector as it relates to animal welfare (Vaarst et al., 2005). Moreover, it is rational to incorporate animal health, animal welfare, and food safety at the production level into one HACCP-based program since (Noordhuizen and Metz, 2005): (1) hazards in any of these areas are predominantly multicausal; (2) focus must be on risk identification and management; (3) HACCP principles may compromise hazard and risk identification; (4) production processes can be brought under control more efficiently; and, (5) final product quality can be assured more effectively than if each of these aspects is approached separately. The HACCP concept is a program based on preventative measures dealing with hazard and risk identification, process analysis, designation of critical control points, monitoring of control points, and documentation and verification of the program (Ropkins and Beck, 2000). The seven principles of HACCP adapted to livestock production are as follows (Cullor, 1997):

1. Draw detailed descriptions of the production process using flow charts.
2. Identify and evaluate potential hazards and risks related to the hazards during the production process.
3. Determine critical control points (CCPs) in the production process where such risks can be controlled.
4. Specify when the CCPs are under control by setting standards, criteria, and tolerances (limits).
5. Design an on-farm monitoring system involving CCPs to check whether all specifications are being met.
6. Determine corrective actions for events where CCPs exceed their tolerances (limits).
7. Verify the plan using additional information or actions.

There have been increasing attempts in recent years to develop HACCP-based approaches looking to integrate animal health and welfare with food safety at the production level (Grandin, 2004; Noordhuizen and Metz, 2005; Sorensen et al., 2006). Quality assurance (QA) programs have been developed in recent years through collaborative efforts by industry and consumers to demonstrate wholesomeness, quality, and safety of food products (Mench, 2003; Webster, 2005). Although these programs focused primarily on food safety, they are considered appropriate for application to other goals such as animal welfare and animal health. However, there is a consensus that there is insufficient information about the interaction of animal health and welfare and food safety to fully implement an HACCP-based system. Yet, there is ongoing research regarding the assessment of risks to animal health and welfare and identification of CCPs where there is high risk and where control may be needed in the production sector. Grandin (2004) has outlined some of the CCPs that may be used in monitoring animal welfare in the development of an HACCP-based animal-welfare auditing scheme that includes, for example: (1) type of housing; (2) quality and functionality of euthanasia equipment; and (3) access to functional and well-maintained water and feeding equipment. These are but a few of the CCPs for animal welfare in the production sector. Using the list of CCPs Grandin (2004) proposed for different production sectors at the farm level, CCPs for animal health and food safety could be integrated to find areas of complementarity as a starting point for a comprehensive HACCP-based program. Sorensen et al. (2006) consider the development of an HACCP-based system by the production sector to be too extensive and the number of necessary control points too numerous to be operable. Undoubtedly, development of an HACCP-based farm management system would be expensive. In addition, Lievaart et al. (2005) states that introduction of HACCP-based programs will not run smoothly for a number of concerns of producers that: (1) are not willing to change routine practices; (2) need explanation of the ultimate goal; (3) need to be convinced that quality control will help reduce quality failure costs and improve market retention instead of increase profits and expand market segments; and (4) need to be convinced that such a system would provide early warning of impending problems and result in saving losses due to disease through risk implementation strategies. These are likely challenges that the livestock industry would encounter outside of the complexity associated with designing such a system, yet the perceived benefit of integrating animal health and welfare and food safety into one monitoring program is evident. Sorensen et al. (2006) proposed a compromise to develop a generic set of hazards and risk factors for the production sectors, and to develop CCPs, critical levels/tolerances, and corrective actions for the specific farm rather than for a generic HACCP-based farm management system. With such a system, the number of hazards will be reduced to a few with controllable risk factors and residual risk factors will be controlled by good manufacturing practices (GMPs) without CCPs. Until more is known about the effect of animal health and welfare on food safety, the implementation of an all-encompassing HACCP-based system seems unfeasible. However, any management system should be based on generic GMPs with a few of the high-risk hazards controlled by HACCP-like CCPs, alarm levels/tolerances, and corrective actions.

SUMMARY AND FUTURE OUTLOOK

Maintaining the safety of the food supply is essential to all countries and the livestock industry spends significant resources in assuring consumers that their products are safe and wholesome. However, industry, government, and consumers are increasingly aware that animal health and animal welfare issues are closely linked to food safety. The processing sectors have adopted quality assurance and process control strategies, such as HACCP, that reduce food safety risks and provide public confidence in product quality. Nevertheless, there are concerns that process quality in the production sector involving rearing of livestock for food does not demonstrate "due diligence" in managing animal health and welfare. The production sector has adopted and consistently applied quality assurance programs and GMPs to address process quality, but without established "standards," such programs are very subjective and their impact is too difficult to measure. The management of food safety and animal health and welfare from farm to table requires coordination and integration that is simply not provided by the current regulatory framework or policies. The priorities of food safety and animal health and welfare are often very different. However, approaches to reduce the conflict between regulations and standards governing food safety and animal health and welfare are essential to development of national and international standards governing animal health and welfare and the quality and safety of their products.

An ideal approach to harmonize animal health, animal welfare, and food safety would be through implementation of an HACCP-based program focused on preventative measures dealing with hazard and risk identification, process analysis, designation of CCPs, monitoring of control points, and documentation and verification of the program. The concept of an HACCP-based program to assure product and process quality in the conversion of food animals to food seems unfeasible due to the complexity in considering all hazards, risk factors, and control points for animal health and welfare and food safety. Thus, the approach should be centered on areas of complementarity, considering fewer hazards with controllable risk factors, while the residual risk factors are controlled by GMPs. The hazards and risk factors should be generic for all types of livestock farms and the CCPs, alarm levels/tolerances, and corrective actions should be developed for each specific farm. The scientific literature should be used to determine CCPs for animal welfare, animal health, and food safety, and areas of overlap should be prioritized for inclusion in an HACCP-based program. Furthermore, the criteria for assessing animal health and welfare should be established based on the knowledge that logistic-based criteria (design and engineering) of housing and transport facilities are often inflexible and unfeasible to replace or change. Therefore, the focus should be on less subjective and more measurable criteria including behavioral, physiological, immunological, animal health, and production output level as indices of animal welfare. Management techniques of logistic-based criteria (breeding, housing, transportation, feeding, and medication) should be based on animal-based criteria. An integrated system based on this approach would provide measurable feedback and improvements via a set of standards in animal health, animal welfare, and food safety. These standards will improve the health and welfare of animals and the safety of their products, as well as consumer confidence in the process of converting animals to food and will provide a framework for international regulation and trade.

For international acceptance of the standards to improve animal health and welfare and food safety, the biggest challenge for implementation will be in development and oversight of the program across the numerous entities involved in production, marketing, monitoring, regulation, and trading of animal food products. A short list of the entities required to implement such a system includes: Codex Alimentarius Commission, World Animal Health Organization, World Trade Organization, Global Food Safety Initiative, representative bodies for livestock industries (i.e., National Cattlemen's Beef Association, National Pork Producers Council, etc.), regulatory bodies (United States Department of Agriculture, Food and Drug Administration, etc.), trade bodies (United States Meat Export Federation, etc.). While not as complex as international trade, the mere implementation of such a system within a country will require input from various segments of the food chain from farm to table and require establishment of a body of experts to champion the concept of regulations and standards that harmonize animal health, animal welfare, and food safety.

REFERENCES

Arthur, T.M., N. Kalchayanand, J.M. Bosilevac, D.M. Brichta-Harhay, S.D. Shackelford, J.L. Bono, T.L. Wheeler, and M. Koohmaraie. 2008. Comparison of effects of antimicrobial interventions on multidrug-resistant *Salmonella*, susceptible *Salmonella*, and *Escherichia coli* O157:H7. *J. Food Prot.* 71: 2177–2181.

Bacon, R.T., and J.N. Sofos. 2003. Food hazards: Biological food; characteristics of biological hazards in foods. In: *Food Safety Handbook*, R.H. Schmidt and G. Rodrick, Eds. New York: Wiley Interscience, pp. 157–195.

Baines, R.N,. and W.P. Davies. 2000. Meeting environmental and animal welfare requirements through on-farm food safety assurance and the implications for international trade. IFAMA Agribusiness Forum, June, 25, Chicago, IL. http://www.ifama.org/events/conferences/2000Congress (accessed December 12, 2010).

Basset-Mens, C., and H.M.G. van der Werf. 2005. Scenario-based environmental assessment of farming systems: The case of pig production in France. *Agric. Ecosyst. Environ.* 105: 127–144.

Bicudo, J.R., and S.M. Goyal. 2003. Pathogens and manure management systems: A review. *Environ. Technol.* 24: 115–130.

Bracke, M.B.M., J.H.M. Metz, A.A. Dijkhuizen, and B.M. Spruijt. 2001. Development of a decision support system for assessing farm animal welfare in relation to husbandry systems: Strategy and prototype. *J. Agric. Environ. Ethics* 14: 321–337.

Bywater, R.J., and M. Casewell. 2000. An assessment of the impact of antibiotic resistance in different bacterial species and of the contribution of animal sources to resistance in human infections. *J. Antimicrob. Chemo.* 46: 643–645.

Cahill, S., K. Morley, and D.A. Powell. 2010. Coverage of organic agriculture in North American newspapers, Media: Linking food safety, the environment, human health and organic agriculture. *Brit. Food J.* 112: 710–722.

Casewell, M., C. Friis, E. Marco, P. McMullin, and I. Phillips. 2003. The European ban on growth-promoting antibiotics and emerging consequences for human and animal health. *J. Antimicrob. Chemother.* 52: 159–161.

Cullor, J.S. 1997. HACCP (hazard analysis critical control points): Is it coming to the dairy? *J. Dairy Sci.* 80: 3449–3452.

Davies, P.R. 2010. Intensive swine production and food safety. *Foodborne Path. Diseas.* 8(2): 189–201. doi:10.1089/fpd.2010.0717.Aheadofprint,http://www.liebertonline.com/doi/abs/10.1089/fpd.2010.0717 (accessed December 20, 2010).

De Passille, A.M., and J. Rushen. 2005. Food safety and environmental issues in animal welfare. *Rev. Sci. Tech. Off. Int. Epiz.* 24: 757–766.

Doyle, M.P., and M.C. Erickson. 2006. Emerging microbiological food safety issues related to meat. *Meat Sci.* 74: 98–112.

EC (Commission of the European Communities). 2000. White paper on food safety. http://ec.europa.eu/dgs/health_consumer/library/pub/pub06_en.pdf (accessed January 4, 2010).

Esteban, J.I., B. Oporto, G. Aduriz, R.A. Juste, and A. Hurtado. 2008. A survey of foodborne pathogens in free-range poultry farms. *Int. J. Food Micro.* 123: 177–182.

Farm Foundation. 2006. Food safety and animal health. In: *Future of Animal Agriculture in North America*, Chapter 5, pp. 1–26. http://www.farmfoundation.org/projects/documents/Foodsafetyanimalhealth.pdf (accessed December 2, 2010).

Field, C.J., I.R. Johnson, and P.D. Schley. 2002. Nutrients and their role in host resistance to infection. *J. Leuk. Biol.* 71: 16–32.

Gauly, M., C. Bauer, R. Preisinger, and G. Erhardt. 2002. Genetic differences of *Ascaridia galli* egg output in laying hens following a single dose infection. *Vet. Parasitol.* 103: 99–107.

Grandin, T. 2004. Animal welfare audits for cattle, pigs, and chickens that use the HACCP principles of critical control points. http://www.grandin.com/welfare.audit.using.haccp.html (accessed December 20, 2010).

Grandin, T. 2006. Progress and challenges in animal handling and slaughter in the U.S. *Appl. Anim. Behav. Sci.* 100: 129–139.

Hermansen, J.E. 2003. Organic livestock production systems and appropriate development in relation to public expectations. *Livest. Prod. Sci.* 80: 3–15.

Hoglund, J., H. Nordenfors, and A. Uggla. 1995. Prevalence of the poultry red mite, *Dermanyssus gallinae*, in different types of production systems for egg layers in Sweden. *Poult. Sci.* 74: 1793–1798.

Hoglund, J., C. Svensson, and A. Hessle. 2001. A field survey on the status of internal parasites in calves on organic dairy farms in southwestern Sweden. *Vet. Parasitol.* 99: 113–128.

Hovi, M., A. Sundrum, and S.M. Thamsborg. 2003. Animal health and welfare in organic livestock production in Europe: Current state and future challenges. *Livest. Prod. Sci.* 80: 41–53.

ICMSF (International Commission for Microbiological Specifications in Foods). 1996. *Microorganisms in Foods 5: Characteristics of Microbial Pathogens*. London: Blackie Academic & Professional.

ICMSF (International Commission for Microbiological Specifications in Foods). 2002. *Microorganisms in Foods 7: Microbiological Testing in Food Safety Management*. New York: Kluwer Academic/Plenum Publishers.

IFT (Institute of Food Technologists). 2006. Antimicrobial resistance—implications for the food system. *Comp. Rev. Food Sci. Food Safety* 5: 71–137.

Jacob, M.E., J.T. Fox, S.L. Reistein, and T.G. Nagaraja. 2008. Antimicrobial susceptibility of foodborne pathogens in organic or natural production systems: An overview. *Foodborne Path. Disease* 5: 721–730.

Kijlstra, A. and I.A.J.M. Eijck. 2006. Animal health in organic livestock production systems: A review. *NJAS Wageningen J. Life Sci.* 54-1: 77–93.

Kijlstra, A., B.G. Meerburg, and M.F. Mul. 2004. Animal-friendly production systems may cause re-emergence of *Toxoplasma gondii*. *NJAS* 52-2: 119–132.

Koski, K.G., and M.E. Scott. 2001. Gastrointestinal nematodes, nutrition and immunity: Breaking the negative spiral. *Ann. Rev. Nutr.* 21: 297–321.

Koutsoumanis, K., and J.N. Sofos. 2004. Microbial contamination of carcasses and cuts. In *Encyclopedia of Meat Sciences*, W.K. Jensen, Ed. Amsterdam, The Netherlands: Elsevier Academic Press, pp. 727–737.

Koutsoumanis, K.P., I. Geornaras, and J.N. Sofos. 2006. Microbiology of land animals. In: *Handbook of Food Science, Technology and Engineering*, Y. H. Hui, Ed. Boca Raton, FL: CRC Press Taylor & Francis Group, pp. 52.1–52.43.

Lievaart, J.J., J.P.T.M. Noordhuizen, E. van Beek, C. van der Beek, A. van Risp, J. Schenkel, and J. van Veersen. 2005. The hazard analysis critical control points (HACCP) concept as applied to some chemical, physical and microbiological contaminants of milk on dairy farms. A prototype. *Vet. Quart.* 27: 21–29.

Lou, Y., and A.E. Yousef. 1997. Adaptation to sublethal environmental stresses protects *Listeria monocytogenes* against lethal preservation factors. *Appl. Environ. Microbiol.* 63: 1252–1255.

Lund, V., and B. Algers. 2003. Research on animal health and welfare in organic farming — a literature review. *Livest. Prod. Sci.* 80: 55–68.

Magnusson, U. 2001. Breeding for improved disease resistance in organic farming — possibilities and constraints. *Acta. Vet. Scand.* 95: 59–61.

Maine, D.C., A.J.F. Webster, and L.E. Green. 2001. Animal welfare assessment in farm assurance schemes. *Acta Agric. Scand.* 51: 108–113.

McMahon, M.A.S., and I.G. Wilson. 2001. The occurrence of enteric pathogens and *Aeromonas* species in organic vegetables. *Int. J. Food Microbiol.* 70: 155–162.

Mead, P.S., L. Slutsker, V. Dietz, L.F. McCaig, J.S. Bresee, C. Shapiro, P.M. Griffin, and R.V. Tauxe. 1999. Food-related illness and death in the United States. *Emerg. Infect. Dis.* 5: 607–625.

Meerburg, B.G., M. Bonde, F.W. A. Brom, S. Endepols, A.N. Jensen, H. Leirs, J. Lodal, G.R. Singleton, H.–J. Pelz, T.B. Rodenburg, and A. Kijlstra. 2004. Towards sustainable management of rodents in organic animal husbandry. *NJAS* 52-2: 195–205.

Meerburg, B.G., G.R. Singleton, and A. Kijlstra. 2009. Rodent-borne diseases and their risks for public health. *Crit. Rev. Microbiol.* 35: 221–270.

Mench, J.A. 2003. Assessing animal welfare at the farm and group level: A United States perspective. *Anim. Welf.* 12: 493–503.

Morris, J.G. Jr. 2011. How safe is our food? *Emerg. Infect. Dis.* Ahead of publication. http://www.cdc.gov/eid/content/17/1/pdfs/10-1821.pdf (accessed December 21, 2010).

Mukherjee, A., D. Speh, E. Dyck, and F. Diez-Gonzalez. 2004. Preharvest evaluation of coliforms, *Escherichia coli*, *Salmonella*, and *Escherichia coli* O157:H7 in organic and conventional produce grown by Minnesota farmers. *J. Food Prot.* 67: 894–900.

NACMCF (National Advisory Committee on Microbiological Criteria for Foods). 1998. Hazard analysis and critical control point principles and application guidelines. *J. Food Prot.* 61: 1246–1259.

Ni, J.Q., C. Vinckier, J. Coenegrachts, and J. Hendriks. 1999. Effect of manure on ammonia emission from a fattening pig house with partly slatted floor. *Livest. Prod. Sci.* 59: 25–31.

Nollet, N., D. Maes, L. De Zutter, L. Duchateau, K. Houf, K. Huysmans, H. Imbrechts, R. Geers, A. de Kruif, and J. van Hoof. 2004. Risk factors for the herd-level bacteriologic prevalence of *Salmonella* in Belgian slaughter pigs. *Prev. Vet. Med.* 65: 63–75.

Noordhuizen, J.P.T.M., and J.H.M. Metz. 2005. Quality control on dairy farms with emphasis on public health, food safety, animal health and welfare. *Livest. Prod. Sci.* 94: 51–59.

Passantino, A. 2009. Welfare of animals at slaughter and killing: A new regulation on the protection of animals at the time of killing. *J. Verbr. Lebensm.* 4: 273–285.

Permin, A., M. Bisgaard, F. Frandsen, M. Pearman, J. Kold, and P. Nansen. 1999. Prevalence of gastrointestinal helminthes in different poultry production systems. *Brit. Poult. Sci.* 40: 439–443.

Regula, G., R. Stephan, J. Danuser, B. Bissig, U. Ledergerber, D.L. Wong, and K.D. Stark. 2003. Reduced antibiotic resistance to fluoroquinolones and streptomycin in "animal-friendly" pig fattening farms in Switzerland. *Vet. Rec.* 152: 80–81.

Rodenburg, T.B., M.C. van der Hulst-van Arkel, and R.P. Kwakkel. 2004. *Campylobacter* and *Salmonella* infections on organic broiler farms. *NJAS* 52-2: 101–108.

Ropkins, K., and A.J. Beck. 2000. Evaluation of worldwide approaches to the use of HACCP to control food safety. *Trends Food Sci. Technol.* 11: 10–21.

Rostagno, M.H. 2009. Can stress in farm animals increase food safety risk? *Foodborne Path. Disease* 6: 767–776.

Sagoo, S.K., C.L. Little, and R.T. Mirchell. 2001. The microbiological examination of ready-to-eat organic vegetables from retail establishments in the United Kingdom. *Lett. Appl. Microbiol.* 33: 434–439.

Samelis, J., and J.N. Sofos. 2003. Strategies to control stress-adapted pathogens and provide safe foods. In: *Microbial Adaptation to Stress and Safety of New-Generation Foods,* A.E. Yousef and V.K. Juneja, Eds. Boca Raton, FL: CRC Press, pp. 303–351.

Scallan, E., R.M. Hoekstra, F.J. Angulo, R.V. Tauxe, M.–A. Widdowson, S.L. Roy, J.L. Jones, and P.M. Griffin. 2011a. Foodborne illness acquired in the United States — major pathogens. *Emerg. Infect. Dis.* Ahead of publication. http://www.cdc.gov/EID/content/17/1/pdfs/09-1101p1.pdf (accessed December 17, 2010).

Scallan, E., P.M. Griffin, F.J. Angulo, R.V. Tauxe, and R.M. Hoekstra. 2011b. Foodborne illness acquired in the United States—unspecified agents. *Emerg. Infect. Dis.* Ahead of publication. http://www.cdc.gov/eid/content/17/1/16.htm (accessed December 17, 2010).

Scharff, R. 2010. Health-related costs from foodborne illness in the United States. http://www.producesafetyproject.org/admin/assets/files/Health-Related-Foodborne-Illness-Costs-Report.pdf-1.pdf (accessed December 18, 2010).

Skandamis, P.N., Y. Yoon, J.D. Stopforth, P.A. Kendall, and J.N. Sofos. 2008. Heat and acid tolerance of *Listeria monocytogenes* after exposure to single and multiple sublethal stresses. *Food Microbiol.* 25: 294–303.

Sofos, J.N. 2002. Approaches to pre-harvest food safety assurance. In: *Food Safety Assurance and Veterinary Public Health; Volume 1, Food Safety Assurance in the Pre-Harvest Phase*, F.J.M. Smulders and J.D. Collins, Eds. Wageningen, The Netherlands: Wageningen Academic Publishers, pp. 23–48.

Sofos, J.N. 2005. *Improving the Safety of Fresh Meat*. Cambridge, England: CRC/Woodhead Publishing Limited.

Sofos, J.N. 2006. Field data availability and needs for use in microbiological risk assessment. In: *Food Safety Assurance and Veterinary Public Health. Vol. 4. Towards a Risk-based Chain Control*, F.J.M. Smulders, Ed. Wageningen, The Netherlands: Wageningen Academic Publishers, pp. 57–74.

Sofos, J.N. 2008. Challenges to meat safety in the 21st century. *Meat Sci.* 78: 3–13.

Sofos, J.N. 2009. ASA centennial paper: Developments and future outlook for postslaughter food safety. *J. Anim. Sci.* 87: 2448–2457.

Sofos, J.N., and I. Geornaras. 2010. Overview of current meat hygiene and safety risks and summary of recent studies on biofilms, and control of *Escherichia coli* O157:H7 in nonintact, and *Listeria monocytogenes* in ready-to-eat, meat products. *Meat Sci.* 86: 2–14.

Sorensen, J.T., P. Sandoe, and N. Halberg. 2001. Animal welfare as one among several values to be considered at farm level: The idea of an ethical account for livestock farming. *Acta Agric. Scand.* 51: 11–16.

Sorensen, J.T., M. Bonde, T. Rousing, S.H. Moller, and L. Hegelund. 2006. Herd health surveillance and management in an integrated HACCP based system. Proceedings of the 11th International Symposium on Veterinary Epidemiology and Economics. http://www.sciquest.org.nz/elibrary/download/64161/T3-5.3.6_-_Herd_health_surveillance_and_management_in_an_integrated_HACCP_based_system (accessed December 11, 2010).

Stopforth, J.D., J. Samelis, J.N. Sofos, P.A. Kendall, and G.C. Smith. 2003. Influence of extended acid stressing in fresh beef decontamination runoff fluids on sanitizer resistance of acid-adapted *Escherichia coli* O157:H7 in biofilms. *J. Food Prot.* 66: 2258–2266.

Stopforth J.D., and J.N. Sofos. 2006. Recent advances in pre- and post-slaughter intervention strategies for control of meat contamination. In: *Advances in Microbial Food Safety, ACS Symposium 931. Recent Advances in Intervention Strategies to Improve Food Safety*, V.K. Juneja, J.P. Cherry, and M.H. Tunick, Eds. Washington, DC: American Chemical Society, Oxford University Press, pp. 66–86.

Stopforth, J.D., R. Shulaim, B. Kottapalli, W.E. Hill, and M. Samadpour. 2008. Thermal inactivation D- and z-values of multidrug-resistant and non-multidrug-resistant *Salmonella* serotypes and survival in ground beef exposed to consumer-style cooking. *J. Food Prot.* 71: 509–515.

Thamsborg, S.M., A. Roepstorff, and M. Larsen. 1999. Integrated and biological control of parasites in organic and conventional production systems. *Vet. Parasitol.* 84: 169–186.

Tuyttens, F., M. Heyndrickx, M. De Boeck, A. Moreels, A. van Nuffel, E. van Pouche, E. van Coillie, S. van Dongen, and L. Lens. 2008. Broiler chicken health, welfare and fluctuating asymmetry in organic versus conventional production systems. *Livest. Sci.* 113: 123–132.

Vaarst, M., S. Padel, M. Hovi, D. Younie, and A. Sundraum. 2005. Sustaining animal health and food safety in European organic livestock farming. *Livest. Prod. Sci.* 94: 61–69.

Von Borell, E. 2000. Assessment of pig housing based on the HACCP concept — critical control points for welfare, health and management. In *Improving Health and Welfare in Animal Production,* EAAP Publication, Volume 2, H. Blokhuis, D. Ekkel, B. Wechsler, Eds. Wageningen, The Netherlands: Wageningen Pers Publ., pp. 75–80.

Waller, P.J., and S.M. Thamsborg. 2004. Nematode control in "green" ruminant production systems. *Trends Parasitol.* 20: 493–497.

Webster, A.F.J. 2001. Farm animal welfare: The five freedoms and the free market. *Vet. J.* 161: 229–237.

Webster, J. 2005. The assessment and implementation of animal welfare: Theory into practice. *Rev. Sci. Tech. Off. Int. Epiz.* 24: 723–734.

Windon, R.G. 1996. Genetic control of resistance to helminthes in sheep. *Vet. Immunol. Immunopath.* 54: 245–254.

Winter, C.K. and S.F. Davis. 2006. Organic foods. *J. Food Sci.* 71: R117–R124.

Young, I., A. Rajic, B.J. Wilhelm, L. Waddell, S. Parker, and S.A. McEwen. 2009. Comparison of the prevalence of bacterial enteropathogens, potentially zoonotic bacteria and bacterial resistance to antimicrobials in organic and conventional poultry, swine and beef production: A systematic review and meta-analysis. *Epidemiol. Infect.* 137: 1217–1232.

CHEMICAL FOOD SAFETY

Steve L. Taylor and Joseph L. Baumert

INTRODUCTION

While the preceding section of this chapter focused on the very important issues surrounding microbial food safety of animal-based food products, chemical hazards are also important. Unlike microbiological hazards, chemical agents do not multiply in foods unless they are associated with microbial growth. With chemical hazards, the focus is on hazard identification and assessment with control efforts focused on the prevention of their entry into the food with various raw materials. However, a few potentially hazardous chemical substances are produced by microorganisms sometimes associated with animal-based foods including botulinum toxin from growth of *Clostridium botulinum*. This section focuses on the nature of various potential chemical hazards and their monitoring and control including a focus on food allergens, which have emerged in recent years as a chemical safety issue that must be controlled through the development and application of allergen control plans.

CHEMICAL HAZARDS ASSOCIATED WITH ANIMAL-BASED FOOD PRODUCTS

Foods can be viewed as complex mixtures of chemicals with many being nutrients essential to sustain life. Nevertheless, non-nutrient chemicals can and do exist in foods. Some of these chemicals can be toxic and hazardous under certain circumstances of exposure, although, fortunately, most are not hazardous under typical circumstances of exposure. Even some nutrients can be toxic under certain circumstances of exposure. The central axiom of toxicology is that the dose makes the poison so the amount of exposure to a given chemical is related to the potential hazard. The focus here is on chemical substances in foods that may pose a risk in animal-based food products under some reasonably expected circumstances of exposure.

Chemicals in foods arise from two principal sources—naturally occurring substances and manufactured chemicals. The naturally occurring substances in foods include the nutrients that have limited toxicological properties when consumed as part of the diet. However, some naturally occurring substances are potentially hazardous including both naturally occurring constituents of certain foodstuffs and naturally occurring contaminants. Fortunately, very few such chemicals exist in animal-based food products beyond the naturally occurring contaminants found in seafood such as ciguatera toxins in fish and various shellfish toxins, all arising from algae consumed as part of the food chain in ocean environments. These toxic contaminants will not be extensively discussed because seafood is not a principal focus of this book.

The major categories of manufactured chemicals that can occur in animal-based food products are feed additives and veterinary drugs, although food additives, chemicals migrating from packaging materials, and inadvertent or accidental contaminants occurring as industrial and environmental pollutants can also be a concern on occasion. Chemicals produced by reactions occurring during the processing, preparation, storage, and handling of foods could also be considered artificial because these processes occur through human intervention.

NATURALLY OCCURRING TOXICANTS IN ANIMAL-BASED FOODS

Few naturally occurring constituents occur in animal-based foods with the exception of certain hazardous species of marine organisms (Table 10.1). The only known exceptions are certain naturally occurring plant toxicants that can be ingested by animals feeding on certain noxious weeds; the toxicants can then be passed through to meat, milk, and eggs (Beier and Nigg, 1994). Such situations happen very rarely but are more likely to occur with livestock grazing on open range in regions where certain noxious weeds are endemic. The levels of alkaloid toxins

---

TABLE 10.1
Naturally Occurring Toxicants in Animal-Based Food Products

Naturally Occurring Constituents

Poisonous animals (puffer fish)

Plant toxicants passed through to meat, milk, and eggs

Constituents causing allergies or intolerances

    Milk allergens

    Egg allergens

      Fish allergens

    Crustacean shellfish allergens

    Molluscan shellfish allergens

      Meat allergens

    Lactose for lactose intolerance

Naturally Occurring Contaminants

Bacterial toxins (botulinum toxin)

Mycotoxins (aflatoxins)

Algal toxins (saxitoxins in paralytic shellfish poisoning)

---

that pass through to meat, milk, and eggs, and the hazards associated with the intake of these animal-based food products have not been studied extensively. Thus, these situations with a couple of rare exceptions would best be described as concerns rather than known hazards. The so-called milk sickness from the ingestion of milk from cows that grazed on white snakeroot is probably the most noteworthy example of such a situation. Tremetone is the identified toxicant present in white snakeroot. Notably, Abraham Lincoln's mother died of milk sickness in Illinois in 1818, but this illness has not been reported in recent years in the United States (Beier and Nigg, 1994).

Naturally occurring contaminants can also enter the food supply from natural sources. With animal-based food products, the principal concerns are bacterial toxins and mycotoxins from molds. Bacterial foodborne diseases are typically caused by viable pathogenic bacteria that invade cells and tissues, multiply, and thereby cause inflammation and injury. However, a few bacteria are toxigenic and produce exogenous toxins in foods before the food is eaten. In these cases, the ingestion of the toxins causes the illness even if the bacteria are destroyed in processing or preparation. The staphylococcal enterotoxins and botulinal toxins are the best examples.

Staphylococcal food poisoning is one of the most common forms of foodborne disease and is caused by ingestion of staphylococcal enterotoxins. The staphylococcal enterotoxins are produced in foods by certain strains of *Staphylococcus aureus,* which grow on foods, including animal-based food products, under certain conditions such as temperatures between 10°C and 45°C (Wong and Bergdoll, 2002). Upon ingestion, the enterotoxins cause nausea and vomiting within 1 to 6 hours. Low microgram amounts of the enterotoxins are sufficient to elicit symptoms (Wong and Bergdoll, 2002). The enterotoxins are small proteins with molecular weights of 25,000 to 29,000 daltons, and nine distinct, but structurally related, enterotoxins have been identified as being produced by various strains of *Staphylococcal aureus* (Wong and Bergdoll, 2002). The enterotoxins are relatively stable to digestion and are quite heat resistant. For this reason, staphylococcal food poisoning is often associated with foods that were cooked after improper storage at elevated temperatures that allowed the proliferation of *S. aureus.* Staphylococcal food poisoning is prevented by food storage conditions that do not allow *S. aureus* to grow and produce the enterotoxin.

Another toxigenic bacterium is *Clostridium botulinum,* which can produce potent neurotoxic botulinal toxins under anaerobic conditions (Parkinson and Ito, 2002). Because of the requirement for anaerobic growth conditions, botulinal toxin formation occurs most frequently in improperly processed (canned), low-acid foods, including meat products. The vegetative cell of *C. botulinum* and the botulinal toxins are easily destroyed by heat. However, the spores of *C. botulinum* are heat-resistant, survive improper thermal processing, and germinate and grow under suitable anaerobic conditions (Parkinson and Ito, 2002). The commercial canning process is predicated on the destruction of spores of *C. botulinum* so that the spores will not germinate, grow, and produce toxin during storage of the canned product. The botulinal toxins are proteins with a molecular mass of approximately 150 kDa. Seven toxin types have been identified as being produced by various strains of *C. botulinum* (Parkinson and Ito, 2002) with types A, B, and E most commonly associated with foodborne illness. The botulinal toxins are extremely potent. Clinical symptoms develop within 12 to 48 hours after ingestion of the implicated food. Symptoms include serious neurological manifestations including blurred vision, inability to swallow, aphasia, and weakness of the skeletal muscles progressing to respiratory paralysis and death. Proper operation of canning equipment is the key to industrial control points to prevent introduction of botulism into canned food.

Mycotoxins are naturally occurring contaminants produced when certain species of molds grow on certain foods (Chu, 2002). Typically, the toxin-producing molds grow on cereal grains and oilseeds. However, in the case of aflatoxin, ingestion of moldy feed by cows can result in the appearance of an aflatoxin metabolite in the milk. The aflatoxins are produced primarily by fungi of the *Aspergillus* genus, namely, *A. flavus* and *A. parasiticus,* which are molds that can contaminate peanuts and corn (Chu, 2002). Aflatoxins B and G are the forms of aflatoxin that have been identified in legumes and cereals. Dairy cows fed aflatoxin-contaminated grains or oilseeds are known to release a related form of aflatoxin, aflatoxin M, into their milk. The aflatoxins are potent hepatocarcinogens. The control of mycotoxin formation in foods is predicated on the control of mold growth in stored grains, oilseeds, and other foods. Regarding aflatoxin M in milk, the most critical measure is to avoid feeding moldy grains to dairy cows.

POTENTIALLY TOXIC MANUFACTURED CHEMICALS IN ANIMAL-BASED FOOD PRODUCTS

Foods may contain a variety of manufactured chemical substances that are either intentionally or unintentionally added (Table 10.2). With the intentionally added chemicals, these substances should be safe under normal circumstances of exposure. However, overuse or inappropriate uses can lead to

TABLE 10.2
Potentially Toxic Manufactured Chemicals in Animal-Based Food Products

Food Additives (with overuse)

Sodium nitrite

Agricultural Chemicals

Feed additives
Veterinary drugs and antibiotics

Industrial Chemicals

Polychlorinated biphenyls
Polybrominated biphenyls

Intentional Adulterants

Melamine and cyanuric acid

hazardous situations. With unintentional manufactured chemicals, the exposure dose is also important, but the mere presence of the substance can be considered as a source of concern.

Food Additives

Food additives are intentionally added to foods to provide a wide variety of technical benefits. Several thousand food additives exist, although many of these chemicals are used in rather small amounts.

The degree of hazard associated with the food additives used in animal-based food products is quite low primarily because the safety of food additives is well established (Taylor, 2005). In many cases, food additives have been subjected to safety evaluations in laboratory animals and use levels are maintained at exposure doses far below any dose that would be hazardous. Furthermore, many food additives have long histories of safe use even if classical toxicological evaluations in laboratory animals have not always been exhaustively performed. Many of these substances are generally recognized as safe (GRAS). Finally, the use of food additives is deliberately controlled in manufacturing operations. As long as additives are used in accordance with good manufacturing practices, hazardous situations can be avoided.

The primary hazard associated with food additives is their misuse. An example relating to the popular processed meat additive, sodium nitrite, will illustrate the consequences of misuse. Sodium nitrite is a white granular substance easily confused with other salts, including sodium chloride, which are much less toxic. In the illustrative incident, a small grocery store was repackaging additives such as sodium chloride, sodium nitrite, and monosodium glutamate (MSG) from bulk containers into home-use packets (Taylor and Hefle, 2002). Somehow, sodium nitrite was erroneously labeled as MSG. The mislabeled product was used in hazardous amounts by consumers, resulting in acute methemoglobinemia and at least one death.

Agricultural Chemicals

An array of various chemicals is used in modern animal agriculture. Residues of these chemicals can sometimes be found in the raw and processed animal-based food products. Public health authorities evaluate the safety of such chemicals and regulate and monitor their use in food-producing animals (Taylor, 2002). Feed ingredients and veterinary drugs, including antibiotics, are the primary concerns with food-producing animals. When properly used, minimal hazards are posed by the residues of these chemicals remaining in foods. Thus, the primary approach to lessen this particular hazard is to use such materials only as recommended.

Feed Additives

Like food additives, substances added to feed do not often cause health-related concerns among consumers of meat, milk, and eggs. Some years ago, concerns were raised when diethylstilbesterol (DES) was allowed and used as a growth promoter in beef cattle. Subsequently, DES was shown to be carcinogenic, and its use as a feed additive was banned. DES is definitely carcinogenic to humans; its use as a drug to prevent miscarriages in pregnant women was linked to certain types of cancer in their offspring. However, there is no evidence that the very low levels of DES in edible beef occurring after the use of DES as a growth promoter pose any carcinogenic risk to humans.

Veterinary Drugs and Antibiotics

A variety of veterinary drugs and antibiotics can be used on food-producing animals. If properly used, residues in foods are typically low and hazards are small. Some concerns have arisen especially when these chemicals are used inappropriately. As an example, penicillin is a common antibiotic used in animal as well as human health. Some consumers are allergic to penicillin primarily because of its use in human medicine. The likelihood of allergic reactions to the very low levels of penicillin residues found in foods is quite remote (Dewdney and Edwards, 1984), but improper use could lead to higher levels of consumer exposure.

Industrial Chemicals

Industrial chemicals enter the food supply principally as environmental pollutants. Typically, the residue levels of industrial chemicals found in foods is rather low, resulting in inconsequential hazards. However, on the rare occasions where hazardous levels of industrial chemicals enter the food supply, devastating consequences can occur from both a health and economic perspective because of the potential magnitude of the contamination.

Polychlorinated Biphenyls (PCBs) and Polybrominated Biphenyls (PBBs)

Animal food products have become contaminated with environmentally persistent chemicals, PCBs and PBBs, on several past occasions (Taylor, 2002). PCBs and PBBs are primarily industrial chemicals with PBBs commonly used as fire retardants and PCBs frequently used in transformer fluid. Residues exist in the food chain as toxic pollutants from industrial practices. PCBs and PBBs are not particularly worrisome as acute toxicants in foods. However, since they are fat-soluble, elimination from the body is slow and the chronic effects of exposure to these contaminants in foods are a concern. Many years ago in Michigan, an incident occurred involving the accidental contamination of dairy feed with PBBs. This episode resulted in the destruction of many cows and their milk. While the health consequences remain uncertain, the economic impact was considerable (Reich, 1983). Leaking heat exchangers or transformers are the principal sources of PCBs. The most famous incident of PCB contamination occurred in Japan when PCBs leaked from a heat exchanger used in the deodorization process for rice bran oil. Ingestion of the oil was responsible for many cases of "yusho" (meaning oil disease) in Japan (Miyata, Murakami, and Kashimoto, 1978). The toxic effects were chronic with symptoms persisting in many of the victims for 8 years or more after exposure. Such incidents continue to occur periodically although fortunately without the large number of human illnesses experienced in the yusho incident. Leaking transformers have contributed to the contamination of feeds with PCBs, which led to the destruction of chickens, eggs, and egg-containing food products (Taylor, 2002). Clearly, this type of environmental pollution with industrial chemicals can and should be prevented.

Intentional Adulterants

Of course, the intentional adulteration of foods can also result in potentially hazardous chemicals entering the food supply. The classic example is melamine, which perhaps together with cyanuric acid was intentionally added to milk and wheat gluten in China to increase apparent protein levels. These chemicals elicit misleading results in some protein assays based upon nitrogen content. However, melamine together with cyanuric acid is a rather potent toxic combination of chemicals that resulted in adverse reactions in infants exposed to the adulterated milk and pets ingesting the contaminated pet foods (Hau, Kwan, and Li, 2009). Of course, in most countries, it is illegal to add intentional adulterants to foods although catching the perpetrators can be problematic unless some knowledge exists to suggest possible analytes for testing.

FOOD ALLERGENS FROM ANIMAL-BASED FOOD PRODUCTS

Certain naturally occurring constituents of animal-based food products are capable of causing food allergies or intolerances. Over the past decade, food allergies and intolerances have been increasingly recognized as serious food safety issues. Food allergies involve abnormal responses of the human immune system usually to naturally occurring substances, primarily certain specific proteins, in foods (Taylor and Hefle, 2001). Food allergies occur only in certain individuals in the population with an overall estimated prevalence of 3.5 to 4.0% in the United States. These individuals have immune systems that respond abnormally to specific naturally occurring proteins in foods that most consumers can ingest with no adverse consequences. Both humoral (antibody- or IgE-mediated) and cell-mediated allergies occur with foods. Food allergies can involve both animal- and plant-based foods. The most common foods involved in IgE-mediated allergic reactions are peanuts, tree nuts, soybeans, and wheat from the plant kingdom and cow's milk, egg, crustacean shellfish, and fish from the animal kingdom. Many other foods can cause allergic reactions on a more infrequent basis. The symptoms of IgE-mediated food allergies are individually variable ranging from very mild skin rashes and itching to life-threatening asthma and anaphylactic shock. Rather low levels of exposure to residues of allergenic foods are sufficient to elicit an allergic reaction in some affected individuals. Thus, food-allergic individuals must follow rather strict avoidance diets in an attempt to eliminate all exposure to those foods that trigger their allergic responses (Taylor, Hefle, and Munoz-Furlong, 1999). In addition to IgE-mediated food allergies, abnormal cell-mediated immunological reactions can also occur with foods. However, allergic reactions of this type have not been well studied especially with respect to animal-based food products.

Milk and eggs will serve as the primary examples of commonly allergenic foods of animal origin. All types of mammalian milks (cow, goat, sheep, etc.) are allergenic and cross-reactions frequently occur between milk from different species (Sicherer, 2001). Eggs from all species of domestic birds (chicken, turkey, duck, goose, etc.) are allergenic and cross-reactions are frequent among eggs from different species (Sicherer, 2001). Despite serving as excellent sources of protein, meats such as beef, pork, chicken, and turkey are not considered as commonly allergenic foods. Milk and eggs are the most common allergenic foods among infants, affecting as many as 2 to 3% of infants and young children under the age of 3 years (Taylor, 2005). Most milk- and egg-allergic infants outgrow these particular food allergies. However, milk and egg allergies persist in some individuals so the development of oral tolerance is not universal (Skripak et al., 2007; Savage et al., 2007). Recent evidence has indicated that young children may become tolerant to heated forms (baked) of milk and egg before becoming tolerant of less well-cooked forms of egg or milk (Lemon-Mule et al., 2008; Nowak-Wegrzyn et al., 2008).

The primary allergens in milk and eggs are naturally occurring proteins. In milk, the major allergenic proteins are casein, β-lactoglobulin, and α-lactalbumin (Besler, Eigenmann, and Schwartz, 2000). These proteins also happen to be the most prominent proteins in milk. For eggs, the major allergenic proteins are ovomucoid, ovalbumin, ovotransferrin, and lysozyme (Besler, 1999). These egg proteins are the most prominent proteins in egg white. Egg yolk also contains known allergens, but they do not appear to be allergenic as frequently. Bovine serum albumin (BSA), a blood protein, is a minor allergen also found in cow's milk. However, BSA appears to be the major allergen in beef. BSA is more heat-labile than other milk allergens, so most allergic reactions to beef can be prevented by eating well-done beef (Nowak-Wegrzyn and Fiocchi, 2009). A similar protein, chicken serum albumin (CSA), is the major allergen present in chicken meat. CSA can also be found in egg yolks and is responsible for bird-egg syndrome, a condition where individuals are allergic to pet or domestic birds and are reactive to some egg products (Quirce et al., 2001).

Some food-allergic subjects react to rather low doses of their offending foods. For these individuals, the implementation of a safe and effective avoidance diet is a major obstacle. Because of these low thresholds, allergen control has become a key concern in food manufacturing facilities where multiple formulations are made on shared equipment and in shared facilities.

Food intolerances are also individualistic adverse reactions to foods or food components but, in this case, they occur through mechanisms that do not involve the immune system (Taylor and Hefle, 2001). Several types of food intolerances are known to occur. However, the metabolic food disorders are the category most frequently associated with animal-based food products. Metabolic food disorders occur either when individuals respond abnormally to a food component because they have a deficiency in an enzyme needed to metabolize that substance or because the substance affects their metabolic processes in an unusual manner. With animal-based foods, lactose intolerance is the best example of a metabolic food disorder (Suarez and Savaiano, 1997). Lactose is a disaccharide found in cow's milk. Lactose-intolerant individuals have low levels of the enzyme, β-galactosidase (lactase), in their small intestine. As a result, the disaccharide cannot be hydrolyzed into its constituent monosaccharides, glucose and galactose. While glucose and galactose can be absorbed and used for energy, lactose is not absorbed from the intestine unless it is hydrolyzed. The undigested, unabsorbed lactose then enters the colon where resident colonic bacteria convert it to CO_2, H_2, and H_2O creating flatulence and frothy diarrhea. A very large number of consumers are affected by lactose intolerance because it is common among Asians, Hispanics, and African-Americans. While these individuals must follow dairy product avoidance diets, most of them can safely ingest some lactose in their diets without experiencing adverse reactions. In this case, the threshold dose is much higher than for IgE-mediated milk allergy.

SUMMARY

Animal-based food products do not frequently present chemical hazards to consumers. The chemicals that are intentionally used in the production of animals or the processing of animal-based products are generally well evaluated for safety and are of limited concern when used according to good agricultural or good manufacturing practices. The most significant hazards involve naturally occurring toxicants, industrial environmental contaminants, and intentional adulterants. Control measures can be implemented to lessen the risks posed by any of the known chemical hazards. Food allergies and intolerances represent a well-known risk to the sensitized segment of the consuming public. However, food-allergic individuals can lessen their risk simply by avoiding products made with certain animal-based components such as milk, egg, or lactose.

REFERENCES

Beier, R.C. and Nigg, H.N. (1994) Toxicology of naturally occurring chemicals in foods. In: *Foodborne Disease Handbook, Volume 3, Diseases Caused by Hazardous Substances,* Hui, Y.H., Gorham, J.R., Murrell, K.D., and Cliver, D.O., Eds. New York: Marcel Dekker, pp. 1–186.

Besler, M. (1999) Allergen data collection: Hen's egg white (*Gallus domesticus*). *Internet Symp. Food Allergens* 1: 13–33.

Besler, M., Eigenmann, P., and Schwartz, R.H. (2000) Allergen data collection: Cow's milk (*Bos domesticus*). *Internet Symp. Food Allergens* 2: 9–74.

Chu, F.S. (2002) Mycotoxins. In: *Foodborne Diseases,* 2nd ed. Cliver, D.O. and Riemann, H.P., Eds. San Diego, CA: Academic Press, pp. 271–303.

Dewdney, J.M. and Edwards, R.G. (1984) Penicillin hypersensitivity — is milk a significant hazard?: A review. *J. Royal Soc. Med.* 77: 866–877.

Hau, A.K., Kwan, T.H., and Li, P.K. (2009) Melamine toxicity and the kidney. *J. Am. Soc. Nephrol.* 20: 245–250.

Lemon-Mule, H., Sampson, H.A., Sicherer, S.H., Shreffler, W.G., Noone, S., and Nowak-Wegrzyn, A. (2008). Immunologic changes in children with egg allergy ingesting extensively heated egg. *J. Allergy Clin. Immunol.* 122: 977–983.

Miyata, H., Murakami, Y., and Kashimoto, T. (1978) Studies on the compounds related to PCB. VI. Determination of polychlorinated quaterphenyls (PCQ) in Kanemi rice oil caused "Yusho" and investigation on the PCQ formation. *J. Food Hyg. Soc.* 19: 417–425.

Nowak-Wegrzyn, A., Bloom, K.A., Sicherer, S.H., Shreffler, W.G., Noone, S., Wanich, N., and Sampson, H.A. (2008). Tolerance to extensively heated milk in children with cow's milk allergy. *J. Allergy Clin. Immunol.* 122: 342–347.

Nowak-Wegrzyn, A. and Fiocchi, A. (2009) Rare, medium, or well done? The effect of heating and food matrix on food protein allergenicity. *Curr. Opinion Allergy Clin. Immunol.* 9: 234–237.

Parkinson, N.G. and Ito, K. (2002) Botulism. In: *Foodborne Diseases,* 2nd ed. Cliver, D.O. and Riemann, H.P., Eds. San Diego, CA: Academic Press, pp. 249–259.

Quirce, S., Maronon, F., Umpierrez, A., de las Heras, M., Fernandez-Caldas, E., and Sastre, J. (2001) Chicken serum albumin (Gal d 5) is a partially heat-labile inhalant and food allergen implicated in the bird-egg syndrome. *Allergy* 56: 754–762.

Reich, M.R. (1983) Environmental politics and science: The case of PBB contamination in Michigan. *Am. J. Public Health* 73: 302–313.

Savage, J.H., Matsui, E.C., Skripak, JM., and Wood, R.A. (2007) The natural history of egg allergy. *J. Allergy Clin. Immunol.* 120: 1411–1417.

Sicherer, S.H. (2001) Clinical implications of cross-reactive food allergens. *J. Allergy Clin. Immunol.* 108: 881–890.

Skripak, J.M., Matsui, E.C., Mudd, K., and Wood, R.A. (2007) The natural history of IgE-mediated cow's milk allergy. *J. Allergy Clin. Immunol.* 120: 1172–1177.

Suarez, F.L. and Savaiano, D.A. (1997) Diet, genetics and lactose intolerance. *Food Technology* 51(3): 74–76.

Taylor, S.L. (2002) Chemical intoxications. In: *Foodborne Diseases,* 2nd ed. Cliver, D.O. and Riemann, H.P., Eds. San Diego, CA: Academic Press, pp. 305–316.

Taylor, S..L (2005) Food additives, contaminants and natural toxins and their risk assessment. In: *Modern Nutrition in Health and Disease,* 10th ed. Shils, M.E., Shike, M., Ross, A.C., Caballero, B., and Cousins, R.J., Eds. Philadelphia: Lippincott Williams & Wilkins, pp. 1809–1826.

Taylor, S.L. and Hefle, S.L. (2001) Food allergies and other food sensitivities. *Food Technology* 55(9): 68–83.

Taylor, S.L. and Hefle, S.L. (2002) Naturally occurring toxicants in foods. In: *Foodborne Diseases,* 2nd ed. Cliver, D.O. and Riemann, H.P., Eds. San Diego, CA: Academic Press, pp. 193–210.

Taylor, S.L., Hefle, S.L. and Munoz-Furlong, A. (1999) Food allergies and avoidance diets. *Nutrition Today* 34: 15–22.

Wong, A.C.L. and Bergdoll, M.S. (2002) Staphylococcal food poisoning. In: *Foodborne Diseases,* 2nd ed. Cliver, D.O. and Riemann, H.P., Eds. San Diego, CA: Academic Press, pp. 231–248.

11 Animal Welfare in the Context of Ecological Sustainability

Frederick Kirschenmann

CONTENTS

INTRODUCTION

Animal welfare has largely been addressed as an insular issue. Often it is addressed as a moral obligation to treat animals "humanely," and sometimes as a matter of animal "rights," which suggests that animals should be treated as non-human persons with certain inalienable rights and therefore should not be treated as property. Important as these conversations have been (reaching back to ancient times), it is now a good time to revisit the issue of animal welfare in a new context. The Pew Commission on Industrial Farm Animal Production recently began that process by addressing the issue of animal welfare in a more complex context that included public health, the environment, and rural economies.

This chapter addresses the issue of animal welfare within the context of future agricultural challenges and ecological farming systems designed to meet those challenges.

PRINCIPLES OF INDUSTRIALIZATION

The principles of industrialization began to be applied to agriculture as early as the second decade of the twentieth century. By that time, F.W. Taylor's "Principles of Scientific Management" had provided much of the economic rationale for industrial enterprises.

Consequently, industrial agriculture adopted essentially the same goal as other sectors of the industrial economy—maximum, efficient production and short-term economic return. The means of achieving that goal also followed the principles adopted by other sectors—specialization, simplification, and concentration. It was widely assumed that these principles could be applied to agriculture as readily as they had been to other sectors of the economy. An additional incentive to urge the adoption of these principles was the need to reduce labor requirements on the farm in order to "free" people to pursue career goals unrelated to animal agriculture and to serve the needs of industrial manufacturing in urban areas.

The application of these industrial principles to agriculture seemed to work relatively well so long as key resources (cheap energy, surplus fresh water, and relatively stable climates) were available, and so long as unintended negative consequences, including environmental degradation, could be ignored. It is becoming increasingly clear that these key resources will soon dissipate and that we can no longer ignore industrial agriculture's unintended negative consequences. Hence, we will be forced to redesign our agriculture in the future, and animal agriculture will be an integral part of this transformation.

The industrial principles that currently shape animal agriculture were initially applied to crop agriculture, and were adopted aggressively after World War II. They were adopted in animal agriculture much later—essentially in the 1980s and 1990s. As one example of this transition in animal agriculture, the number of hogs produced between 1982 and 2002 increased in the United States by 10%, while the number of hog *farms* decreased by 76% (USDA/NASS, 2006). This is but one indication of the increased concentration in animal agriculture during this period.

Throughout the industrial era, modern farming systems have become increasingly specialized, simplified, and dependent on technological innovation. The principle objectives of this innovation were to increase labor efficiency, increase the yields of a few crop and animal species, and rely on control management to solve production problems. Like all other industrial economies, this strategy was heavily dependent on the availability of two primary reserves—inexpensive natural resources (principally cheap fossil fuels, abundant fresh water, and rich virgin soils) to sustain the system, and adequate natural sinks to absorb the wastes from the system. Since both of these resources are now in decline, agriculture is coming under pressure to design alternative systems.

NEW STRATEGIES: DIVERSITY

One of the new strategies under consideration is to re-introduce more diversity into future farming systems but to do so with creative new designs that incorporate the principles of ecology, evolutionary biology, and the science of networks. The primary question confronting this post-modern agriculture is whether a technology based on new synergies between crops, livestock, and other organisms can now replace fossil fuels and other one-dimensional technologies to increase productivity while at the same time beginning the necessary task of restoring the ecology of primary resources (soil, water, and biodiversity) essential to future productivity.

In an especially poignant essay, Masae Shiyomi and Hiroshi Koizumi state the issues succinctly.

> The development of agriculture in advanced countries from the 1950s to the 1970s occurred largely because of enormous increases in the use of fossil fuel energy. Specifically, it was supported by the increased use of fertilizers and agrochemicals, which are produced with fossil fuels, agro machinery that burns large amounts of fuel, and the breeding of new varieties of crops that are responsive to and compatible with such chemical inputs and cultural practices. ... The use of intra- and interspecific interactions and interactions between organisms and the environment, such as climatic factors and soils, are given little consideration in the current agricultural system. ...Modern agricultural practice has viewed these interactions as production constraints that must be overcome to make high production possible. Because the direct effects of fossil fuel energy and its products on agricultural production have been so powerful, reliable, and dramatic, little attention has been paid to the complex networks of biological interactions. ... [But] the present system of agriculture, which depends on consumption of tremendous quantities of fossil fuel energy, is now being forced to change to a system where the interactions between organisms and the environment are properly used. There are two reasons for this transformation. The first is the depletion of readily obtainable fossil fuel resources. The second is that consumption of fossil fuels has induced deterioration of the environment. (Shiyomi and Koizumi, 2001, pp. 1–2)

The environmental degradation to which Shiyomi and Koizumi refer will increasingly impose itself as a central issue for farmers. Farmers in the Mississippi Basin, for example, can no longer ignore the fact that one of the largest hypoxic zones, in the Gulf of Mexico, is largely due to the specialized, input-dependent farming practices in the basin. A significant increase in diversification will be essential to sufficiently scale back nitrogen releases in order to begin shrinking the dead zone. In addition, shrinking the zone is essential to the future health of aquatic life in the Gulf of Mexico. Recent studies indicate that in order to meet the target of reducing the hypoxic zone to 5000 km^2, set by the federal government at the turn of the twenty-first century, nutrient releases would have to be cut back by 40 to 45% (Raloff, 2004). No one believes that such a reduction is possible under current specialized monoculture farming systems.

David Tilman proposes another reason why post-modern farmers may find it necessary to shift from highly specialized monoculture farming systems to more diverse farming operations.

> Although owners of the businesses were probably shocked, I doubt if epidemiologists were surprised that Hong Kong chicken operations, housing up to a million genetically similar chickens, were susceptible to a rapid and devastating outbreak of disease last year. When those running massive livestock operations realize that chronic disease and catastrophic epidemics are the expected result of high densities and low diversity, and when society restricts the release of pollutants from such operations, it may again be profitable for individual farms, or neighborhood consortia, to have mixed cropping and livestock operations tied together in a system that gives an efficient, sustainable, locally closed nitrogen cycle. (Tilman, 1998, p. 212)

In 1946, Aldo Leopold had already articulated the ecological principles embedded in Tilman's perception. Leopold observed that

> The trend of animal ecology shows, with increasing clarity, that all animal behavior-patterns, as well as most environmental and social relationships, are conditioned and controlled by *density*. It seems improbable that man is any exception . . . I have studied animal populations for twenty years, and I have yet to find a species devoid of maximum density controls. . . . In all species one is impressed by one common character: If one means of reduction fails, another takes over. (Leopold, 1946, p. 225; emphasis added)

In other words, nature functions by the ecological principle of diversity and synergy. Through a long process of natural selection, this is the principle by which nature has developed the capacity for self-renewal. Moreover, the evolutionary dynamics of nature always seek to re-establish that synergy when the density of any one species threatens it. Leopold understood that modern industrial agriculture, which purposely introduced such species densities, was at odds with this ecological principle and that the principles of ecology needed to be introduced into agriculture, as well as conservation, if the self-renewing capacity of the biotic community was to be sustained.

> . . . there is urgent need of predictable ecology at this moment. The reason is that our new physical and chemical tools are so powerful and so widely used that they threaten to disrupt the capacity for self-renewal in the biota. This capacity I will call land-health. (Leopold, 1946, p. 219)

SYMPTOMS OF DISORGANIZATION

> The symptoms of disorganization, or land sickness, are well known. They include abnormal erosion, abnormal intensity of floods, decline of yields in crops and forests, decline of carrying capacity in pastures and ranges, outbreak of some species as pests and the disappearance of others without visible cause, a general tendency toward the shortening of species lists and of food chains, and a world-wide dominance of plant and animal weeds. With hardly a single exception, these phenomena of disorganization are only superficially understood. (Leopold, 1946)

Highly specialized, species-dense monocultures are, in other words, very brittle and very vulnerable to environmental perturbations. As the capacity for self-renewal of such systems has diminished, and as farmers rely on one-dimensional, single-tactic technologies to maintain productivity, the system has become increasingly costly to operate. Even with intense, relatively cheap fossil fuel-based technologies to sustain the system, farmers on average retain little to no net income (see Figure 11.1).

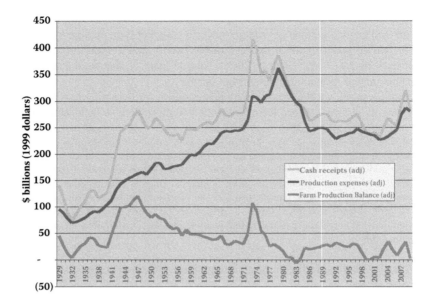

FIGURE 11.1 (See color insert)   Farm production balance in the United States, 1929–2007. (From USDA/ ERS. Chart by Ken Meter, 2009.)

Joe Lewis, pest management specialist, and his colleagues, with the Agriculture Research Service, have recognized the core problem with this approach, not only with respect to pest management, but also with virtually all human enterprises—"the attempted solution becomes the problem" (Lewis, van Lenteren, Phatak, and Tumlinson, 1997). Lewis points out that in pest management, as in other systems, the basic principle for managing undesirable variables is one of applying a direct external counterforce against it. However, that approach, he argues, will only secure short-term relief because within diverse, dynamic ecosystems such strategies are always met by "countermoves that 'neutralize' their effectiveness."

SOLUTION

The solution, according to Lewis, is to develop "farming practices that are compatible with ecological systems" and to design "cropping systems that naturally limit the elevation of an organism to pest status." Lewis suggests that we have ignored the inherent capacity of nature to keep pests in check and to make farming more profitable for farmers.

> We historically have sold nature short, both in its ability to neutralize the effectiveness of ecologically unsound methods as well as its array of inherent strengths that can be used to keep pest organisms within bounds. If we will but understand and work more in harmony with nature's checks and balances we will be able to enjoy sustainable and profitable pest management strategies, which are beneficial to all participants in the ecosystem, including humans. (Lewis et. al. 1997, p. 12248)

Lewis goes on to point out that such alternative farming practices will require the introduction of more diversity into the system. At a minimum, farming systems must include the habitat that can "provide the important refugia for developing natural enemy/pest balances" (Lewis et. al., 1997).

Research that is more recent has confirmed Lewis' observations. The results of two studies reported in the July 1, 2010 issue of *Nature* magazine demonstrate the advantage of increased diversity for achieving more effective pest control. Single-tactic, therapeutic intervention strategies to control pests tend to "disrupt the communities of those natural enemies—which, in turn, provide less effective pest control" (Turnbull and Hector, 2010). Furthermore, "intensification of farming can drastically distort the relative-abundance distributions of natural enemy communities in favour of a few dominant species" (Crowder, Northfield, Strand, and Snyder, 2010), which increase pest pressures.

Economic data now confirm that the highly specialized farming systems so endemic to industrial farming systems have failed farmers economically. Despite the initial appeal of seemingly quick-fix solutions to pest and other production problems, and despite the obvious labor efficiency achieved through highly specialized systems, farmers find themselves on technology (Cochrane, 1979) and pesticide (van den Bosh, 1978) treadmills that have contributed to their economic malaise. Due to the rapidly increasing expenses of these monoculture systems, net farm income is now lower in both Canada and the United States than it was in 1929 despite a sevenfold increase in gross income (USDA/ERS). As farmers are driven out of business, the rural communities that depend heavily on local agriculturally related economies also decline. Subsequently, the public services on which farmers depend for their own economic health—public roads, schools, and other services—begin to deteriorate, placing additional economic burdens on farmers (Ettner, 2010).

As the cost of fossil fuels increases (due to the increased expenses of extracting such depleting resources), as climate change causes greater instability, and as agriculturally related environmental degradation becomes increasingly visible, and therefore intolerable to the public at large, the pressure to develop an alternative to specialized, industrial agriculture will increase.

NEW FOCUS: REPLACE CURRENT FOSSIL ENERGY TECHNOLOGIES TO ENHANCE AGRICULTURAL PRODUCTION

It may be critical, therefore, that we now focus a sufficient portion of our agricultural research agenda on answering the crucial question that Shiyomi and Koizumi raised: "Is it possible to replace current technologies based on fossil energy with proper interactions operating between crops/livestock and other organisms to enhance agricultural production?" (2001, p. 6).

The future of animal agriculture and the issue of animal welfare now need to be explored in this new context. Perhaps the prescient wisdom of Sir Albert Howard, urging us to use nature as the model for our agriculture, will now finally appeal to us as never before. As our reserves of cheap energy, surplus water, and stable climates disappear, we will have to look for new models to redesign agriculture. Highly specialized, simplified, concentrated forms of agriculture that require excessive quantities of cheap energy, fresh water, and stable climates will become increasingly dysfunctional in our new world. Consequently, adhering to the "main characteristics of Nature's farming" may serve as a useful guide to design a future sustainable agriculture which will incorporate a more humane and essential animal component in the system. As Howard put it so succinctly:

> Mother earth never attempts to farm without live stock; she always raises mixed crops; great pains are taken to preserve the soil and to prevent erosion; the mixed vegetable and animal wastes are converted into humus; there is no waste; the processes of growth and the processes of decay balance one another; ample provision is made to maintain large reserves of fertility; the greatest care is taken to store the rainfall; both plants and animals are left to protect themselves against disease." (Howard, 1943, p. 4)

ETHICAL PRINCIPLES

There are obvious ethical principles involved in this vision for agriculture. This transformation would require what Leopold called the development of an "ecological conscience," which "reflects a conviction of individual responsibility for the health of the land. Health is the capacity of the land for self-renewal" (Leopold, 1949). A new agriculture designed along these principles would be more diverse, animals would be integrated into the landscape in numbers that are appropriate to the self-renewing capacity of the land, there would be more perennials, and animals would be able to perform their normal functions out on the landscape. Such a redesigned agriculture would likely go a long way toward achieving animal welfare objectives that have been articulated for centuries.

SUMMARY

One suspects that most of us have not yet fully comprehended the scope of the changes that are in store for our food and agriculture systems as we transition from an industrial economy to an ecological economy. The end of cheap energy, climate destabilization, and the depletion of fresh water resources are but three of numerous changes that will likely drive that transformation. How we manage animal agriculture will be part of that more comprehensive transition. Biological synergies will likely replace many of our current energy intensive inputs in the new designs of future agriculture, and animals, integrated into creative new designs that simultaneously address issues of energy conservation, resource depletion, environmental degradation, and animal welfare. Numerous models already exist (Kirschenmann, 2007). In her recent book, *The End of the Long Summer*, Dianne Dumanoski suggests that we are, in fact, at "a fundamental turning point in the relationship between humans and the Earth, arguably the biggest step since human mastery of fire" (Dumanoski, 2009).

It is not hard to imagine some of the transformations that these shifts will have on animal agriculture. It will be difficult to maintain large numbers of animals in concentrated, confinement facilities when crude oil reaches $200 or $300 per barrel. By most estimates, when crude oil hit $147 per barrel in 2007, confinement hog operations were reportedly losing over $20 per hog due to increased feed and other costs. Animals that were managed in multi-species, intensive rotational grazing systems, in which creative biological synergies, like those on Joel Salatin's farm, were the principal management strategy, had a clear competitive advantage in such high-energy input cost circumstances. More diverse, integrated, crop/livestock systems will likely move toward more complex, smaller operations or neighborhood consortia rather than uniform, single species, concentrated operations, since diverse systems tend to be more knowledge intensive and require more on-site management, all of which will likely transition livestock operations to more decentralized, grass-based systems. While such transitions will not automatically guarantee humane animal treatment, they lend themselves much more to systems where animals can perform their natural functions.

REFERENCES

Cochrane, W. 1979. *The Development of American Agriculture*. Minneapolis: University of Minnesota Press.
Crowder, D.W., T.D. Northfield, M.R. Strand, and W.E. Snyder. 2010. Organic agriculture promotes evenness and natural pest control, *Nature*, 466: 109–112.
Dumanoski, D. 2009. *The End of the Long Summer*. New York: Three Rivers Press.
Ettner, L. 2010. Roads to ruin: Towns rip up the pavement: Asphalt is replaced by cheaper gravel, *The Wall Street Journal*, July 17.
Howard, A. 1943. *An Agricultural Testament*. New York: Oxford University Press.
Kirschenmann, F. 2007. Potential for a new generation of biodiversity in agroecosystems of the future, *Agronomy Journal*, 99: 373–376.
Leopold, A. 1946. The land-health concept and conservation. In: *For the Health of the Land*. J.B. Callicott and E.T. Freyfogle, Eds. Washington, DC: Island Press, pp. 218–226.
Leopold, A. 1949. *A Sand County Almanac*. New York: Oxford University Press.
Lewis, W.J., J.C. van Lenteren, S.C. Phatak, and J.H. Tumlinson, III. 1997. A total system approach to sustainable pest management, *Proceedings of the National Academy of Sciences*, 94: 122243–12248.
Raloff, J. 2004. Massive oxygen-starved zones are developing along the world's coasts, *Sciences News Online*, 165(23): 6.
Shiyomi, M., and H. Koizumi, Eds. 2001. *Structure and Function in Agroecosystem Design and Management*. New York: CRC Press.

Tilman, D. 1998. The greening of the green revolution, *Nature*, 396: 212.

Trumbull, L.A., and A. Hector. 2010. How to get even with pests, *Nature*, 466: 36–37.

USDA/Economic Research Service/ U.S. and State Farm income Data. Available on the ERS Web page at: http://www.ers.usda.gov/Data/FarmIncome/finfidmu.htm#farmnos

USDA/NASS. 2006. Census of Agriculture and Historical Highlights. Washington, DC.

van den Bosch, R. 1978. *The Pesticide Conspiracy*. New York: Doubleday.

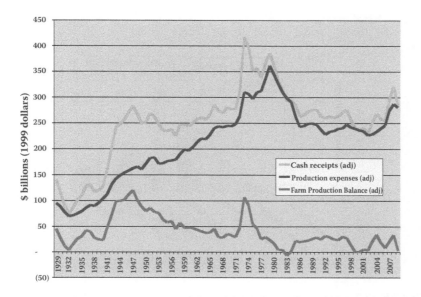

COLOR FIGURE 11.1 Farm production balance in the United States, 1929–2007. (From USDA/ERS. Chart by Ken Meter, 2009.)

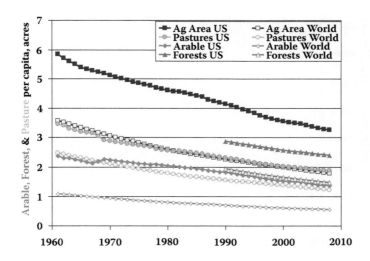

COLOR FIGURE 12.1 Per capita availability of arable, pasture, and forestland in the United States and the world during the past century. (From Food and Agriculture Organization, 2010.)

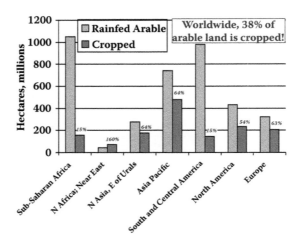

COLOR FIGURE 12.2 Arable and crop land worldwide and by region. (From Food and Agriculture Organization, 2000. *World Soil Resources Report 90.*)

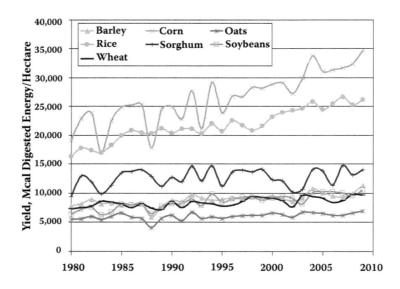

COLOR FIGURE 12.3 Capture of digestible energy per hectare in crops harvested from various cereal grains and soybeans in the United States from 1980 to 2009.

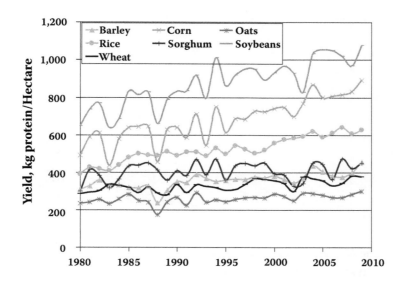

COLOR FIGURE 12.4 Protein yields per hectare from various crops in the United States from 1980 to 2009.

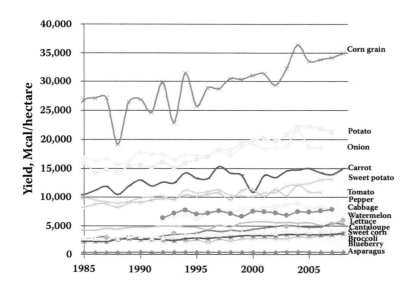

COLOR FIGURE 12.5 Energy capture in edible megacalories per hectare by corn grain, various fruits and vegetables, and berries from 1985 to 2008. (From USDA/ERS, 2010; USDA/ARS, 2010.)

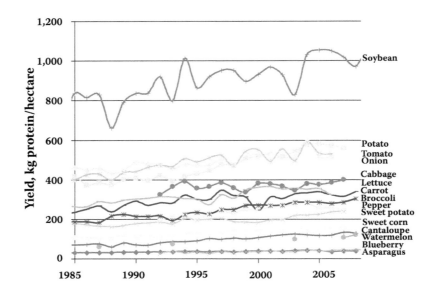

COLOR FIGURE 12.6 Protein yields per hectare for food crops and soybeans from 1985 to 2008. (From USDA/ERS, 2010; USDA/ARS, 2010.)

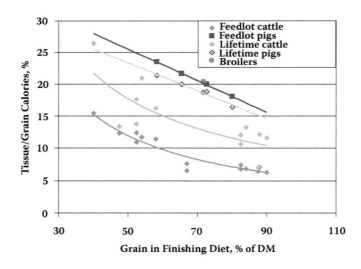

COLOR FIGURE 12.7 Feedlot phase versus lifecycle caloric efficiencies for various animal products. (Based on data from Whitney et al. (2006), *J. Anim. Sci.* 84: 3356–3363; Bremer et al. (2007), *2008 Nebraska Feeders Report*, pp. 39–40; Dozier et al. (2007), *J. Appl. Poult. Res.* 16: 206–218; and Klopfenstein et al. (2008), *J. Anim. Sci.* 86: 1223–1231.

12 Competition between Animals and Humans for Cultivated Crops
Livestock Production and our Food Supply

Fred Owens and Christa Hanson

CONTENTS

INTRODUCTION

> Worldwide, people die because affluent individuals consume foods of animal origin (meat, milk). Feeding animals is wasteful; using food and our scarce land resources that could be used to produce food for people. Ultimately, consumption of animal products constitutes misuse of the earth's resources and leads to abuse not only of animals, but also of starving humans worldwide.

Slightly paraphrased, that is the message advocated by Francis Lappe in her widely marketed 1971 (updated in 1991) book *Diet for a Small Planet* that is repeatedly iterated by critics of production and consumption of livestock products. Surprisingly, these concepts have gone largely unchallenged.

Yet, this scenario is based on several inherent assumptions that need closer attention and scientific scrutiny. Does the worldwide supply of arable land limit food production? Is livestock production limited by the amount of arable land? Can land used for production of livestock feeds be converted readily to produce food for humans? Do all the calories fed to livestock come from cereal grains and oilseeds? Are products unsuitable for human consumption fed to livestock? Are food and feed crops equally efficient in production (yield per hectare) of calories, protein, and other essential nutrients? How efficiently do livestock convert dietary calories and protein from feeds into edible products? Are large-scale livestock production units inefficient and irresponsible? In this chapter, these questions will be addressed in an attempt to appraise whether increases in food prices and worldwide starvation should be blamed on production and consumption of livestock products.

DOES THE WORLDWIDE SUPPLY OF ARABLE LAND LIMIT FOOD PRODUCTION?

Millions of hectares of land worldwide are too hot, too cold, too wet, too dry, too steep, too rocky, or inaccessible for raising crops. An additional one-half of the land area on Earth consists of meadows, pastures, forests, and woodlands (Table 12.1). This leaves only about 13% of the total land area on Earth available for crop production. Land suitable for crop production is called "arable." As a percentage of the total land area, arable land varies among countries from under 0.01% (Iceland, Djibouti) to over 55% in the Ukraine, Moldova, and Bangladesh (Nationmaster, 2005).

Food is needed to support a world population that is increasing at a cumulative rate of 1.1% per year; fortunately, this rate is decreasing and a continued decrease is projected (U.S. Census Bureau, 2010). Within the 48 contiguous states of the United States, special uses (roads, railroads, parks, defense, and urbanization) occupied 11% of the total land area, a fraction increasing at a rate of 0.1% per year (Lubowski et al., 2002). Additional cropland is being lost due to erosion and other forms of land degradation. As outlined by Malthus centuries ago, growth in the human population decreases the per capita availability of land for crop production, pasture, and forests (Figure 12.1).

IS ALL OF THE ARABLE LAND AVAILABLE CURRENTLY BEING USED FOR CROP PRODUCTION?

Based on data from 1994 (FAO, 2000) and shown in Figure 12.2, cropland actually exceeds arable land in North Africa and in the Near East thanks to irrigation. Averaged across the world, only 38% of the arable land currently is cropped, and arable land in more tropical regions of the world

TABLE 12.1
Land Use in the United States and Worldwide in 2005

Land use	Area (hectares/person)		Percentage of land	
	U.S.	World	U.S.	World
Total land area	3.02	2.01	—	—
Arable land	0.59	0.27	19.51	13.18
Forest land	0.74	0.59	24.70	29.40
Meadows, pastures	0.72	0.52	24.00	25.90
Permanent crops	0.02	0.03	0.56	1.53
Irrigated land	0.08	0.06	2.50	2.94

Source: Central Intelligence Agency. (2009). *The World Factbook.* https://www.cia.gov/library/publications/the-world-factbook/fields/2097.html. Accessed April 2010.

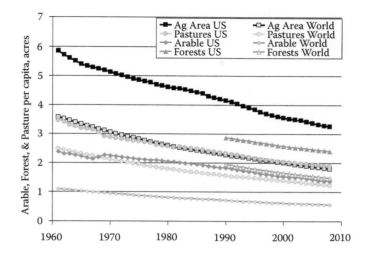

FIGURE 12.1 (See color insert) Per capita availability of arable, pasture, and forestland in the United States and the world during the past century. (From Food and Agriculture Organization, 2010.)

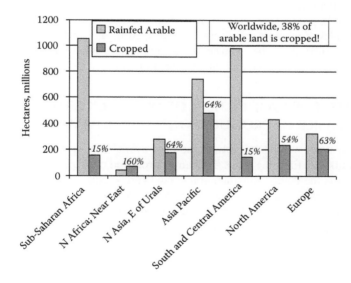

FIGURE 12.2 (See color insert) Arable and crop land worldwide and by region. (From Food and Agriculture Organization, 2000. *World Soil Resources Report 90.*)

will yield more than one crop each year. Based on these estimates of the arable land supply, crop production could be increased markedly in sub-Saharan Africa and South and Central America, where only 15% of arable land currently is being cropped. In other regions of the world, one-half to two-thirds of arable land is cropped. Arable land may not be cropped for several reasons. These include inadequacy of an infrastructure to transport and market crops, insufficient financial return on land and crop investments, trade restrictions or tariffs, lack of agronomic inputs necessary to enhance crop production, and political instability. As such limitations are resolved and the demand for and price of food and feed products increase, this untapped production potential certainly can expand the worldwide supply of food and feed. Nevertheless, based on these data, worldwide crop production at present is not limited by the amount of arable land available for crop production.

IS LIVESTOCK PRODUCTION LIMITED BY THE AMOUNT OF ARABLE LAND?

Because non-ruminant animals (poultry, swine, and humans) are incapable of digesting feeds rich in fiber, non-ruminants in developed countries generally are fed diets composed of cereal grains and protein supplements that, in turn, are derived from arable land. In developing countries, non-ruminants often are fed diverse by-products and wastes unsuitable or not desired for human consumption. In contrast with poultry and swine, herbivores (non-ruminants and ruminants that include cattle, sheep, and goats) in both developing and developed nations harvest grass and weeds while foraging on large expanses of non-arable land. Meadows and pastures that comprise approximately 26% of the global land area and a portion of the forested land, an additional 29% of the worldwide land area (Table 12.1), usually is considered suitable for grazing livestock. Although forests must be clear-cut for planting crops, forestland need not be cleared for grazing; retaining some trees can enhance pasture and livestock productivity as reviewed by Belsky, Mwonga, and Duxbury (1993). In addition, the fibrous residues from cereal grains and oilseed crops produced from arable land can be fed to and digested by domesticated herbivores. Through converting unused or underutilized fiber-rich resources into milk and meat and through grazing forages from land areas inaccessible for harvest, herbivores can enhance both the energy and nutrient supply for humans. Similarly, feeding food residues and wastes to non-ruminants results in a net increase in the supply of calories and protein for humans. As an example, in New Zealand in 2009, pastures of non-arable land grazed by cattle served as the primary source of both calories and protein for 5.8 million dairy cows that yielded milk and milk products to not only feed their population of 4.3 million, but also to export to other countries (New Zealand Agricultural Statistics, 2010). Through grazing, herbivores remove surface forage and brush that, when allowed to accumulate, provide fuel for devastating wildfires that often destroy scenery, property, and human and animal life.

CAN LAND USED FOR PRODUCTION OF LIVESTOCK FEEDS BE CONVERTED READILY AND EASILY TO PRODUCE FOOD FOR HUMANS?

Season length, soil type, weather conditions, and the availability of water, equipment, labor, and markets all can markedly limit the crops that can be produced on a specific plot of land or in a specific area. This limits the degree to which arable land can be converted from one crop to another. For example, all rice produced within the United States is grown on irrigated land (Smith, 2001). Vast tracts of non-irrigated land in the "corn belt" currently producing cereal grains could not be converted to grow rice productively without extensive land modification and costly investments even if such a change were feasible agronomically. Crops differ markedly in their tolerance of environmental temperatures and season length, so the feasibility of converting arable land from one crop to another is limited.

Traditionally, family farms produced a variety of crops and livestock species. Because of the factors listed previously, land type and environmental conditions must be matched with crop requirements if productivity is to be maximized. Land used to grow crops unsuited for the soil or environment reflects an inefficient use of available resources just as land that stands idle and does not produce a crop. Similarly, using arable land as pasture for grazing cattle represents inefficient use of valuable land resources. Efficiency of food production is enhanced by trade on a local, national, and international basis. Consumption of foods produced locally, although espoused by individuals attempting to reduce the amount of energy used to transport goods and materials, often results in inefficient use of available resources. A century ago when the United States was settled, diets for the winter months in temperate areas consisted of dried fish or meats, potatoes, and home-canned vegetables produced in the garden. All foods were obtained at home or nearby. Although such diets may have a nostalgic appeal, returning to the use of only locally produced foods and products is unlikely to satisfy today's consumers in developed countries where markets currently provide consumers with a very wide choice of out-of-season fruits and vegetables, specialty breads, coffees, and livestock products amassed from distant parts of the globe. Today products are designed, trimmed, pre-processed, and packaged to meet the desires of consumers for taste, composition, quality, and convenience, all at an additional cost to consumers.

DO ALL THE CALORIES FED TO LIVESTOCK COME FROM CEREAL GRAINS AND OILSEEDS OR FROM PRODUCTS UNSUITABLE FOR HUMAN CONSUMPTION?

Beyond the surpluses of crops produced on arable land and the pastures and forages produced on non-arable land, vast quantities of by-products are fed to or consumed by livestock. With most crops, less than one-half of the aboveground biomass produced is suitable for consumption by humans. This leaves a substantial amount of residues that can be harvested by livestock, fed to livestock, or returned to the soil. By-products also are generated during grain processing and conversion of crops into foods, beverages, and fiber for human use. During production of corn grain, for example, only one-half of the dry matter of the mature plant is grain. The remaining half, consisting of stalks, leaves, and husks, can be grazed or harvested and fed to livestock as silage when the grain is included to yield corn silage or it can be stored and fed separately from grain as stover silage. During industrial conversion of grain to starch, corn sweeteners, or alcohol and other products for human use or production of fuel, an additional diverse stream of by-products (hominy feed, corn gluten feed, corn bran, corn gluten meal, brewers' grains, and distillers' grains) is generated that typically comprises over 30% of the processed grain. When grains are used for making chips or baking, additional by-products are created. Foods and baked goods beyond their expiration date also are fed to livestock. DePeters et al. (2000) published a list of 17 by-products commonly fed to dairy cattle in California. Likewise, Bath et al. (2001) compiled nutrient composition data for dozens of by-product feeds commonly fed to livestock. Were such products not fed to livestock, some other means of handling and disposal would be needed. Finally, food wastes are fed to animals. In 2002, 49% of the pigs on earth were in China (FAO, 2002) thriving largely on crop and industrial by-products and food waste. The fact that the population of pigs in China is equal to 75% of the human population illustrates how a large animal population can co-exist viably with a large human population. Within developed countries, the population density of domestic animals varies regionally depending on the availability of feed resources and market demand. For example, averaged across the United States, only one pig (*Suis domestica*) exists for every 4.6 people, but within Iowa, pigs outnumber people by 4.6 to 1 (as often becomes apparent downwind). The dependence of livestock on by-products and on other waste materials would be expected to increase in the future as competition for grain for other purposes (e.g., biofuels, combustion to generate electricity) increases the cost of grain.

Widespread availability of economical by-products increases their use in livestock diets. Most high concentrate feedlot diets a decade ago contained 80% grain, but today up to one-half of that grain has been displaced with by-products of biofuel production (e.g., distillers' grains plus solubles). Through converting pastures, forages, and by-products that otherwise would be wasted, livestock production increases the supply of food available for humankind.

ARE FOOD AND FEED CROPS EQUALLY EFFICIENT IN PRODUCTION (YIELD PER HECTARE) OF CALORIES, PROTEIN, AND OTHER ESSENTIAL NUTRIENTS?

Life on earth relies on photosynthesis of the past, present, and future. Efficiency of converting light energy to biomass energy for most cereal crops falls between 1 and 2%. The C4 type of crops (sugarcane, corn, and pineapple) trap nearly two-thirds more solar calories than C3 plants. Energetic efficiency of photosynthesis (conversion of solar energy to plant energy) as high as 8% has been reported for cane plants although only a portion of that energy is retained as sucrose (Govindjee and Govindjee, 2000). Efforts are under way to genetically modify wheat and rice to increase the photosynthetic efficiency of those crops.

Return in terms of calories per hectare also differs among crops due to differences in inherent genetic potential, weather conditions, water availability, and crop management (fertilization, irrigation, and control of pests, weeds, and diseases). These management factors often limit crop production in developing countries. Yield of a crop grown under ideal agronomic conditions should provide an index of the relative genetic capability of that crop available currently.

For comparison among cereal grains and soybeans in terms of yield of calories and protein, one might presume that crops grown in the United States in past years should be produced under agronomic and management conditions that should approach being "ideal." As an estimate of potential yields from various cereal crops, yields in the United States were compiled from the USDA-ERS (2010) database for the past 30 years. These crop yields from various cereal grains and soybeans were converted to megacalories of metabolizable energy by multiplying yield per hectare by the caloric content of products based on available (metabolizable) energy content of various grains (NRC, 1998). Values are shown in Figure 12.3.

Note that for corn grain, the yield of calories per hectare was more than twice that of other cereals with the exception of rice. Furthermore, the increase in the yield of calories alone during the past 30 years from rice and corn exceeds the total calorie yield from most other crops! Linear regression of the yield of digestible energy from 1980 to 2009 provides an estimate of the percentage increase each year. For these crops, the average annual increase has been positive: Corn (2.17%), rice (1.75%), soybeans (1.75%), barley (1.12%), wheat (0.81%), oats (0.69%), and sorghum (0.42%). The substantial yield increases for corn and soybeans, feed resources for livestock, bode well for the future of animal production. However, the slow increase in the yield of calories from sorghum grain is disconcerting considering that among these cereal crops, sorghum is most resistant to drought and thrives in regions with very limited rainfall and water availability. In addition to differences among cereal crops in their need for water (less for sorghum and wheat than for corn or rice), these crops also differ in their need for supplemental nitrogen (N) fertilizer, being much less for soybeans and other legumes due to the capacity of synergistic bacteria associated with legume roots to fix N from the air. When selecting a crop to plant, farmers must consider not only

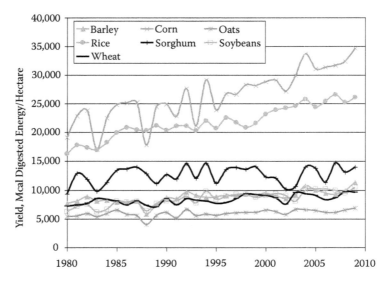

FIGURE 12.3 (See color insert)   Capture of digestible energy per hectare in crops harvested from various cereal grains and soybeans in the United States from 1980 to 2009.

crop yields (amount and consistency) and crop value, but also total economic return per hectare to pay input costs and support a family and laborers. Return will differ with numerous input costs as well as adaptation of the crop to regional and temporal environmental conditions. Relative risk of crop failure also differs among crops. Biotechnological advances have been achieved through traditional plant breeding and through genetic modifications that reduce insect damage, the cost of weed control, and the plant's need for water and fertilizer. These modifications cannot only reduce input costs, but also allow crops to be produced in regions or under conditions previously unsuited for that crop, potentially increasing the amount of land suitable for production of that crop. On a worldwide basis, productivity of cereal grains also has been increasing steadily. Except for four crops (corn grain, sorghum grain, peanuts, and rice), crop yields averaged across all countries in the world are currently surprisingly similar (70 to 139%) to production rates within the United States (USDA-FAS, 2011). For the four specific crops noted, however, worldwide production over the past 5 years has averaged only 39, 37, 43, and 53%, respectively, of that in the United States, probably due to greater application of genetics and biotechnology and additional agronomic inputs within the United States and the inherent responsiveness of these crops to selection and agronomic inputs.

In addition to calories, protein components (essential amino acids) are required for growth and maintenance of animals and humans. Protein return per hectare from various crops and soybeans was calculated in the same manner as for calories (Figure 12.4).

Protein yield per unit of arable land is greater for soybeans than for cereal grains primarily due to the high protein content of soybeans. Linear regression of yield of protein against year (from 1980 to 2009) gives an estimate of the yearly increase. Averaged across this 29-year period, yearly increases in protein yields were positive for soybeans (1.75%), corn (2.17%), rice (1.75%), sorghum (0.42%), barley (1.11%), wheat (0.81%), and oats (0.69%). These increases are almost identical to the changes in yield of calories. Parallel increases in yield of calories and protein are expected unless protein content of the crop changes.

Among these cereal grains, protein content is higher for wheat and barley than for other cereal grains. The value of a protein source must consider not only its protein content, but also its quality (limiting amino acids and balance among amino acids). Shortages of essential amino acids (often

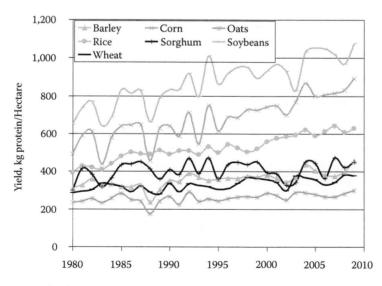

FIGURE 12.4 (See color insert) Protein yields per hectare from various crops in the United States from 1980 to 2009.

lysine and tryptophan) limit the quality of proteins from commercially produced cereal grains. In contrast, proteins present in animal products and soybeans provide a well-balanced complement of amino acids. Based on the Protein Digestibility Corrected Amino Acid Score (PDCAAS), values for humans for the protein from casein, egg, whey, and soybeans are all near 100, the maximum value possible, whereas values of proteins from other crops and foods is lower (legumes, 70; beef, 92; fruits, 76; vegetables, 73; , cereals, 59; wheat, 42) (Schaafsma, 2000). Because of shortages of specific essential amino acids or total protein, diets for humans and animals based on cereal grains must be supplemented with proteins from other sources in order to compensate for their amino acid shortages for optimum growth, performance, and health.

For growth and maintenance, a minimum quantity of each of the individual essential amino acids is required (grams per day). The PDCAAS of a source of protein or a diet serves as an index of how well the ratios of amino acids within a protein or diet match the ratio of individual amino acids required. Amino acid requirements can be met by providing a very large excess of protein from a source with a low biological value or by a small amount of protein with a high biological value. Because protein sources can differ in their first limiting amino acid, the PDCAAS for different protein sources is not additive, but can be synergistic. By combining protein sources, one protein source can complement another so that the combination has a higher PDCAAS than the average for each of the two protein sources. Because the proteins present in cereal grains, leafy vegetables, and fruits are low in lysine and tryptophan as well as total protein, protein sources rich in lysine and tryptophan and richer in total protein content are needed to complement such foods or feeds. Protein sources that are particularly rich in lysine and tryptophan as well as in total protein content include animal products (meat, milk) and certain legumes (various beans). Although animal source proteins are convenient and have been used for centuries to complement plant protein sources, beans can be substituted for animal proteins in vegetarian diets. Indeed, proteins derived from soybeans including isolated soy protein, tofu (a soy precipitate), and fermented soy products (meso, tempeh) are used widely as a substitute for proteins of animal origin in infant formulas for milk in developed countries and as a supplement for grain-based diets in developing countries. The efficiency of converting solar energy directly to these well-balanced plant proteins theoretically should be considerably greater than expecting animals to convert various plant proteins to animal products with a high PDCAAS. On this basis, precipitated or fermented soy products have been touted widely as being ideal for displacing animal protein sources in diets for humans. Unfortunately, several factors currently limit the acceptance of soy-based protein sources by humans. These include the presence of anti-nutritional compounds, adverse taste components, and high commercial cost. The organoleptic issues and other limitations (presence of estrogenic isoflavones, protease inhibitors, and phytic acid) in soy products can be alleviated largely through selection of specific soy cultivars and modified industrial processes. For example, fermentation of soy products can reduce or eliminate protease inhibitors and hemaglutinin. Marketed tofu and its derivatives typically contain over 87% water and on a wet matter basis equivalent have only about one-third of the protein in ground beef or pork. Costs per unit of protein, not per unit of food, need to be considered when comparing various sources of protein. In 2011, the cost per unit of protein from available tofu and soy products in supermarkets in the United States was about 1.7 times the cost of protein from ground beef or pork. Whether this high price differential between soy protein products and protein sources of animal origin will decrease in the future if tofu is produced on a larger scale is uncertain. Tofu and soy products are more extensively marketed and consumed in countries where protectionist trade policies restrict the importation of animal products so that the availability of animal products is limited and higher in price.

HOW EFFICIENT IS PRODUCTION OF VARIOUS FOOD CROPS?

Some of the arable land currently used for crop production could be used to produce vegetables, fruits, and other food products that are edible at harvest or readily converted to products to be consumed directly by humans; hereafter, designated "food crops." When arable land is used to produce various food crops frequently consumed by humans, what return in calories can one expect per hectare? Again, yields of calories and protein from various food crops within the United States were calculated. Considering that agronomic conditions should be optimum under production conditions in the United States, these relative values should reflect the current genetic merit of various crops. Crop yields, calculated from acreage and production of various food crops in the United States (USDA-ERS, 2010), were combined with calorie content for each crop derived from USDA-ARS (2010) data tables for energy and nutrient content to calculate yields in terms of megacalories of metabolizable energy as illustrated in Figure 12.5. Corn grain energy yields from Figure 12.3 provide a comparison.

Only two crops, potatoes and onions, retained as much as one-half of the energy in food for human consumption as the corn plant deposits merely in corn grain. Among the vegetables, the root crops (potatoes, onions, carrots, and sweet potatoes) had considerably greater retention of edible energy per hectare of arable land than vegetable crops, sweet corn, or blueberries. Like these root crops for which information is available, other root crops (cassava, beets) might be expected to retain more solar energy in edible products than food crops derived from the aerial portion of plants. The yield of calories is considerably lower from sweet corn than field corn due to harvest at a very immature stage and low grain yields. The fact that yields of calories for food crops are all considerably lower than for corn grain illustrates how diversion of cropland from grain production to vegetable production would markedly reduce the amount of solar energy captured in edible products. Expansion of the land area that is used for production of either feed or food crops would also be expected to decrease productivity because the land areas that are added likely will be less suited climatically or agronomically for production than land in regions currently being farmed. Regression of yield per hectare since 1985 was used as an index of the rate of increase in productivity of these crops. Again, productivity changes over time for various crops were positive (see Figure 12.5). That yields of all crops showed an upward trend is encouraging. However, despite these substantial yearly increases in food crop production, yield of total calories per hectare remains much lower for these crops consumed directly by humans than for most cereal grains.

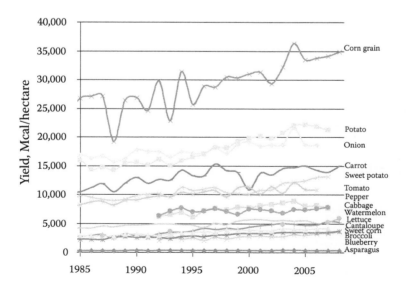

FIGURE 12.5 (See color insert)   Energy capture in edible megacalories per hectare by corn grain, various fruits and vegetables, and berries from 1985 to 2008. (From USDA/ERS, 2010; USDA/ARS, 2010.)

In contrast to management and harvest procedures for cereal grains, most steps in production and harvest of vegetables, fruits, and nut crops are not readily mechanized. Instead, management and harvest of vegetables and fruits typically require extensive and expensive seasonal manual labor. Further, in contrast with cereal grains that are readily stored for later use, vegetables and fruits high in moisture content typically are fragile and have a short "shelf life." Certainly, food crops at local and seasonal farmer's markets have consumer appeal in terms of freshness and credence attributes. However, in areas of the world where rainfall is seasonal and winter temperatures are cold, either long-term storage or importation is required if vegetables and fruits are to be consumed beyond their season of production. Most vegetables require dehydration or special handling and conditions for storage (e.g., refrigeration, canning, and freezing). Even in developing countries located in tropical climates, particularly in isolated rural areas, lack of appropriate storage facilities obviates dependence on fresh food crops as a yearlong source of calories and nutrients. These same factors complicate and increase the cost of handling and transporting food crops, particularly for produce that is exported. Distance and time from markets often limits the physical locations where vegetable crop farms can be sited. The concept of basing a diet on the minimum "food miles" to reduce the carbon footprint may be more of an illusion than a fact (Capper et al., 2009). As compared with fruits and vegetables, dry cereal grains are readily handled, transported, and stored for years or even centuries when protected from insects and rodents. Hence, distance, storage conditions, and time limitations are less restrictive for dry cereal grains than for most food crops. Although grazing of livestock in temperate regions also is seasonal, feeding of harvested forages and grains to livestock allows milk and meat production to continue uninterrupted throughout the year and avoids seasonal fluctuations in the supply and price of foods for humans. Yet, the need for refrigeration or freezing for longer-term storage of animal products can limit availability of meat and milk in developing countries. However, dehydration, preservatives, and, for milk, ultra-high temperature treatment can help to reduce the need for specialized storage facilities. Through converting cereal grains to milk and meat, the impact of local crop failures on availability of feed for livestock is cushioned readily by transport of grains or by-products from areas of surplus to areas of shortage.

Protein yields from various food crops relative to soybeans calculated as above for the past 23 years are presented in Figure 12.6.

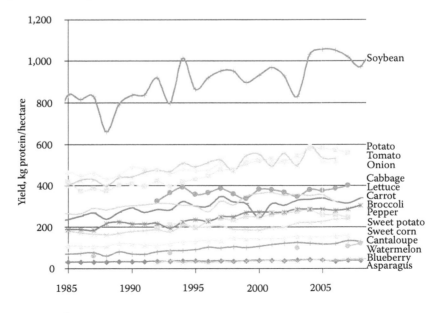

FIGURE 12.6 (See color insert)   Protein yields per hectare for food crops and soybeans from 1985 to 2008. (From USDA/ERS, 2010; USDA/ARS, 2010.)

Relative to soybeans, protein yields from food crops ranged from only 4% for asparagus to 55% for root crops and tomatoes (Figure 12.6). As with calories, yields of protein from crops fed to livestock typically are double those of food crops, so conversion of arable land used for production of grain and oilseed to production of food crops consumed directly by humans would markedly reduce both protein and calorie output from crops. A portion of this greater yield of calories and protein can be ascribed to successes by plant breeders in enhancing productivity of specific cereal grains and oilseeds, but inherent differences in photosynthetic capacity and efficiency and in plant productivity probably are involved, as well.

HOW EFFICIENTLY DO LIVESTOCK CONVERT DIETARY CALORIES AND PROTEIN FROM FEEDS INTO EDIBLE PRODUCTS?

In developed countries, approximately one-third of calories in the human diet are derived from animal products (FAO, 1995). Substantial amounts of certain cereal crops (e.g., sorghum, corn, barley, oats, millet, and rye) are fed to animals, but feeding of several "food" grains (wheat and rice) to livestock is rare unless other "feed" grains cannot be readily or economically sourced. Yet, in certain countries and regions, like the United Kingdom, poultry diets often contain wheat due to its availability and trade restrictions. Selection of specific feeds for formulation of animal diets is based first on the availability of various homegrown materials (crop by-products, pasture, and forages). These are supplemented as needed with additional energy, amino acids, minerals, and vitamins. Livestock producers who grow crops must determine whether homegrown products should be fed to livestock or if those feeds and livestock should be sold to markets outside the farm or ranch. Forage-based diets usually require supplemental energy, protein, vitamins, and minerals. For poultry and swine in the United States, 50 to 60% of the total cost of production is attributed to the cost of dietary ingredients. For non-ruminants, diets must be low in fiber content. Consequently, diets for poultry and swine in developed countries typically are composed primarily of cereal grains supplemented with various oilseed by-products (e.g., soybean meal, peanut meal, cottonseed meal, and canola meal). For cattle, sheep, and goats, the cost of supplemental feeds will vary depending on the farming system employed and stage of growth of livestock. Of the total cost of production, the cost of feed for ruminants will range from under 5% for supplementing grass-based diets with needed nutrients to over 80% for finishing cattle in feedlots (OSU, 2003).

Selection of the individual feeds to be included in diets for animals generally is based on computerized least-cost diet formulation programs. This method of formulation minimizes the total cost of a diet by optimizing synergies of proteins to provide amino acids with addition of other nutrients from various feed ingredients and mineral and vitamin supplements. When the cost of one feed relative to a second with similar nutrient composition increases markedly, the second feed will displace the first as an ingredient in a livestock diet. In contrast, when all feed costs increase, the amount of feed that is fed to livestock decreases because livestock production is no longer economically viable. Only when the commercial value of livestock products increases simultaneously with feed prices will feed use for livestock continue unabated. Economic survival of any enterprise is possible only when product value exceeds the total input cost. Consequently, whenever the cost of grain relative to that of animal products increases, the amount of grain that is fed to livestock automatically decreases sparing grain for other uses. Except for hobby farmers, livestock production is a business that can survive only through converting commodities or feed ingredients that are low in cost into products that are valued by consumers. Perturbations in this relationship (e.g., crop failures, competition with biofuel producers for grain, taxation or subsidies for specific commodities, and reduced consumer demand) force livestock producers to modify their formulated diets and, in some cases, to fail economically. A decrease in product value, induced by consumer concerns or fears (e.g., animal abuse concerns, healthfulness perceptions like "swine flu," hoof and mouth disease, zoonoses) also can reduce the demand for animal products, the economic return for producers, and ultimately livestock production.

In countries that are more affluent where grain is abundant, grain is and will continue to be fed to animals as long as the economic return generated through conversion of grain to animal products is adequate to cover all costs. The degree that composition of a diet for a specific animal will change in response to a change in price of a feed is much smaller for diets for pets and horses than for diets used for other livestock. Production economics dictates the sustainability of a livestock production system, but not the sustainability of pets in developed countries. Cereal grains, when refined and processed, also are consumed by humans. When feeds that could be milled into food for humans are fed to animals for production of animal products, calories are lost. However, as discussed previously, grain is not the single source of calories for livestock when forages, by-products, and gleaned residues are available. Hence, when calculating the efficiency of conversion of calories and protein from grain products to animal products (meat, milk, and eggs), one must consider the wide diversity among animal species in their dietary requirements, in feed availability and cost, and in animal production practices.

Estimates of caloric and protein efficiencies when converting dietary components into foods of animal origin will vary widely, both temporally and regionally, due to the wide spectrum of livestock production systems used in developed and developing countries. Efficiency values generated by CAST (1999) for conversion of weight, calories, and protein from grains to edible animal products averaged within developed and developing countries were derived from extensive comparisons and are provided in Table 12.2.

As shown in Table 12.2, the efficiency with which energy and protein are converted from cereal grains into livestock products differs markedly both among animal species and between developed and developing countries. This diversity usually is not recognized or understood by critics of livestock production. The efficiency of converting calories from cereal grains into calories present in edible livestock products in developed countries ranges from a mean of only 12% for pork production to more than 50% for milk production. In developing countries, caloric efficiency ranges from a low of 22% for egg production (although values should be much greater for layers that scavenge for food) to over 175% for beef harvested largely after being finished on grass. An efficiency that exceeds 100% indicates that animals are obtaining energy and nutrients from non-grain resources inedible for humans so that the output of calories or protein in animal products is exceeding the

TABLE 12.2
Efficiency of Converting Grain to Edible Animal Products

	Weight Efficiency[a]	Caloric Efficiency[b]	Protein Efficiency[b]
Developed Countries			
Beef	0.38	0.21	0.49
Pork	0.27	0.12	0.24
Lamb or mutton	1.28	0.27	1.26
Poultry	0.47	0.20	0.32
Milk	3.03	0.51	0.56
Eggs	0.46	0.16	0.32
Developing Countries			
Beef	3.23	1.75	4.51
Pork	0.57	0.25	0.50
Lamb or mutton	3.03	0.65	2.97
Poultry	0.64	0.27	0.44
Milk	4.55	0.76	0.84
Eggs	0.64	0.22	0.44

[a] Calculated from CAST (1999) based on typical animal diets.
[b] Calculated from weight efficiencies from CAST (1999) and calorie and protein content of individual animal products from USDA-ARS (2010).

input of calories or protein from grain. However, an efficiency that exceeds 100% should not be mis-interpreted to mean that feeding more grain to livestock would increase efficiency. Including more grain in a forage diet often increases the *rate* of growth or milk production, but additional dietary grain generally decreases caloric *efficiency* of production because added grain displaces energy that otherwise is obtained from non-grain (by-products and forage) portions of the diet. High rates of production of milk and meat are desired in order to dilute the maintenance costs of animals and to increase cash flow. Therefore, the economic incentive for feeding diets rich in grain is greatest when overhead, investment cost, and interest rates are high. In contrast, low land cost encourages grazing and, with improved types of forage, can markedly reduce or fully eliminate the need to feed cereal grains to ruminants. The fact that caloric efficiency is greater for animal products in developing than in developed countries reflects a heavier reliance on non-grain feed resources in developing than in developed countries.

The efficiency values in Table 12.2 were calculated in 1999 before the rapid expansion of the biofuel industry within the United States for conversion of feed grains to ethanol and oilseeds to biodiesel. Fermentation of grain to produce ethanol has markedly increased the supply of distillers' grains and corn gluten feed, by-products inedible by humans but able to replace one-half or more of the grain in diets for dairy and beef cattle. If an inedible by-product like distillers' grain displaces one-half of the cereal grain of a diet and production per unit of diet remains unchanged, efficiency of grain use is doubled! The degree to which by-products of biofuel production can displace dietary cereal grains differs among livestock species. Thanks to microbial digestion in the rumen, grow-ing and adult ruminants can digest fiber-rich by-products that growing poultry and swine cannot. Through modifying the biofuel manufacturing process to separate components either before or after fermentation, by-products that are sufficiently low in fiber content to have greater feeding values for poultry and swine can be generated. Alternative uses for feeds and by-products, including combus-tion to generate electricity as currently used in Brazil, fermentation of silages to generate methane and electricity in Europe, and the production of ethanol from cellulosic by-products, likely will increase the competition for and the price of these ethanol by-products and other feeds that are cur-rently fed widely to livestock.

Efficiency values cited in Table 12.2 markedly exceed those suggested by certain sources (Lappe, 1991). This difference can be ascribed largely to differences in the database used for calculations. Efficiency values reported in the past generally were calculated on a weight conversion basis, not on a caloric basis, and typically for only a single phase of livestock production, often the feedlot phase of production when grain intakes are highest. More logically, efficiencies should be cal-culated through total life-cycle assessment methods. Furthermore, they should consider only the dietary products that are edible by humans. Caloric efficiencies during the feedlot phase of produc-tion for pigs, cattle, and broilers are contrasted with those over the total life cycle in Figure 12.7. In these experiments, corn grain calories were displaced to various degrees by calories from distillers' grains to decrease the grain content of the diet.

In all cases, displacing corn grain with distillers' grains increased the tissue-to-grain caloric ratio (efficiency of converting the remaining corn grain calories to tissue calories). Caloric effi-ciency based on the total life cycle can be either greater or less than the efficiency calculated for the grain-feeding or feedlot phase of production. With pigs, lifetime efficiency was slightly less than feedlot efficiency because grain typically is fed to sows during gestation. In contrast, the caloric efficiency for cattle is considerably greater over the life span than during the feedlot phase. This reflects the fact that cows usually graze pasture or consume forages during gestation. In addition, for about one-half of their growth, calves obtain energy by nursing and from grazed forages, and are not fed supplemental grain during this pre-feedlot growth interval of 6 to 9 months. In addition, many feedlot cattle are "backgrounded" on pasture for several months after being weaned. Because of this extended period of growth without supplemental grain, caloric efficiency for beef produced on grass or fed diets low in grain content can exceed the caloric efficiency of converting grain to poultry and swine products because non-ruminants typically are fed diets rich in grain for their full

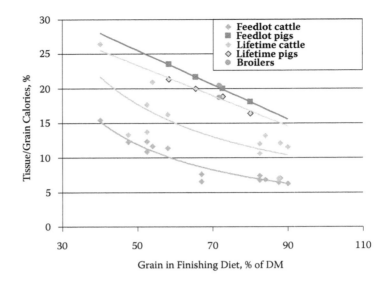

FIGURE 12.7 (See color insert) Feedlot phase versus lifecycle caloric efficiencies for various animal products. (Based on data from Whitney et al. (2006), *J. Anim. Sci.* 84: 3356–3363; Bremer et al. (2007), *2008 Nebraska Feeders Report*, pp. 39–40; Dozier et al. (2007), *J. Appl. Poult. Res.* 16: 206–218; and Klopfenstein et al. (2008), *J. Anim. Sci.* 86: 1223–1231.)

life span. These differences among livestock species, in both the extent and the efficiency of grain use, although seldom recognized, can markedly alter the efficiencies of converting calories from cereal grains to livestock products.

Compared with energetic efficiencies noted in Table 12.2, protein efficiencies from feeds fed to animals (including soybean meal, a by-product of production of soy oil) generally is greater. Again, values within developed and developing countries were averaged across a wide variety of production conditions. Despite having low rates of production, livestock that glean waste products and consume crop residues typically have higher efficiencies of caloric and protein use from the feed grains than do livestock raised under confinement conditions being fed harvested grains and forages. As noted previously, economic viability of large-scale confinement units (industrial animal agriculture or "factory farms") relies heavily on least-cost sources of energy from grain or grain by-products, as well as economies of scale.

Additional estimates of efficiency were derived by Oltjen and Beckett (1996) based on conversions of energy and protein from feeds edible by humans to edible products produced by cattle. Competition for food on a "humanly edible" basis inherently seems more logical than on a "grain" basis. These authors provided extensive details about their methods for calculating "human edible" returns across regions within the United States that differ in availability of feed resources. Their estimates are shown in Table 12.3.

Values for milk produced at Dairy I in Table 12.3 represent a typical California production system where by-products comprised a substantial portion of the diet. Values for Dairy II were derived for milk production by cows fed alfalfa, corn silage, corn grain, and soybean meal. Here, concentrations of dietary grain were greater, leading to lower efficiencies of both energy and protein use. Efficiency of beef production differed with region depending largely on the resources available for maintenance of the cowherd (grazed forages versus harvested corn silage), the duration of the feeding period, and the choice of grain used in the feedlot (wheat versus corn). The marked discrepancy between efficiency values shown here and those calculated by others (Gerbens-Leenes, Nonhebel, and Ivens, 2002; Peters, Wilkins, and Fick, 2007) can be ascribed to regional differences in available

TABLE 12.3
Humanly Edible Returns from Cattle Production in the United States

	Digestible Energy (%)	Digestible Protein (%)
Dairy I	128	276
Dairy II	57	96
Beef		
Colorado	37	65
Iowa	28	52
Texas	59	104

Source: Oltjen, J.W., and J.L. Beckett. 1996. *J. Anim. Sci.* 74: 1406–1409.

Note: Values are calculated as output energy or protein divided by input energy or protein multiplied by 100.

resources, inherent assumptions related to the edibility by humans of various feed sources, calculation of efficiency values in a given segment of growth versus the full life span, and the assumed productivity of crops that might be grown on arable land currently used for grain production.

In addition to feed, water is used to grow crops fed to livestock and water is consumed directly by animals during growth and production. One early estimate that 20,000 l of water were required to produce 1 kg of boneless beef was challenged in an extensive study by Beckett and Oltjen (1993). Under current production practices, they estimated that 3682 l of water were required to produce 1 kg of boneless beef. Of this, less than 4% (145 l) was consumed by the animals, 96% was used for production of feed crops, and the remainder (<1%) was used for beef processing. Based on their estimates, production of 1 kg of alfalfa, wheat, sorghum, barley, and corn require 911, 1417, 1689, 1665, and 900 l of water, respectively. Note that these amounts of water needed per kilogram of feed do not match estimates of the amount of water needed per hectare for various crops because crop yields per hectare differ among these crops. In regions where crops are produced without irrigation, water usage is of limited concern. For irrigation, however, the energy used for and the cost of pumping water as well as the depletion of aquifers are ongoing concerns. Whether crops are produced for export or for domestic use should not alter water use. Water is needed for growth and calorie production by crops. The amount of water as well as energy used for processing generally is greater for food crops than for cereal grains, partly due to the low productivity of food crops.

ARE LARGE-SCALE LIVESTOCK PRODUCTION UNITS INEFFICIENT AND IRRESPONSIBLE?

Benefits from specialization in crop production and in animal care have caused large-scale production systems to gradually displace the less efficient "family farms" that often grew crops not ideally suited for climatic and soil conditions for feeding multiple livestock species. Increased attention to economics of livestock production has led to housing systems and precisely balanced diet formulas based on least-cost resources that maximize economic returns in terms of rate and efficiency of growth, health, reproduction, and activity. Although growth might be criticized as an index of well-being, similar techniques are used with infants (APGAR scores) and children (growth charts) to assess health, development, and well-being in the human population. Housing or management systems or diets that fail to consider the basic needs of livestock for space, health, and nutrients usually result in depressed rates or efficiencies of growth. Economics, productivity, and market demands and anthropomorphic concerns have led to the livestock and crop management practices typically used today.

To reduce cost of production or to meet consumer desires, certain practices have evolved within large-scale commercial livestock units that may have adverse effects on the well-being of animals. Several of these involve alterations in the diets available to or provided for livestock. For example, the meat color of growing calves fed milk as their only dietary ingredient is white or light gray. Based on tradition, restaurants and consumers voice a preference for veal that is white or gray-white in color. To produce veal of that color, myoglobin concentrations in muscle tissue must be low. With formulated feeds, light meat color can be achieved through feeding diets very low in iron content. Although such diets may not retard growth, veal calves fed such diets have lower blood hemoglobin concentrations than found in calves with access to forage to graze. Consequently, in order to meet consumer desires, special diets and confinement production units are used for veal production. Agronomic advancements to increase productivity and nutritional value of forages also have complicated livestock production. Development and widespread use of legumes that fix nitrogen from the air and of highly productive winter wheat for grazing cattle has increased the incidence of bloat among grazing cattle beyond that seen among cattle grazing slower growing, unimproved forages or native pastures. Productivity of grass crops and forages generally is increased by applying organic or inorganic fertilizers. However, with certain plant species, particularly drought-resistant sorghums and sudans, a high level of nitrogen fertility in soil when combined with weather or frost stress causes these plants to accumulate nitrate which is toxic to ruminants. Because the cost of net energy typically is lower for concentrates than forages, formulated diets for livestock and poultry generally are rich in concentrate feeds and low in fiber content. Grains often are processed (finely ground, pelleted, steam-flaked, extruded, or fermented) to increase the availability of energy and nutrients even further. High rates of digestion, fine particle size, and low concentrations of dietary fiber are associated with an increased incidence of gastric ulcers in swine and of ruminal acidosis in dairy cows and cattle and sheep in feed yards. To maintain production and reduce subclinical disease problems of animals and poultry, feed additives (probiotics, yeast, enzymes) often are included in diets for livestock to increase nutrient availability and stabilize or supplement the digestion process. The prevalence of respiratory disease among cattle is increased markedly by assembling groups of livestock from multiple origins. Thus, large-scale production systems typically employ preventive and treatment measures to reduce microbial infections to a greater degree than smaller-scale production units do where livestock never leave their farm of origin.

WHAT IS THE ROOT CAUSE OF STARVATION IN TODAY'S WORLD?

It is difficult to rationalize that production or consumption of animal products is directly or indirectly contributing to human starvation when one considers that:

1. A large fraction of arable land in the world currently is not being used for crop production
2. Many animal species thrive on diets composed of forages and inedible by-products
3. Feed grains yield substantially more energy and protein than do food crops
4. Land used to grow cereal crops is not readily or easily converted to grow food crops
5. The efficiency of converting dietary calories and protein to animal products can exceed 100% when composed of by-products and waste materials

Instead, imbalanced economics both among and within nations is considered to be the primary cause of food deprivation and starvation. Economics drives production and distribution of nutrients, goods, and services worldwide. Current food production in the world is sufficient to provide calories for 7 billion people while the current world population is approximately 6 billion (Oracle, 2000). Caloric distribution is inequitable both among and within nations. Most food is produced in more developed countries, but individuals in need of food usually cannot afford to pay the price that farmers in developed or developing countries need to produce additional food. In addition, weather, insects, and disease can reduce crop and livestock production temporarily and regionally.

Most experts attribute food imbalances and starvation to lack of purchasing power of citizens and to restrictive political policies, not to the worldwide supply of food (Oracle, 2000). Certain countries (e.g., the United States, Brazil, Argentina, and Australia) are blessed with a surplus of arable land, an infrastructure for food transport, economic power to obtain materials that increase crop productivity (fertilizers, pesticides, and irrigation), aggressive and economically savvy crop and livestock producers, and a relatively low population density. Consequently, these countries export both crop and animal products to other nations.

Local and regional factors always have and will continue to dictate the economic feasibility of production of livestock, as well as food and feed crops. Feeding of grain to livestock will continue as long as economic return is generated, for example, local grain prices remain low relative to the demand-driven prices for livestock products. The choice among crops that landholders produce to pay their expenses is driven by value of the crop produced minus all input costs. Whenever competition for or a shortage of grain results in an increased price of grain, economic return to livestock producers is reduced and producers are forced by economics to decrease or cease production. Alternatively, producers can rely to a greater degree on calories supplied from non-arable land available for grazing and locally available crop and industrial by-products that are unsuitable for human consumption or for other uses (e.g., fermentation or combustion to generate electricity). Because diets of poultry, swine, and feedlot cattle are strongly dependent on grain for calories, such enterprises respond more abruptly and drastically to changes in grain prices than livestock enterprises that exist on diets composed of by-products, waste products, and forages. Production of fruit and vegetable crops that can be marketed directly to consumers likely will increase in regions where producers have nearby access to urban markets and low labor costs. Because each of these factors, as well as the demand for specific animal products, varies internationally and regionally, production of livestock would be expected to increase in certain regions of the world and to decrease in other areas depending on the availability and cost of inputs, environmental restrictions, and consumer values of specific products. Although global marketing of agricultural products should improve the nutrient balance among nations, two factors are expected to limit food availability. First, persistent nationalism to either restrict or encourage imports or exports biases worldwide prices and inhibits trade, often reducing production and availability of food in other countries or at home. Second, consumer attitudes and governmental regulations that restrict access to or production of more highly productive or efficient hybrids or varieties of both crops and livestock serves to reduce the worldwide food supply.

WHAT MIGHT ONE EXPECT AS THE HUMAN POPULATION INCREASES IN THE 21ST CENTURY?

As is apparent across nations (Figure 12.8), an increase in per capita income in developing countries generally results in an increase in the demand for and the consumption of animal products. Per capita consumption of meat and milk is expected to increase during the current 23-year evaluation period (1997 to 2020) by 13 and 16 kg (16 and 8%, respectively) in developed countries and by 11 and 19 kg (44 and 44%, respectively) in developing countries according to Delgado (2003). When combined with the increase in the global population, this means that to meet demands, total meat and milk production will need to increase by 62 and 49%, respectively, during this 23-year period. During this period, per capita use of cereals for feeding livestock is projected to increase to 375 kg in developing countries and to 72 kg in developed countries. Over 70% of this increase in the meat supply is expected to come from pork and poultry. To feed these animals and meet this demand, exports of grain and meat from developed countries is expected to increase while interchange of both crops and livestock among developing countries will increase. How long such increases in crop and meat production and exports can continue to meet the worldwide demand for livestock

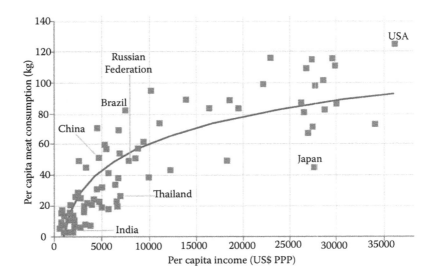

FIGURE 12.8 Meat consumption versus per capita income for various nations. (From FAO. 2006a. Livestock's long shadow: Environmental issues and options. ftp://ftp.fao.org/docrep/fao/010/a0701e/a0701e01.pdf)

products has been questioned. In those countries where arable land is not fully used, crop production can be increased readily in response to market forces. Yet, the expansion of crop production on marginal, erodible land and the conversion of forested to arable land are of concern because of deleterious environmental effects and degradation of arable and potentially arable land (FAO, 2000).

Despite public concerns in developing and developed countries, the demand for and the price of foods and feeds is expected to increase. In response to these economic incentives, crop and livestock producers can be expected to begin cropping large expanses of potentially arable land not currently being used for crops in sub-Saharan Africa and South and Central America. As fossil fuel prices increase, the use of cereal crops, cellulosic products, and oilseeds for biofuel production and for generation of electricity (via combustion or biogas production) also should grow. Increased stringency of environmental regulations in developed countries likely will restrict development or expansion of crop and livestock production units, increase inter-dependency among countries, and expand international trade. Private and public investments in research dealing with genetics and management of specific plant species will continue to increase efficiencies of production by reducing environmental limitations (insect damage, insufficient water, low fertility, heat, and soil salinity). Where economic incentives to convert feed grains to animal products exist, production of livestock will increase, with expansion being greatest in localities where grain is abundant, grain prices are low, and underutilized resources and by-products for livestock feed are readily available.

CURRENT AND FUTURE CONCERNS

Numerous food issues are widely discussed and debated.

1. According to a 2009 report from the FAO, 19% of the people on Earth (more than 1 billion people) are undernourished with prevalence being greatest in Asia and Africa, and world hunger is increasing (FAO, 2006b; 2009). While politicians discuss the potential effects of global warming on species diversity and the future food supply for humans, humans are starving today.
2. Within developed countries, some individuals question whether food products should be distributed to people in other parts of the world while poor and needy people exist within

their own community. Which has greater lasting impact: Supplying fish to feed the needy or teaching the needy to fish? Investment in an Agricultural Peace Corps to increase efficiency of food production internationally should help to decrease armed conflicts created by food insecurities in developing nations.

3. Some individuals adopt specific lifestyles in an attempt to reduce global warming and to spare food for humans that, without a full understanding of economic and trade issues, are likely to prove vain and futile. That locally produced foods and a simple reduction in food miles can reduce the carbon footprint of food, though seemingly logical, has been soundly refuted by Capper et al. (2009). Likewise, that the carbon footprint is less from beef finished on grass rather than on grain has been debunked (Capper et al., 2009; Peters et al., 2010) even though grass-finished beef can reduce competition for arable land. This illustrates how actions to achieve different environmental goals can conflict with each other. Public understanding about nutrition, food production, and food safety needs to be enhanced and misinformation needs to be challenged promptly. Efforts by policymakers and the public to save the environment and to feed the world may prove erroneous and deleterious unless all economic aspects are analyzed carefully as outlined by Simmons (2009).

4. Activist groups that condemn animal production ignore the critical role that animal production plays in providing a livelihood for producers, processors, and retailers as well as calories and protein sorely needed for rural communities of both the developing and the developed world (IFAD, 1993; Speedy, 2003; Randolph et al., 2007). Beneficial effects of animal production on rural development internationally are mirrored by the fact that a 2010 World Food Prize was awarded to Heifer International. This organization provides chickens, pigs, or pregnant heifers to rural communities around the world in order to stimulate food production and an animal industry.

5. Traditional commodity food products are designed to be economical. Specialty products advertised to have added nutrient values (fortified) or consumer appeal (preparation ease) typically have added costs. Specifications and regulations related to food origin, processing, labeling, and handling to assure consumers that their food is wholesome, nutrient-fortified, safe, and environmentally friendly necessarily add cost to our food supply.

6. Concerns about food production methods and food safety have increased interest in "organic products." Such products do provide additional options for consumers. Because plant and animal foods produced to match the specifications for an "organic" label have either lower productivity or higher input costs (or both), they typically cost much more than commodity foods. Yet, no benefits based on scientific measurements in terms of nutrient content, nutrient availability, presence of pesticide and herbicides, and food safety have been demonstrated.

7. Elected legislators strive to maintain or reduce food prices for the public. These often thwart food production. Such efforts must not compromise food quality, wholesomeness, and safety. Governmental programs designed to provide economic protection for crop and livestock producers in their own nation through subsidies (e.g., for biofuel production) and import restrictions of seed and food (e.g., genetically altered products) by reducing trade are limiting the supply of food, increasing its cost, and modifying farming practices internationally by altering commodity prices.

8. The impact of developments that continually remove arable land from production (estimated at 1 million hectares annually in the United States; FAO, 2000) and the increased cost of energy for local and international transport are presumed to have limited impact on food costs. These topics require more attention and extensive production and economic modeling.

9. To meet the projected increase in demand for animal products worldwide, greater integration of feeding systems to fully use available feeds and biomass as well as strategic supplementation of animal diets with nitrogen or non-protein nitrogen sources and with required

minerals, vitamins, and energy during certain critical periods should be encouraged. Similarly, increased input of "best practices" in crop production, including use of genetically improved crops and application of adequate, but not excessive amounts of fertilizer and water, can improve crop productivity while reducing their environmental footprint.

10. Most of the major world religions espouse sympathy for the sick, the sorrowful, the poor, and the needy. Certainly, individuals deficient in calories and specific nutrients fall into this group. Agricultural development teams are needed to serve around the globe to assure long-term food security and enhance productivity of foods of plant and animal origin.

SUMMARY

The population of the world is increasing at a cumulative rate of 1.2% annually. In the next 50 years, the need for food will double, with much of that need being met by advances in science and biotechnology. Of the total land on our globe, only about 13% is arable and readily suited for crop production. Yet, only 39% of this arable land in the world currently is used for crop production with the largest potential for expansion being in Central and South America and sub-Saharan Africa. Various livestock species differ markedly in their capacity to consume and digest by-products and forages and in the degree to which they compete with humans for nutrients and calories. Swine and poultry in developed countries typically are fed diets composed of grain and oilseed by-products, but herbivores can obtain a large fraction of their calories and protein needs through grazing and consumption of fiber-rich crop by-products and industrial wastes. When fed crops suitable for direct human consumption, animals compete with humans for calories and nutrients. However, when fed by-products not suited for human consumption or when grazing forage produced on non-arable land, animals do not compete with humans for calories and nutrients, but instead can add to the supply of food for humans (see Chapter 13 by Lardy and Caton in this book). This is particularly evident in developing countries. Based on measured yields, crops differ widely in their capacity to convert solar energy to food or feed energy. Calorie yields consistently are much greater for cereal grains than for food crops suited for more immediate human consumption. The highest caloric yields per hectare come from corn and rice, while the highest protein yields are obtained from soybeans. Conversion of cropland from growing cereal crops to growing vegetables is limited by agronomic (crop), environmental, and economic (labor, market) factors. In developing countries, crop yields also are limited by crop management and inputs (fertilizer, insect and weed control), as well as the environmental conditions (seasonal rainfall, temperature). The current supply of food is adequate to feed all people on the globe. Nevertheless, inequitable distribution of food, crop failures, war, pestilence, political barriers, and poverty are causing starvation among the poor and needy worldwide. Economics of supply and demand for calories and nutrients will continue to drive feed and food production where production, trade, and distribution are not restricted by governmental policies, religious practices, and personal food preferences or concerns. Political pressures and protectionist trade policies intended to obtain or maintain low food prices have failed to provide safe and abundant food for consumers. Because technological advances in crop and animal productivity in a free market system ultimately serve to benefit food consumers, not food producers, financial support of agricultural research and extension practices to increase plant and animal productivity are critical public investments for maintaining a safe, affordable, and abundant supply of food for the expanding human population.

REFERENCES

Bath, D., J. Dunbar, J. King, S. Berry, and S. Olbrich. 2001. Byproducts and unusual feedstuffs. *Feedstuffs* 73: 30–37.

Beckett, J.L., and J.W. Oltjen. 1993. Estimation of the water requirement for beef production in the United States. *J. Anim. Sci.* 818–826.

Belsky, A.J., S.M. Mwonga, and J.M, Duxbury. 1993. Effects of widely spaced trees and livestock grazing on understory environments in tropical savannas. *Agroforestry Systems* 24: 10–20.

Bremer, V.R., G.E. Erickson, and T.J. Klopfenstein. 2007. Meta-analysis of UNL feedlot trials replacing corn with WDGS. *2008 Nebraska Feeders Report*, pp. 39–40. http://beef.unl.edu/beefreports/2008.pdf

Capper, J.L., R.A. Cady, and D.E. Bauman. 2009. Demystifying the environmental sustainability of food production. *Cornell Nutrition Conf.* pp. 174–190.

CAST (Council on Agricultural Science and Technology). 1999. Animal Agriculture and Global Food Supply. Ames, IA: CAST. http://www.cast-science.org/publicationsInfo.asp

CIA (Central Intelligence Agency). 2009. *The World Factbook.* https://www.cia.gov/library/publications/the-world-factbook/fields/2097.html Accessed April 2010.

Delgado, C.L. 2003. Rising consumption of meat and milk in developing countries has created a new food revolution. *J. Nutr.* 133: 3907S–3910S.

DePeters, E.J., J.G. Fadel, M.J. Arana, N. Ohanesian, M.A. Etchebarne, C.A. Hamilton, R.G. Hinders, M.D. Maloney, C.A. Old, T.J. Riordan, H. Perez-Monti, and J.W. Pareas. 2000. Variability in the chemical composition of seventeen selected by-product feedstuffs used by the California Dairy Industry. *Prof. Anim. Sci.* 16: 69–99.

Dozier, W.A. III, J.L. Purswell, M.T. Kidd, A. Corzo, and S.L. Branton. 2007. Apparent metabolizable energy needs of broilers from two to four kilograms as influenced by ambient temperature. *J. Appl. Poult. Res.* 16: 206–218.

FAO (Food and Agriculture Organization of the United Nations). 1995. Staple foods: What do people eat? http://www.fao.org/docrep/u8480e/u8480e07.htm

FAO. 2000. *Land Resource Potential and Constraints at Regional and Country Levels.* World Soil Resources Report 90. ftp://ftp.fao.org/agl/agll/docs/wsr.pdf

FAO. 2002. Global pig numbers — World hog population. http://www.thepigsite.com/articles/7/markets-andeconomics/858/global-pig-numbers-world-hog-population-2002

FAO. 2006a. Livestock's long shadow: Environmental issues and options. ftp://ftp.fao.org/docrep/fao/010/a0701e/a0701e01.pdf

FAO. 2006b. World hunger increasing. http://www.fao.org/newsroom/en/news/2006/1000433/index.html

FAO. 2009. The state of food insecurity in the world. ftp://ftp.fao.org/docrep/fao/012/i0876e/i0876e02.pdf

Gerbens-Leenes, P.W., S. Nonhebel, and W.P.M.F. Ivens. 2002. A method to determine land requirements relating to food consumption patterns. *Agriculture, Ecosystems and Environment* 90: 47–58.

Govindjee, and R. Govindjee. 2000. What is photosynthesis? http://www.life.illinois.edu/govindjee/whatisit.htm

IFAD (International Fund for Agricultural Development). 1993. Institutional and economic framework conditions for livestock development in developing countries and their interrelationships. http://www.ifad.org/lrkm/theme/husbandry/framework/framework_bib.htm

Klopfenstein, T.J., G.E. Erickson, and V.R. Bremer. 2008. Use of distillers by-products in the beef cattle feeding industry. *J. Anim. Sci.* 86: 1223–1231.

Lappe, F.M. 1991. *Diet for a Small Planet.* 20th Anniversary Edition. New York: Ballantine Books.

Lubowski, R.N., M. Vesterby, S. Bucholtz, A. Baez, and M.J. Roberts. 2002. Major uses of land in the United States. USDA-ERS. http://www.ers.usda.gov/publications/EIB14/eib14a.pdf

Nationmaster. 2005. World Development Indicators Database. http://www.nationmaster.com/graph/agr_ara_lan_hec-agriculture-arable-land-hectares&date=2005#definition

New Zealand Agricultural Statistics. 2010. Agricultural Production Statistics: June 2009. www.stats.govt.nz/browse_for_stats/industry_sectors/agriculture-horticulture-forestry/AgriculturalProduction_MRJune10final.aspx

NRC (National Research Council). 1998. *Nutrient Requirements of Swine*, 10th Revised Edition. Washington, DC: National Academy of Sciences.

Oltjen, J.W., and J.L. Beckett. 1996. Role of livestock in sustainable agricultural systems. *J. Anim. Sci.* 74: 1406–1409.

Oracle. 2000. An end to world hunger: Hope for the future. Team C002291. http://library.thinkquest.org/C002291/high/present/stats.htm

OSU (Oklahoma State University). 2003. OSU feedlot performance program. http://www.beefextension.com/files/FEEDLOT%20CALC.xls

Peters, C.J., J.L. Wilkins, and G.W. Fick. 2007. Testing a complete-diet model for estimating the land resource requirements of food consumption and agricultural carrying capacity: The New York State example. *Renewable Agriculture and Food Systems* 22: 145–153.

Peters, G.M., H.V. Rowley, S. Wiedemann, R.Tucker, M.D. Short, and M. Schulz. 2010. Red meat production in Australia: Life cycle assessment and comparison with overseas studies. *Environ. Sci. Technol.* 44: 1327–1332.

Randolph, T.F., E. Schelling, D. Grace, C.F. Nicholson, J.L. Leroy, D.C. Cole, M.W. Demment, A. Omore, J. Zinstag, and M. Ruel. 2007. Role of livestock in human nutrition and health for poverty reduction in developing countries. *J. Anim. Sci.* 85: 2788–2800.

Schaafsma, G. 2000. The protein digestibility–corrected amino acid score. *J. Nutr.* 130: 1865S–1867S.

Simmons, J. 2009. Technology's role in the 21st century: Food economics and consumer choice. Why agriculture needs technology to help meet a growing demand for safe, nutritious and affordable food. *Cornell Nutr. Conf.* pp. 159–173.

Smith, C.W. 2001. Rice. http://cwaynesmith.tamu.edu/agro306-500/materials/fall2010/topics/rice.10.1.evol.types.ppt#274,1,Slide 1

Speedy, A.W. 2003. Global production and consumption of animal source foods. *J. Nutr.* 133: 4048S–4053S.

US Census Bureau. 2010. International Data Base: World population growth rates 1950–2050. http://www.census.gov/ipc/www/idb/worldgrgraph.php

USDA-ARS (Agricultural Research Service). 2010. USDA National Nutrient Database for Standard Reference, Release 23. Nutrient Data Laboratory Home Page, http://www.ars.usda.gov/main/site_main.htm?modecode=12-35-45-00

USDA-ERS (Economic Research Service). 2010. Feed Grains Database. http://www.ers.usda.gov/Data/FeedGrains/download.htm

USDA-FAS. 2011. World agricultural production archives. http://www.fas.usda.gov/wap_arc.asp

Whitney, M.H., G.C. Shurson, L.J. Johnston, D.M. Wulf, and B.C. Shanks. 2006. Growth performance and carcass characteristics of grower-finisher pigs fed high-quality corn distillers dried grain with solubles originating from a modern Midwestern ethanol plant. *J. Anim. Sci.* 84: 3356–3363.

13 Crop Residues and Other Feed Resources

Inedible for Humans but Valuable for Animals

Gregory Lardy and J.S. Caton

CONTENTS

INTRODUCTION

Domesticated livestock have been an essential integrated component of agriculture and human food systems for thousands of years (Bradford, 1999). Ecosystems of both arable and non-arable lands have developed in the presence of animals and consequently ecological succession is clearly shaped by the plant–animal interface. Sustainable coexistence of plant and animal agriculture is under-pinned by the health and well-being of each component. This is well illustrated by the "Ancient Cow Contract," which simply states that the herdsman contracts with his animals to provide hous-ing, feed, safety, and care in exchange for milk, meat, fiber, and other products (Anderson, 2000). Ruminant animals have and will continue to play a unique and essential role in the human–animal–plant interface because of their diversity, adaptability, and ability to consume feedstuffs that are inedible by humans. These types of feedstuffs are abundant and include forages, crop residues, native grasslands, food-processing by-products, and other feed resources. A commonality shared by these feedstuffs is that the microorganisms in the ruminant foregut can utilize them and the microorganisms subsequently provide the animal with a source of nutrition. This unique symbiotic relationship, which is, in fact, analogous to the Ancient Cow Contract, has played a pivotal and sus-tainable role in providing food, fiber, and clothing for almost every civilization in recorded history. The world population is expected to reach 7.7 billion by 2020 and nearly 9.2 billion by 2050 (United Nations, 2008; medium variant). The majority of the projected population growth is expected to occur in developing countries where the recent trends include increased per capita consumption of animal products (CAST, 1999). Consequently, the demand for more efficient and effective pro-duction of high-quality animal products will increase. Because of their unique ability to convert non-edible products into high-quality, nutrient-dense foods for human consumption, animals, and in particular ruminants, are expected to fill an increasingly significant and sustainable role in the United States in the coming decades. The contribution of enormous quantities of inedible sources of fibrous plants and other waste products on a global basis is addressed by Owens and Hanson in Chapter 12 of this book.

HISTORICAL PERSPECTIVE

Throughout history, ruminants have provided meat, milk, fiber, and a source of draft power to almost every civilization. Even before animals were domesticated, hunting tribes relied on ruminants to play a critical role in providing food, fiber, and clothing for tribe members and subsequently for civilizations on almost every continent. Ruminants are able to exist in climates that range from bitter cold Arctic environments to the tropics of Africa, Asia, and South America. Evidence of human dependence upon animals abounds in the archeological records. Approximately 10,000 years ago, livestock domestication began and was a critical component of the transition from hunter and gatherer to cultivating and shepherding foundations for societal life (Cambell and Lasley, 1975). Sheep and goats (both ruminant species) were among the earliest livestock species domesticated. For early human civilizations, having a ready supply of fresh meat and milk was a competitive advantage. The domestication of cattle provided the added benefit of draft power, which added tremendously to the ability to do physical work, including that of cultivation for agronomic purposes as well as transportation of these crops and other goods to market.

With the advent of cultivation and the domestication of livestock, which were mutually complementary through draft power, and livestock's consumption of inedible waste products, villages flourished and a slow march toward urbanization began. The development of large cities was a slow process with a major limiting factor being a steady supply of food. Consequently, during most of history the majority of the population of most civilizations was engaged in agricultural activities. However, through the industrialization of agriculture and the application of technological breakthroughs, progressively less of the population has been needed for the production of food supplies. In developed countries, the rate of off-farm migration has rapidly accelerated during the last 50 to 100 years to a point where less than 1.5% of the population is directly involved in production agriculture. Currently, an ever-growing emphasis is being placed upon sustainability, efficiency, and integration of agricultural practices to ensure a consistent and safe food supply with minimal ecological impact while optimizing the use of scarce resources including land, water, and other inputs.

ECOLOGICAL BALANCE AND SUSTAINABILITY

Ecosystems of both arable and non-arable lands have developed in the presence of animals and consequently ecological succession is clearly shaped by the plant–animal interface. Ecological balance is the concept of a dynamic equilibrium within a community of organisms where species and ecological diversity remain relatively stable. In theory, good ecological balance and sustainable agricultural practices should be mutually inclusive events. Sustainable coexistence of plant and animal agriculture is underpinned by the health and well-being of each component. Specifically within agro-ecosystems, the utilization of materials inedible by humans as nutrient sources for livestock helps provide ecological balance by providing mutual benefits for the animal, environment, and human population alike. The concept of this mutually beneficial symbiotic relationship between animals and humans within this context is historically present as the Ancient Cow Contract. While most would say that the cow can never call the herdsman into account for a breached contract, animals certainly "communicate" their acceptance of existing contract conditions in various ways. These include increased productivity, behavior that reflects contentment, and improved health and well-being.

From a broad perspective, animal agriculture contributes to human health and well-being in numerous ways. These include the production of nutrient-dense foods, the conversion of inedible plant materials from non-arable land, crop residues, and food processing by-products, by provision of draft power for cultivation and transport, vegetation management through grazing, waste disposal and nutrient recycling functions, and several others (Bradford, 1999). Animals also add tremendous value to the cropping sector. In fact, recent estimates (Harris et al., 2009) indicate that livestock and cereal crops used for feed grains represent nearly 60% of the total agricultural receipts in the United States. Critics of the use of livestock for human food production or the use of various feedstuffs for feeding livestock often state that inefficiencies associated with feeding livestock for food production coupled with extra environmental burdens posed by livestock operations should preclude the long-term use of animal products as major food sources by developed modern societies. While this argument appears reasonable to some at first glance, a more in-depth investigation of the underlying assumptions provides a different picture regarding the use of livestock to meet growing human food demands.

Animal products (meat, milk, and eggs) are nutrient-dense and provide an excellent source of essential amino acids, vitamin A, thiamin, riboflavin, niacin, B12, iron, calcium, zinc, and other essential nutrients (CAST, 1997). Additionally, animal protein has a high biological value for humans and, on a per unit basis, it is difficult if not impossible to match the biological value of animal protein with proteins from plant food sources. While consumption of high-quality protein diets from plant sources may be achievable on an individual or small-scale basis, the issue becomes even more pointed when attempting to supply a high-quality, nutrient-dense food supply to large populations. In addition, critics of the current and likely expanding role of animals in the provision of human foods make the false assumption that all animals compete directly with humans for foods on a per unit basis. This assumption, of course, is incorrect particularly in the case of ruminants, which derive nutritional benefits from fibrous feedstuffs. It is also an incorrect assumption for non-ruminant species as they have historically made and will continue to make extensive use of human inedible resources (Westendorf, 2000) especially in the area of food processing by-products. When calculating whole-animal efficiencies of protein production, non-ruminant animals are more efficient compared with ruminants. However, when adjusting the calculations of efficiency of animal protein production per unit of human edible food consumed by livestock, the calculations show a distinct and unique advantage for ruminants for supplying high-quality, nutrient-dense foods for the human population (CAST, 1999).

Feeding forages, crop residues, and agricultural by-products to ruminants in conjunction with judiciously grazing non-arable lands is a component of good management and stewardship that has health and well-being benefits for both humans and animals, while concomitantly providing opportunities to maintain a desirable ecological balance. A non-exhaustive list of potential human, animal, and ecological benefits of the human–animal partnership and of providing nutrients to livestock from crop residues and other resources that are inedible for humans but valuable for animals is provided in Table 13.1. Unfortunately, as with any tool or practice placed in human hands, the potential for misuse or abuse is always present. Therefore, in the practice of animal husbandry, care must be taken to ensure that good stewardship principles and sustainable management practices that are mutually beneficial are provided. If not, the risk for potential harm to one or more members of the partnership increases. Concerning providing crop residues and other feed resources (including grazing of non-arable lands) that are inedible for humans but valuable for animals, herbivores in general and ruminants in particular maintain a distinct and competitive advantage among domestic animals for these practices.

RUMINANT ANIMALS

Ruminant animals are herbivores that possess a unique arrangement of compartments in their gastrointestinal tract which allows for pre-gastric fermentation of ingested materials (Van Soest,

TABLE 13.1

The Human–Animal Partnership and the Feeding of Crop Residues and Other Human Inedible Feed Resources to Animals: Potentially Mutual Benefits to Human Well-Being, Animal Welfare, and Ecological Balance

Human Health and Well-Being	Animal Health and Well-Being	Potential Improvements in the Ecological Balance
Production of high-quality, nutrient-dense foods, which are highly palatable and add diversity to the diet.	The provision of shelter and the protection from predators.	Improved rangeland ecology though proper grazing management.
Conversion of plant materials from non-arable land, crop residues and food processing by-products, and certain waste products, i.e., materials that humans cannot eat or choose not to eat, into high-quality food products thereby adding value.	Access to a more reliable food source and moderation of seasonal extremes in nutrient supply.	Reductions in biomass that would need to be burned, buried, or disposed of in some other way that could potentially have negative impacts on the environment.
Production of fibers, leathers, pharmaceuticals, and a wide array of other products useful to humans.	Improved animal health and care through good feeding and management practices.	Potential reductions in environmental nutrient loads.
Provision of waste disposal and nutrient recycling functions, which contribute added value to production systems.	Increased gastrointestinal health through feeding of high-fiber feeds.	Production of manures, which are a valuable source of organic plant nutrients and which reduce the need for chemical fertilizers.
Provision of draft power for cultivation and transportation, relieving humans of some of the heavy physical labor associated with crop production, and permitting more timely planting and harvesting of crops.	Possible access to veterinary care when needed.	Reduction and control of non-desirable or invasive plant species to maintain a more desirable plant community.
Assisted vegetation management through grazing, which can have important environmental benefits, e.g., reduction of fire hazard and maintenance of desired plant communities.	Access to a more readily available water source.	Control of excess plant biomass resulting in reduced fire hazard.
Providing a means of savings and a food reserve in times of crop scarcity, for agriculturalists not part of a monetary economy.	Additional protection for offspring, including improved postnatal care.	Positive movement toward ecological balance within rangelands and agro-ecosystems.
Contributing to the flexibility and thus stability of food producing systems, and to the total agricultural economy.	Thriving and sustaining livestock communities that are supported by the consumption of feedstuffs that do not directly compete with the human population.	

Source: Adapted and modified from Bradford, G.E. 1999. Contributions of animal agriculture to meeting global human food demand. *Livestock Production Science* 59: 95–112.

1982; Hofmann 1988). Of all the herbivorous species of animals, ruminants are by far the most numerous and important (Church, 1988). Ruminants are in the subclass referred to as ungulates (hooved mammals) and in the order of Arteriodactyla (an even number of toes) and the suborder of Ruminantia. The word ruminant is derived from the Latin *ruminare*, which means to chew over again. Thus, ruminants are even-toed, cud-chewing mammals with hooves (Church, 1988). Crop residues, forages, and food processing by-products are generally fibrous feedstuffs that contain large amounts of cellulose, hemicellulose, and lignin. Ruminants can make excellent use of forages, crop residues, and other feedstuffs inedible by humans because of the symbiotic relationship with the ruminal microbial population and the resulting fermentation that occurs in this compartment of the foregut. This ruminal fermentation produces short-chain fatty acids (volatile fatty acids; VFA) through anaerobic action on primarily the cellulose and hemicellulose fractions of crop residues (Owens and Goetsch, 1988). These fermentable fibers are associated with the plant cell wall and represent structural components of the plants. The microbial population within the rumen thrives on the fermentable fibrous fractions of dietary ingredients, including forages, crop residues, and other high-fiber feedstuffs. The VFAs produced from microbial actions are a waste end-product of ruminal fermentation; however, the VFAs are absorbed across the rumen wall and make their way to the liver where they are metabolized to compounds useful to the host animal. Approximately 70 to 80% of the energy used by ruminants is derived from these VFAs (Fahey and Berger, 1988). In addition to the beneficial use of these microbial waste products, the microbial cells that undergo digestion in the gastric and intestinal segments of the digestive tract provide a ready and highly digestible source of essential nutrients to the host animal, including protein and water-soluble vitamins. Protein provided from the digestion of ruminal microbial cells in the small intestine provides the vast majority of protein used by ruminant animals. Therefore, the ruminant animal receives a double benefit from its symbiotic relationship with the resident microbial population: (1) the provision of VFAs and (2) the provision of nutrients (protein, vitamins, and others) from digested microbial cells. While this is the case for ruminant animals, it is not true for post-gastric fermenting herbivores such as horses or rabbits. Those species represent post-gastric fermenters and only benefit from the microbial production of VFA.

This ability to ferment high-fiber feedstuffs that are inedible for humans is a tremendous benefit to ruminant animals, and has been capitalized on by humans for thousands of years. In addition, this partnership has mutually beneficial attributes for both humans and animals while potentially improving ecological balance (Table 13.1).

FEED RESOURCES: INEDIBLE FOR HUMANS, BUT VALUABLE FOR ANIMALS

Many grasses and other high-fiber plants, as well as by-products of cereal grains, soybeans, and other crops processed for food or industrial use (e.g., bio-fuels) are inedible for humans, but are well utilized by some animals, particularly ruminants. For the purposes of this chapter, we will discuss two main categories of human inedible animal feed resources: (1) forages from non-arable lands, and (2) crop residues and food processing by-products. Animals, and in particular, ruminants, are able to utilize these feedstuffs as a source of nutrition. In almost all cases, these feed resources are unfit for human consumption in their present form. Barriers to human utilization may be related to nutritional composition (e.g., cellulosic, poor amino acid profile), access (not located in close proximity to human populations; for example, seeds from native grasses), sheer quantity (some fractions of crop processing by-products may be edible but the volume of the material requires feeding to livestock in order to utilize them), or economic barriers (the economic cost of further processing outweighs the current value as a human food ingredient).

Non-Arable Lands

Non-arable lands make up the majority of the land types on the Earth's surface. Estimates indicate that approximately 11% of the Earth's land area is cultivated, 24% is in permanent pasture, 31% is forest or woodlands, and the remaining 34% is comprised of glaciated areas, mountain ranges, and urbanized or industrialized land (Holechek, Pieper, and Herbel, 1989). Approximately 75% of the Earth's land surface has some sort of soil constraint, which limits or restricts its use as arable land (FAO, 2000). These constraints include things such as terrain, drainage, or shallowness of soils. Some of these constraints overlap, for example, some shallow soils may occur in areas with steep terrain. Nonetheless, much of the Earth's land surface is non-arable. Consequently, the sustainable

TABLE 13.2
Means and Standard Deviations of Laboratory Analysis of Upland Range Diet Samples Collected at Gudmundsen Sandhills Laboratory in 1992 and 1994 (Organic Matter Basis)[a]

Sample Date	CP (%)	UIP (%)	DIP (%)	NDF (%)	ADF (%)	IVOMD (%)
JAN	6.3	0.8	5.5	83.6	52.5	58.0
MAR	6.0	1.0	5.0	82.5	53.3	54.8
APR	11.4	1.2	10.2	77.5	43.2	67.6
JUN	13.8	2.5	11.3	72.4	40.6	67.6
JUL	12.3	2.2	10.1	79.8	43.6	67.5
AUG	11.3	1.8	9.5	77.9	46.4	63.7
SEPT	7.4	1.1	6.4	79.7	48.8	60.7
NOV	5.9	0.7	5.2	84.4	56.1	48.3
DEC	6.5	1.2	5.4	86.0	54.5	53.9

Source: Modified and adapted from Lardy, G.P., and J.S. Caton. In press. Beef cattle nutrition in commercial ranching systems. In: *Ruminant Nutrition Production Systems.* Paris: UNESCO Books.

[a]   Each observation represents 4 to 7 diets collected from esophageal fistulated cows or steers with ruminal cannulae. CP, crude protein; UIP, undegraded intake protein; DIP, degraded intake protein; NDF, neutral detergent fiber; ADF, acid detergent fiber; IVOMD, *in vitro* organic matter disappearance. Dominant grass species included little bluestem (*Schizachyrium scoparium*), prairie sandreed (*Calamovilfa longifolia*), sand bluestem (*Andropogon gerardii*), switchgrass (*Panicum virgatum*), sand lovegrass (*Eragrostis trichodes*), Indiangrass (*Sorghastrum nutans*), and blue grama (*Bouteloua gracilis*). Common forbs and shrubs include western ragweed (*Ambrosia psilostachya*) and leadplant (*Amorpha canescens*).

use of these lands will require sustainable grazing practices using ruminant livestock to produce food and fiber for a growing world population. Much of the non-arable land in the world consists of native rangelands or forested areas. Human inedible forage resources in these areas can be used by ruminant livestock in a diversity of production systems. Because the grasses and forbs that grow on these lands are composed primarily of cellulosic materials, the use of ruminant livestock is required to convert them into a usable human food resource.

Grazed native forages exhibit wide variations in nutrient quality and quantity (Table 13.2). Consequently, careful nutrient supplementation during periods of extensive grazing is often needed and practiced to enhance livestock production efficiency and improve the utilization of grazed forages from non-arable grazing lands (Caton, Freeman, and Galyean, 1988; Johnson et al., 1998). Numerous food-processing by-products have played and will continue to play a major role in providing supplemental nutrients to animals in extensive grazing situations. These include soybean meal, cottonseed meal, distiller's grain by-products and other by-products from the ethanol and brewing industry, wheat middlings, and many others. The rise of soybean meal from a waste product of the soybean oil industry to one of the primary protein supplements used in the livestock industry is an amazing story. This rise from waste material to high-value feedstuff is a common story for other crop processing by-products having been repeated in numerous instances in the past and will most certainly occur again in the future.

CROP RESIDUES AND FOOD-PROCESSING BY-PRODUCTS

Crop residues include straws, stovers, and other plant materials associated with the production of food crops. Large amounts of crop residues are produced annually and are generally low in energy and protein and inedible by humans. During the processing of crops into foods and other products, large amounts of by-products are produced. Examples of these include sugar beet pulp, distiller's grain by-products, oilseed meals, hulls and screenings, fiber and bran fractions, and many other residues, which are generally inedible for humans, but are usable sources of nutrients for livestock. Crop residues and food-processing by-products have widely variable nutrient compositions (Table 13.3) and care needs to be taken when including them in livestock diets to ensure that expected and actual nutrient compositions are in agreement. Crop residues are usually high-fiber and low-protein products that are well suited for use in a variety of ruminant production systems. On the other hand, many food-processing by-products, such as oil seed meals, are quite high in

TABLE 13.3
Nutrient Composition for Ruminants of By-Product Feedstuffs Expressed on a Dry Matter Basis

By-Product	Metabolizable Energy (MJ/kg)	Crude Protein (g/100 g)
Miscellaneous		
Almond hulls	7.7	2.1
Bagasse	6.3	1.5
Beet pulp	12.6	9.7
Brans	12.8	17.2
Brewers grains	10.4	25.4
Citrus pulp	12.6	7.3
Whole cottonseed	16.0	23.0
Molasses[a]	11.8	7.2
Cakes		
Soybean	13.8	49.9
Ground nut	12.5	52.3
Sunflower seed	10.3	49.8
Rape and mustard seed	11.0	40.6
Cottonseed	12.3	45.6
Palm kernel[b]	11.0	40.6
Copra	12.1	23.4
Sesame seed	12.5	49.1
Miscellaneous cakes[b]	11.0	40.6
Corn germ meal	11.9	22.3
Corn gluten feed and meal[c]	13.8	33.9
Soap stock oils	30.6	0.0
Crop residues		
Wheat	6.3	3.6
Rice[d]	6.2	4.3
Barley	7.2	4.3
Maize	7.4	5.9

Source: From National Research Council. 1989. *Nutrient Requirements of Dairy Cattle*, 6th ed. Washington, DC: National Academy Press; CAST. 1999. *Animal Agriculture and Global Food Supply*. Ames, IA: Council for Agricultural Science and Technology. Task Force Report No. 135.

[a] Average of low-quality sugar cane molasses and molasses from sugar beets.

[b] Same as rape and mustard seed composition.

[c] Assume 20% corn gluten meal and 80% corn gluten feed.

[d] National Research Council (1984) used for rice crop residue.

amino acids and other nutrients and low in fiber, making them major components of non-ruminant livestock diets.

Crop residues have been a staple ingredient in a variety of ruminant production systems worldwide. In particular, cereal grain straws, corn stover, and other fibrous residues have been widely used in ruminant production systems as both grazed and harvested feedstuffs. The choice to harvest or graze these materials depends on a variety of factors related to availability, cost, climate conditions, and mechanization. Grazing these materials has the advantage of delivering nutrients back to the field in the form of manure without fossil fuel inputs required in more intensive production practices. However, in many production environments, snow cover or other weather conditions may limit grazing activities. Furthermore, the need to provide properly formulated diets to optimize production may also require that crop residues, straws, and stover be mixed with other dietary ingredients to provide a well-balanced diet.

A variety of methods for improving utilization of these fibrous residues has been tested over the years. Application of various physical and chemical treatments can increase utilization of straws and stover in particular, but in many cases, they are not cost effective under current production practices. In the future, it may be necessary to explore these and other methods of improving utilization of these high-fiber materials because provision of high-quality animal protein products may require improved utilization of such materials.

Many different types of food-processing by-products have been used in ruminant diets either as supplements or as staple dietary ingredients. While the exact time when this practice began is not known, it likely dates to civilizations that undertook various attempts at food processing as the need to make use of these materials led those cultures to readily use them as feedstuffs.

Almost every food that is processed for use as a human food ingredient produces a corresponding food-processing by-product. For example, the use of wheat for bread flour or durum to produce semolina (used in pasta production) results in the production of a variety of wheat processing by-products, which are commonly referred to as wheat middlings. This by-product finds widespread use in a variety of ruminant and non-ruminant diets as a source of protein and energy. Processing vegetables such as sweet corn for human use results in cannery waste, a high fiber, high-moisture by-product commonly used in beef and dairy cattle production systems located near these canneries and factories. The processing of citrus crops such as oranges for juice production results in the production of citrus pulp, which is commonly fed to beef and dairy cattle in tropical locales.

There are many reasons why these materials are readily used in ruminant production systems as opposed to use as human foods. In some cases, these materials are quite fibrous (e.g., sugar beet pulp, rice bran), while in others poor protein quality may be a factor (e.g., corn processing by-products such as corn gluten feed). In other situations, the sheer volume of the material precludes its use in human food applications except in small quantities (e.g., by-products of ethanol production from corn or other cereal grains, wheat middlings). In many cases, a combination of these and other factors preclude wide-scale use of these materials in human food applications.

Food-processing by-products play an important role in many different ruminant production systems worldwide. As an example, with the advent of the fuel ethanol industry, the use of distiller's grain by-products from ethanol production as a feedstuff in beef and dairy cattle production systems in the United States has risen from essentially nothing to volumes that rival the use of cereal grains. Other products such as sugar beet pulp find widespread use in a variety of dairy cattle diets worldwide because it is a relatively digestible source of fiber.

WORLD POPULATION GROWTH AND FOOD SUPPLY: THE ROLE OF ANIMALS

World population projections estimate a population of 7.7 billion by 2020 and nearly 9.2 billion by 2050 (United Nations, 2008, medium variant; Figure 13.1), which is roughly equivalent to an average annual compound growth rate of 1.2%. The majority of this population growth (95%) is expected to occur in developing countries, where over 75% of the world population lives and where

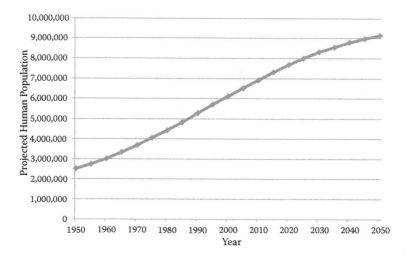

FIGURE 13.1 Past records of world population growth to 2008 and projections for world population growth from 2008 to 2050 (Data from United Nations. 2008. *World Population Prospects: The 2008 Revision.* Population Division, Department for Economic and Social Information and Policy Analysis.)

TABLE 13.4
Past and Projected Trends for Consumption of Meat to the Year 2020

	Total Meat Consumption (mt)			Per Capita Meat Consumption (kg)		
Region	**1983**	**1993**	**2020**	**1983**	**1993**	**2020**
China	17	39	89	16	33	63
India	3	4	8	4	4	7
Other East Asia	2	4	8	22	44	70
Other South Asia	1	2	5	6	7	10
Southeast Asia	4	7	18	11	15	28
Latin America	15	21	38	40	46	57
WANA	5	7	15	20	20	23
Sub-Saharan Africa	4	5	11	10	9	11
Developing world	50	89	194	15	21	31
Developed world	88	99	113	74	78	81
United States	25	31	37	107	118	114
World	139	188	306	30	34	40

WANA = Western Asia and North Africa.
Source: Adapted from CAST. 1999. *Animal Agriculture and Global Food Supply.* Ames, IA: Council
for Agricultural Science and Technology. Task Force Report No. 135.

recent trends are for rapid increases in the consumption of animal products (CAST, 1999). In fact, from the early 1970s into the mid-1990s, the increase in meat consumption in developing countries was nearly three times the increase in developed countries (Delgado, 2003). Consequently, the demand for animal products is projected to increase dramatically (Table 13.4) by 2020 and beyond. In fact, increases in the demand for animal products by world populations have been clearly documented for several years (Bradford 1999; CAST 1999; Delgado, 2003). Delgado states that

by 2020, the share of developing countries in total world meat consumption will expand from 52% currently to 63%. By 2020, developing countries will consume 107 million metric tons (mmt) more meat and 177 mmt more milk than they did in 1996/1998, dwarfing developed country increases of 19 mmt for meat and 32 mmt for milk. The projected increase in livestock production will require annual feed consumption of cereals to rise by nearly 300 mmt by 2020. (Delgado, 2003)

The rapid rise in world populations, increased growth of developing countries, and the concomitant rise in demand for high-quality animal products has begun what Delgado (2003) has described as a "livestock revolution," which, in terms of economic impact and nutrient supply to the human population, will likely far outpace the "green revolution" of a few decades ago. The projected increase in cereal grain production to fuel expanding livestock production for human foods is based on existing proportions of cereal usages to produce animal products. Alternatively, if a greater proportion of crop residues and food processing products were used as animal feeds in conjunction with a larger proportional increase in ruminant livestock production, then projected increases in cereal grain production for livestock feeds would be somewhat muted. Another important fact to consider when viewing the overall importance of animal agriculture to nutrient supply is the impact on economies. In the United States alone, the total economic value added to the economy for all crops and livestock is approximately $281.6 billion with the combined impact of livestock and feed grains representing 59% of the total (Harris et al., 2009; Wilson, Dahl, and Reynolds, personal communication).

SUMMARY AND CONCLUSIONS

Ecosystems of both arable and non-arable lands have developed in the presence of animals and consequently ecological succession is clearly shaped by the plant–animal interface. Sustainable coexistence of plant and animal agriculture is underpinned by the health and well-being of each component. The utilization of inedible feed resources for food animal production contributes to both human and animal welfare by improving ecological balance and expanding animal and human food resources. As we look to the future, it is critically important that society understands and appreciates the unique ability of animals, particularly ruminants, to convert fibrous materials to usable protein and fiber. Ruminant animals play an essential role in the well-being of societies throughout the world. In fact, throughout history, ruminants have provided meat, milk, fiber, and a source of draft power to almost every civilization and they will play a vital role in meeting the needs for food, fiber, and clothing in the future. Since ruminants are able to utilize a variety of forages and feedstuffs that other animals, including humans, cannot use, they will play an increasingly valuable role in the provision of meat, milk, and fiber in the years to come. The variety of forages and feedstuffs that can be utilized in ruminant diets allows them to exist in every corner of the globe. Ruminants will increase in importance because of the roles that they will play in solving many of the complex problems associated with world population growth. As global population increases dramatically in the next 50 years, ruminant production systems will play a pivotal role in providing a sustainable source of nutrients for much of the world's population.

REFERENCES

Anderson, N. 2000. The ancient cow contract—ergonomics, health and welfare issues in dairy cattle housing. *National Mastitis Council Regional Meeting Proceedings.* pp. 17–24.
Bradford, G.E. 1999. Contributions of animal agriculture to meeting global human food demand. *Livestock Production Science* 59: 95–112.
Campbell, J.R. and J.F. Lasley. 1975. *The Science of Animals that Serve Mankind.* New York: McGraw-Hill.
CAST. 1997. *Contributions of Animal Products to Healthful Diets.* Ames, IA: Council for Agricultural Science and Technology. Task Force Report No. 119.
CAST. 1999. *Animal Agriculture and Global Food Supply.* Ames, IA: Council for Agricultural Science and Technology. Task Force Report No. 135.

Caton, J.S., A.S. Freeman, and M.L. Galyean. 1988. Influence of protein supplementation on forage intake, in situ forage disappearance, ruminal fermentation, and digesta passage rates in steers grazing dormant blue grama rangeland. *J. Anim. Sci.* 66: 2262–2271.

Church, D.C. 1988. The classification and importance of ruminant animal. In: *The Ruminant Animal: Digestive Physiology and Nutrition*. D.C. Church, Ed. Englewood Cliffs, NJ: Prentice Hall, pp. 1–13.

Delgado, C.L. 2003. Rising consumption of meat and milk in developing countries has created a new food revolution. *J. Nutr.* 133: 3907S–3910S.

Fahey, G.C. Jr. and L.L. Berger. 1988. Carbohydrate nutrition of ruminants. In: *The Ruminant Animal: Digestive Physiology and Nutrition*. D.C. Church, Ed. Englewood Cliffs, NJ: Prentice Hall, pp. 269–297.

FAO (Food and Agriculture Organization of the United Nations). 2000. *Land Resource Potential and Constraints at Regional and Country Levels*. FAO Report #90. Available at: ftp://ftp.fao.org/agl/agll/docs/wsr.pdf Accessed November 30, 2010.

Harris, M., K. Erickson, J. Johnson, M. Morehart, R. Strickland, T. Covey, C. McGath, M. Ahearn, T. Parker, S. Vogel, R. Williams, and R. Dubman. 2009. Agricultural Income and Finance Report, U.S. Department of Agriculture, Economic Research Service, AIS-88, December 2009.

Holechek, J.L., R.D. Pieper, and C.H. Herbel. 1989. *Range Management Principles and Practices*. Englewood Cliffs, NJ: Prentice Hall.

Hofmann, R.R. 1988. Anatomy of the gastrointestinal tract. In: *The Ruminant Animal: Digestive Physiology and Nutrition*. D.C. Church, Ed. Englewood Cliffs, NJ: Prentice Hall.

Johnson, J.A., J.S. Caton, W.W. Poland, D.R. Kirby, and D.V. Dhuyvetter. 1998. Influence of season dietary composition, intake, and digestion by beef steers grazing mixed-grass prairie in western North Dakota. *J. Anim. Sci.* 76: 1683–1690.

National Research Council. 1984. *Nutrient Requirements of Beef Cattle*, 6th ed. Washington, DC: National Academy Press.

Owens, F.N. and A.L. Goetsch. 1988. Ruminal fermentation. In: *The Ruminant Animal: Digestive Physiology and Nutrition*. D.C. Church, Ed. Englewood Cliffs, NJ: Prentice Hall, pp. 145–171.

United Nations. 2008. World Population Prospects: The 2008 Revision. Population Division, Department for Economic and Social Information and Policy Analysis, United Nations, New York.

Van Soest, P.J. 1982. *Nutritional Ecology of the Ruminant*. New York: Cornell University Press.

Westendorf, M.L. 2000. Food waste as animal feeds: An introduction. In: *Food Waste to Animal Feed*. M.L. Westendorf, Ed. Ames, IA: Iowa State University Press, pp. 3–15.

Wilson, W., B. Dahl and L. Reynolds. Personal communication, North Dakota State University.

14 Welfare, Health, and Biological Efficiency of Animals through Genetics and Biotechnology

Fuller W. Bazer, Duane C. Kraemer, and Alan McHughen

CONTENTS

INTRODUCTION

Dramatic advances in systems biology, including genomics and other new knowledge of gene function and interactions with the environment are constructive forces in improving the health and well-being of both humans and animals. The impact of these genetic effects and advances in biotechnology on animal welfare are explored in this chapter. Long-term food security for animals and humans will depend on advances in efficiency of animal and crop production resulting from biotechnological breakthroughs. Without these efforts, food security for humans and animals will be compromised due to the rapidly increasing human population and a diminishing amount of arable land for production of agriculturally important species of plants and animals.

There is an urgent need for deep sequencing, mapping, and analyses of genomes of animals, plants, and microbes of importance to both the agricultural and biomedical communities. The human genome project has provided revolutionary insights for human health; however, an abundant and safe food supply is fundamental to human health and quality of life. Thus, we must embrace genomics, proteomics, and bioinformatics to explore genomes of animals, crops, and microbes that affect human and animal health. Knowledge of the genome is the beginning, but the next challenge is to use genomic information to understand biology by using genomic sequences to decipher the structure and function of proteins encoded by the genome. This will be achieved by fulfilling the promise of systems biology.

SYSTEMS BIOLOGY

Systems biology requires an interdisciplinary/multidisciplinary approach to understand complex biological systems by interrogating the genome to understand how genome-based molecular processes of cells are linked to higher biological functions. The goal is to understand the genomic bases for low- and high-level function, normal versus abnormal, and disease-susceptible versus disease-resistant specified by gene networks. With that knowledge, one can predict how a change in expression of one gene or a few genes can change the whole organism. The extreme complexity of biological systems requires that reliable predictions be based on outcomes of sophisticated interrogation of the genome of a given species using supercomputing models. Using a combination of statistics, supercomputing, and genome-based molecular biology, biologists, statisticians, and engineers interact to define ways to predict how changes in the cellular environment alter the flow of information through the multiple genome-based pathways controlled by various cell-signaling molecules. Using this approach, biology is expected to move beyond the descriptive stage to join the quantitative sciences where prediction is based on understanding. The vision is to develop and advance antecedent sciences and to transfer technology emanating from the interrogation and translation of genomics biology to improve human and animal health, production agriculture, food safety, and biosecurity.

A primary goal of systems biology is to develop disease signatures that decode patterns of genome-level activity that indicate the presence of a particular disease in humans and animals. Translational genomics will then focus on discoveries for prevention, diagnosis, and therapeutic strategies for treatment of inherited and pathogen-based diseases. Genome signal processing is an outcome of a multidisciplinary approach to integrate understanding of genomics with theory and methods of traditional signal processing that represent disease signatures or production animal agriculture signatures. These signatures are based on decoding patterns of genomic-level activity that signal desired genome signal processing for desired production characteristics of animals used for food, food biosecurity, and food safety for healthy animals and a healthy society.

Systems biology provides a key link between animal agriculture and medicine. As noted previously, knowledge of the human genome has provided revolutionary insights for improved human health, but knowledge of genomes of animals through designed experiments will reveal the genetic bases for resistance to disease and parasites and efficient animal production systems. Rapid advances in understanding the genomic bases for extreme variation among individuals with respect to resistance or susceptibility to disease and parasites and desired phenotype will be accelerated by orders of magnitude using the power inherent in technologies of genomic signal processing. The use of laboratory and domestic animals to do experiments that we are not able to do in humans will allow scientists to unravel genomics and cellular networks responsible for such conditions as natural resistance to diseases and parasites, cancer, cardiovascular diseases, neurological disorders, reproductive health, and nutritional and metabolic diseases. Functional genomics and proteomics will then be used to develop therapeutic strategies to enhance the health and well-being of both humans and animals, as well as strategies for ensuring a safe, abundant, and affordable supply of food to our global society.

GENETICS AND GENOMICS IN CONVENTIONAL ANIMAL BREEDING PROGRAMS

The value of animal agriculture enterprises (poultry, livestock, and fish) in the United States was estimated at $173 billion in 2007 (USDA, 2008a), with the value expected to increase together with increases in both world population and standard of living (Pinstrup-Andersen and Pandya-Lorch, 1999). Although techniques for cloning and producing transgenic animals are becoming more efficient, only commercial production of transgenic fish is poised to affect availability of food animal protein in the near future.

Modern breeds of livestock have achieved high production efficiencies because of traditional animal breeding programs. Between 1945 and 1995, for example, milk production increased threefold; egg production by laying hens increased from 134 to 254 per year; production time of broiler chickens to 3 lb, 15.4 oz (1.8 kg) body weight decreased from 84 to 43 days on one-half the feed; and growth of pigs, sheep, goats, and cattle was faster and resulted in leaner meat (National Research Council, 2002, 2004). These increases can be attributed to various factors depending on species and production systems, including:

1. The use of statistical models to predict breeding values of bulls coupled with sire testing and selection
2. Cross-breeding and artificial insemination (AI) to capture the best genetics from males
3. Synchronization of estrus and ovulation to enhance use of AI
4. Superovulation, AI, and embryo transfer to take advantage of desired genetics from females
5. Artificial incubation of eggs of poultry species to increase hatching rates
6. Improved nutrition
7. Effective disease control through improved animal health
8. Control of seasonality or photo-period to enhance production efficiencies in specific species such as poultry
9. Improved housing to avoid stress resulting from adverse effects of weather
10. Sex reversal in fish to either all female or all male to achieve desired production efficiencies in farm-raised fish

Since the 1960s, more advanced biotechnologies have been used to a limited extent. These biotechnologies include assisted reproductive technologies (*in vitro* maturation of oöcytes and *in vitro* fertilization), embryo splitting to achieve identical twins (clones), sexing sperm, and blastomere nuclear transfer cloning (Norman et al., 2004). Recombinant bovine growth hormone is also a product of biotechnology adopted by some in the dairy industry to sustain lactation performance of cows (see National Research Council, 2004). The value of animal agriculture enterprises in the United States and globally is expected to increase in concert with increases in both world population and standard of living (Pinstrup-Andersen and Pandya-Lorch, 1999).

BIOTECHNOLOGIES FOR ENHANCING ANIMAL HEALTH AND ANIMAL PRODUCTION

Sequencing and mapping genomes of livestock allows scientists to identify genes and understand their regulation in the context of improving production characteristics and health of animals. One outcome is establishment of linkages between inheritances of a desirable trait, for example, milk yield, and segregation of specific genetic markers coupled to that trait. The use of marker-assisted selection can be based on quantitative trait loci (QTL) or, within a QTL, identification of a single-nucleotide polymorphism (SNP) associated with desired traits. A QTL is a marker associated with the presence of a gene of interest, for example for resistance to disease or growth and development characteristics. An SNP is a change in a single nucleotide in DNA that results in a change in the protein for which it encodes. A recent example was the discovery that the difference in size of dogs (e.g., Great Danes and Chihuahuas) is due to differences in frequency of the SNP 5A allele of insulin-like growth factor 1 (IFG1) (Sutter et al., 2007).

There are also QTL for production traits in cattle and swine. A growth hormone receptor variant on bovine chromosome 20 affects milk yield and composition (see http://www.foodproductiondaily. com/Supply-Chain/Gene-identified-to-regulate-milk-content-and-yield) and predicts an increase in milk production of 200 kg per lactation and a decrease in milk fat from 4.4 to 3.4%. A QTL on the long arm of pig chromosome 8, identified as secreted phosphoprotein 1 (SPP1 or osteopontin) is associated with increased litter size (two pigs per litter) and increased prenatal survival of piglets (King et al., 2003). An SNP in SPP1 has also been associated with growth traits and twinning in a population of beef cows selected for high twinning rate (Allan et al., 2007). It is likely that identification of additional QTL and SNPs will have a great impact on the livestock industry. This technology can be coupled with biopsy and genetic analyses of pre-implantation embryos to allow one to choose embryos with the desired genotype to enhance genetic progress in breeding programs. In addition, embryos can be sexed to benefit the animal production enterprise, for example, all females for dairy farms, or semen can be sorted as X chromosome and Y chromosome sperm to achieve the desired sex of offspring.

Current estimates are that there are some 2500 unique genetic phenotypes in animals, excluding humans and mice (Online Mendelian Inheritance in Animals, http://omia.angis.org.au/) with about one-half of them mimicking clinical features of known human genetic disorders. Advances in genomic technologies continue to accelerate so that the current cost to sequence a genome is now tens of thousands rather than millions of dollars and this can be done in days or weeks rather than in months and years. The goal now is to sequence a genome for $1,000 in hours. Scanning genomes for mutations, deletions, and duplications associated with disease in hours is now state of the art technology to interrogate the genomes of animals.

Genomics biology has moved beyond sequencing the genome to defining gene products at the transcriptional level using RNA-Seq or so-called next generation sequencing of the entire transcriptome of organisms. The transcriptome refers to all of the messenger RNAs encoded by genes in the genome of an organism. In principle, RNA-Seq allows determination of the absolute quantity of every molecule in an animal or population of cells. A powerful advantage of RNA-Seq is that it can capture transcriptome dynamics across different tissues or conditions without sophisticated normalization of data sets (Mortazavi et al., 2008). Therefore, it is used to monitor gene expression cell growth and differentiation, track gene expression changes during development, and provide a "digital measurement" of gene expression difference between different tissues. RNA-Seq is invaluable in advancing understanding of transcriptomic dynamics during development and normal physiological changes, and in allowing for robust comparison between diseased and normal tissues, as well as the subclassification of disease states. RNA-Seq allows study of complex transcriptomes to identify and monitor changes in expression of rare RNA isoforms from all genes with greater coverage and depth of sequencing while determining structure and dynamics of the transcriptome (Wang et al., 2008).

Copy number variations (CNVs) in alleles of genes is another focus in genomics because the number of alleles of genes can vary due to deletions (no copy or fewer copies) or duplications (multiple copies), which are associated with mutational mechanisms. CNVs may influence the evolution of gene regulation with the potential to be the basis for causal variants on trait-associated haplotypes. Available results suggest that resources be targeted to identifying genetic variation underlying the "missing" heritability for complex traits that remains unexplained. Although common CNVs are unlikely to account for much of this missing heritability, the strength of selection acting on exonic and intronic deletions suggests that CNVs contribute to rare variants involved in common and rare diseases, as well as variation in other aspects of the phenotypes of animals (Conrad et al., 2010).

The use of genome-based biotechnologies to enhance both animal health and animal production characteristics is desirable in that one can capitalize on normal biological variation in a population. This is done by comparing genomics of animals that are resistant and susceptible to disease or that have high versus low production traits (e.g., milk yield) to identify genetic markers associated with the desired phenotype and then using genomic markers in selection of genotypes that will favor the desired phenotype (health status or production traits). Any controversy associated with cloning and transgenic animals is avoided and the methodology can be applied to large populations of animals used in food production enterprises.

THE UTILITY OF GENETICALLY MODIFIED AND CLONED ANIMALS

Cloning and genetic engineering are very different biotechnologies, but they are often combined in discussions of ethics and animal welfare. Perhaps this is because they can be combined to produce individual animals; however, they will now be discussed individually and then jointly.

Cloning

Cloning of animals is the production of genetically identical individuals. This can occur naturally in the birth of identical (monozygotic) twins (or in the case of nine-banded Armadillos, quadruplets). Up to four identical cattle offspring have been produced by placing single blastomeres from a 4-cell embryo into empty zona pellucidae (a membrane that normally surrounds early embryos), developing them *in vitro* and then transferring them to different recipient females who carried the pregnancies to term (Johnson et al., 1995). The most common current use of the term cloning refers to the biotechnology often referred to as nuclear transfer. When the donor cells are from early stage embryos (Willadsen 1986), it is referred to as embryonic cell nuclear transfer (ECNT). When the cells are from fetuses, or juvenile or mature animals (Campbell et al. 1996; Wilmut et al. 1997), it is called somatic cell nuclear transfer (SCNT). This involves the transfer of a nucleus from a cell of the animal to be cloned into a mature oöcyte (ovum or egg) from which the nucleus has been removed or possibly inactivated. Clones produced in this manner are less identical than those produced naturally, or by embryo splitting, because the mitochondrial DNA (primarily maternal from the ovum) of the offspring is usually that of the recipient ovum, which is usually obtained from a different female of the same or closely related species. Theoretically, it is possible to obtain both the nucleus and the ovum from the same female, but the efficiency of the process is very low and the numbers of ova available from a single female are too few to make this practical. It is important to realize that the sperm mitochondria generally are not transmitted to the offspring. Therefore, even though mitochondria of a male clone are not identical to the original nucleus donor, this difference would not be a factor in the genetics of his offspring.

There are several methods for introduction of the donor cell nucleus into the enucleated (recipient) ovum. They include fusion of the donor cell with the recipient ovum. This can be performed using micromanipulators to introduce the cell into the perivitelline space inside the zona pellucida, followed by electrofusion of the two cells (Willadsen 1986; Wilmut et al., 2002). Alternatively, this can be achieved by microjection of the nucleus, the nucleus plus part of the cytoplasm, or even the entire donor cell directly into the recipient ovum (Lacham-Kaplan et al., 2000). Yet another approach is to remove the zona pellucida from the recipient ovum by either micromanipulation or enzymatic methods, followed by fusion of the donor cell by either chemical or electrical methods (Vajta et al., 2001). These various approaches to nuclear transfer cloning vary between species and laboratories. The main animal welfare concern is the relative inefficiency of the process that influences the large numbers of animals needed to produce the cloned offspring.

Numerous other factors influence the efficiency of the cloning process and the health of the offspring. They include reprogramming of the nuclear DNA, initiation of cell division (activation), *in vitro* culture of the resulting embryo prior to transfer to the recipient uterus, transfer of the embryo to a recipient female (if mammals), plus delivery and postnatal care of the offspring and recipient. Some of these, such as activation and embryo transfer, are fairly well developed for the livestock species. However, others, such as reprogramming and embryo culture, are in need of further improvements. Incomplete reprogramming of the donor cell DNA is probably responsible for much of the embryonic and fetal loss during pregnancies and the developmental abnormalities that are observed in approximately 20% of the offspring (Cibelli et al., 2002; Hill et al., 2002; Panarace et al., 2007). It is important to realize these are the same abnormalities that occur naturally and they are generally not passed on to the offspring of clones. From an animal welfare standpoint, it is essential to have appropriate veterinary care available, or to euthanatize the abnormal offspring humanely, just as should be done for naturally produced abnormal offspring.

One of the risks often mentioned in the use of cloning biotechnology is the reduction of genetic diversity. However, this risk is a function of how cloning technology is used. If cloning is used to produce offspring from valuable animals that cannot otherwise reproduce, it can actually increase genetic diversity (Wells et al., 1998; Westhusin et al., 2007). Another example of this is the use of donor cells from castrated cattle that exhibit outstanding carcass and meat characteristics such as tenderness and flavor that are difficult to measure in live animals. Somatic cells can be collected from these animals several days after their death (if the carcasses are cooled) and used to produce breeding bulls that could transmit the desired genetic components of these traits to their offspring.

It is important to realize that genetically identical animals are not phenotypically identical. This is true of naturally born animals and those animals that are produced using biotechnologies such as cloning. This is mainly due to epigenetic factors that influence how the genes are expressed (i.e., when and in which cells they are turned on or turned off).

In January 2008, the U.S. Department of Health and Human Services, Food and Drug Administration (FDA), Center for Veterinary Medicine (CVM) published Guidance for Industry (179); Use of Animal Clones and Clone Progeny for Human Food and Animal Feed (USD HHS, 2008). The FDA's CVM had previously published a risk assessment titled "Animal Cloning: A Risk Assessment" (US FDA, 2008a). This publication addressed the impact of SCNT on the health of animals involved in the process, and on humans and animals that consume the products of animal clones and their progeny. The risks were evaluated in the context of the use of other artificial reproductive technologies (ARTs) and conventional animal agriculture. This publication includes over 800 references and is available online (http:www.fda.gov/AnimalVeterinary/SafetyHealth/AnimalCloning/ucm124840.htm). Also in January 2008, the FDA published an article on its Consumer Health Information Web page (www.fda.gov/consumer) stating that "meat and milk from clones of cattle, swine (pigs), and goats, and the offspring of clones from any species traditionally consumed as food, are as safe to eat as food from conventionally bred animals." It also points out that "the main use of clones is to produce breeding stock, not food" (US FDA, 2008b).

Most international governmental agencies that have considered the safety of human consumption of food products from animal clones and their offspring agree that their products are as safe as products produced naturally by members of their species. However, there is reluctance to approve the marketing of these products for a variety of reasons other than food safety. The USDA has continued its voluntary moratorium on marketing cloned animal food products (USDA, 2008b), primarily over concerns for the impact on foreign markets for U.S. products.

On October 19, 2010, a "Report from the Commission to the European Parliament and The Council on Animal Cloning for Food Production" was published (http://www.euractiv.com/eu/cap/parliament-calls-eu-ban-cloning-food-news-496089). It indicates that the Commission will propose temporary (5 years) suspension of the use of the cloning techniques in the European Union (EU) "for the reproduction of all food producing animals; the use of clones of these animals; and the marketing of food from clones," and to "Establish the traceability of imports of semen and embryos to allow farmers and industry to set up data bank(s) of offspring in the EU." Cloning would remain possible for all purposes other than for food production such as research, pharmaceutical production, and animal conservation. This recommendation is not based on information that food products of clones and their offspring are unsafe, but that there is a need for more information on the subject. Their main concerns are apparently animal welfare and ethics.

A major welfare issue about the cloning procedure is that only 20 to 30% of pregnancies continue to term; approximately 25% of the pregnant recipients develop hydrops; 20 to 25% of offspring have developmental abnormalities; and 30 to 40% of the calves die before 150 days of age (Panarace et al., 2007) Most of these problems are thought to be due to errors in reprogramming of the DNA of the donor cell after its incorporation into the cytoplasm of the recipient ovum. This is an area undergoing considerable research, and improvements in efficiency of SCNT will probably be made during the next five years. The International Embryo Transfer Society has published recommendations titled "Health Assessment and Care for Animals Involved in the Cloning Process" (IETS, 2008). It recommends that the pregnancies be monitored carefully and that abnormal pregnancies be terminated as early as possible to minimize the adverse health effects on the recipient female and to prevent the birth of unhealthy offspring.

One of the ethical concerns is consumer awareness and labeling. The major cloning companies in the United States addressed this issue by developing a Livestock Cloning Supply Chain Management Program. It complies with the continuing USDA voluntary moratorium on marketing of food products from cloned cattle and pigs. The program includes an animal cloning registry and an incentive program to encourage participation by their clients.

GENETIC ENGINEERING

Recombinant DNA technologies provide a broad array of opportunities for improvement of human health and animal health and well-being. Among those opportunities is the production of transgenic animals, which are among a variety of genetically modified organisms often referred to as GMOs. This activity is sometimes identified as genetic engineering (GE) or gene (DNA) transfer. The objective is usually to add, remove, or rearrange DNA to modify its function or that of its products. An advantage of this technology is that one can modify a single gene locus without perturbing the remainder of the genome. This topic has been extensively reviewed by scientists working in this area (Robl et al., 2007; Laible, 2009; Wall et al., 2009; Thompson et al., 2010), review panels established by agencies of the U.S. government (National Research Council, 2002, 2004), and industry (Council for Agricultural Science and Technology, 2010).

The five basic methods for gene transfer in animal production are described briefly as follows:

1. Pronuclear injection: This was the first method for production of transgenic animals. It was first used in mice and then applied to livestock (Pursel et al., 1989). It involves injection of a DNA construct into the sperm (male) pronucleus of the recently fertilized ovum (zygote). A disadvantage of this method is that there is limited control of the site, or number of copies of the construct integrated into nuclear DNA.
2. Viral vectors: Replication defective viral vectors, most recently lentiviral (Lois et al., 2002; Hofmann et al., 2004), are incorporated into the construct. These constructs may be introduced into the perivitelline space of the ovum or zygote, and they deliver the DNA to the nucleus of the zygote. These viral vector systems are being intensively studied to improve their effectiveness and to evaluate their safety.
3. Sperm-mediated: Because spermatozoa bind exogenous DNA, they can be used to mediate gene transfer at the time of fertilization (Perry et al., 1999). The efficiency of this approach may be increased by intracytoplasmic sperm injection (ICSI) following disruption of the sperm membrane prior to incubation with the DNA to be transferred. Other approaches such as electroporation, liposomes, monoclonal antibodies, and restriction enzymes are being explored to improve the efficiency of this method of gene transfer (Lavitrano et al., 2006).
4. Combining gene transfer and cloning: The donor somatic cells can be genetically engineered using a variety of methods including electroporation or viral vectors. Transgenetic offspring are then produced by SCNT (Lai et al., 2002). The advantage of this method is

that multiple genetic modifications can be made and validated before the nucleus is transferred. A disadvantage is that SCNT is not very efficient.

5. Ectopic DNA transfer: Direct administration of gene constructs, or transgenic stem cells into non-reproductive tissues of fetuses or living animals will result in transgenic animals, but they are not germ line transgenic (Draghia-Akli et al., 2002; Khan et al., 2002). Although these procedures may have profound effects on the individuals, the transgenic traits are not passed on to future generations via the gametes, as is the case for the other methods.

For all of these methods, DNA constructs are prepared *in vitro*. These constructs vary considerably, but most contain the gene of interest (which can be from any species), a promoter segment that influences the location in the body and timing of expression (function) of the gene, enhancer sequences that amplify gene function, and often a marker gene that can be used to detect incorporation of the DNA into the genome of the animal. A recent addition to construct design is the use of RNA interference (RNAi) technology to control gene function (Long, 2010).

BENEFITS FROM PRODUCING TRANSGENIC ANIMALS

There are numerous benefits from producing transgenic livestock. In some cases, there is benefit to both the animals and humans. In others, the benefit may be to animals only, or to humans only.

GE That Is Beneficial for Livestock and Humans

Genetically engineering animals with resistance to zoonotic diseases is expected to reduce the incidence of these diseases in animals, thereby improving their health and well-being, as well as reducing the possibility of those diseases, if zoonotic, being transmitted to humans. Likewise, production of recombinant vaccines against the zoonotic diseases for use in livestock would reduce animal suffering and reduce human exposure, as well as serve as models for the development of vaccines for use in humans. Of the known 69 zoonotic diseases in 28 species of animals, at least 60 of those occur in livestock and poultry (see FASS, 2010). Notable examples of these diseases are brucellosis, tuberculosis, salmonellosis, bovine spongiform encephalopathy, cholera, and rabies (Golding et al., 2006; Richt et al., 2007). An example of a GE modification that fits into this category is mastitis resistance in dairy cattle (Wall et al., 2005).

Genetically engineering animals to produce food and fiber more efficiently might reduce the total number of animals used in animal agricultural enterprises, thereby reducing their negative impact on the environment. This impact can also be reduced by genetically modifying the environmentally damaging waste products produced by livestock and other animals (Golovan et al., 2001). Animal transgenesis can also be used to produce animal models for biomedical research that provides basic information for improving human and animal health and well-being.

GE That Is Beneficial Mainly for Livestock and Other Animals

Genetically engineering animals for resistance to non-zoonotic diseases would benefit animals primarily, but would give comfort and economic benefits to their owners (see Council for Agricultural Science and Technology, 2010). An example of this type of modification is pseudorabies virus-resistant pigs. This is also true for genetically engineering animals to be more adaptive to their environment, especially if global warming continues.

GE of Animals Mainly to Benefit Humans

Animals that are genetically engineered to improve food products, such as decreased saturated fats in meat and milk would benefit the consumer. Livestock producers and the consuming public would be the main beneficiaries from GE changes that increase feed efficiency and product quantity and quality. Animals that are genetically modified to produce medicines in their milk or organs for transplantation to humans would clearly benefit humans more than the animals (USDA, 2008b). However, some of the pharmaceuticals are likely to be beneficial for treating animals.

REGULATIONS FOR GENETICALLY ENGINEERED ANIMALS

Clearly, there are numerous situations and circumstances in which GE would improve the well-being of animals, both domestic and wild. If livestock producers are to meet the demands for live-stock products in the future while maximizing the well-being of livestock and other animals, it will be important to use the biotechnologies of GE and cloning wisely. In January 2009, the FDA-CVM published Guidance for Industry Number 187 titled "Regulation of Genetically Engineered Animals Containing Heritable Recombinant DNA Constructs" (FDA-CVM, 2009). This publica-tion contains non-binding recommendations to guide the industry regarding heritable constructs. It states that the FDA intends to regulate non-heritable constructs in much the same manner as for heritable constructs. It points out that the USDA Animal Health Inspection Service (APHIS) regu-lates DNA constructs that are veterinary biologics, and that the EPA may assert jurisdiction over certain GE animals as well. In addition to the regulation of cloning and genetic engineering by the FDA, USDA, and other federal agencies, all research on the biotechnologies must have approval of local Institutional Animal Care and Use Committees, which evaluate the protocols, inspect the facilities and animals, and monitor compliance of the investigators (FASS, 2010).

BIOTECHNOLOGY FOR PRODUCTION OF IMPROVED ANIMAL FEEDS

GE for the ultimate benefit of animals started with genetic transformation of microbes and plants. That is, feed plants or microbes to produce improved inoculants, vaccines, and so forth were geneti-cally modified with the intention of enhancing the well-being of the consuming animal. Here we describe some examples of the use of GE to improve animal welfare without directly modifying the animal. GE is used to introduce new genes, giving rise to novel traits due to the presence of the associated proteins, and to reduce the presence of pathogens or metabolites responsible for diseases or undesired traits, respectively.

INTRODUCING NEW TRAITS

In theory, GE might be used to improve animal feeds by augmenting or enhancing content of certain nutrients, such as amino acids (e.g., increased lysine in cereals and cysteine/methionine in legumes). Such nutritional enhancements would reduce the need for feed supplementation with the respective amino acids. However, as supplementation is routine for deficient feeds any-way, the net result on the animal is likely to be minimal, with the benefit accruing mainly in the form of cost savings and convenience to the producer, who would no longer have to supple-ment the bulk feed. An early example of using biotechnology methods to modify feed, from the 1980s, was the proposed development of transgenic tannin synthesizing "bloat free" alfalfa to benefit ruminants consuming fresh alfalfa. However, although the biochemistry and physiology of bloating is reasonably well understood, the means to achieve the goal remains elusive. Several groups have developed transgenic alfalfa with modulated protein digestibility, especially via the implicated tannins in various ways, but none of these has proved sufficiently efficacious as to warrant commercialization.

A more recent and more promising example of feed improved through adding genetic mate-rial addresses the phytate, problem. Feed for monogastric animals (e.g., pigs, chickens, and fish) contains the elemental nutrient phosphorus, but about one-half to three-fourths of the phosphorus is bound in indigestible phytate, making it nutritionally unavailable. The undigested phosphorus is eliminated in the feces, which then can end up polluting agricultural lands and waterways. This problem is exacerbated with the excess phosphorus leaching into ground water, causing fish kills, algal blooms, and other undesirable environmental effects. However, the problem can be addressed using phytase, an enzyme that breaks down phytate, allowing the phosphorus to be digested by the animal. In conventional practice, phytase or more readily digestible phosphorus is often added as a supplement to the monogastric feed to make up for the deficiency caused by the nonavailability of phosphorus in the phytate. Now, however, several transgenic plants, including rice and corn, have been developed that produce phytase directly. These are now undergoing agronomic testing, but have already been awarded biosafety certificates in China, where they may be released for cultivation soon, in spite of activist opposition (Hepeng, 2010). Having phytase present in the feed will render superfluous the need to add supplemental phytase.

Similar approaches to improving animal feed qualities are being applied to, for example, improving digestibility of glucans in barley, oats, and rye (Zhang et al., 1999). Alternatively, poor quality feed might be improved by modifying cellulose (Hall et al., 1993). However, modified feeds using these approaches have yet to reach the marketplace.

LIMITING PATHOGENS OR REDUCING ANTINUTRIENTS

In addition to adding useful genes to improve feeds directly, GE can restrict certain pathogens or anti-nutrients. In this mode, instead of inserting a gene construct coding for a protein/trait of interest, GE inserts nucleic acid segments to interact with and metabolize or otherwise interfere with the expression of the undesirable feature. The most promising group of these methods include various forms of RNAi, using specific RNA sequences to interfere (hence, "RNA interference") with endogenous protein synthesis (Rana, 2007). This approach to reducing or removing undesirable features might target anti-nutritional factors, such as excessive trypsin inhibitors, saponins, phytate, oxalates, or complex carbohydrates. One example of using RNAi attacks pathogen development in mice by designing an RNA sequence designed to interfere with a specific gene on the pathogenic genome required for disease establishment (Pfeifer et al., 2006). This strategy is being used in offering protection from several pathogens, especially viral pathogens and even bovine spongiform encephalopathy (BSE) (Richt et al., 2007).

UNPLANNED BENEFITS OF GE CROPS

One unplanned animal health benefit of GE crops has been the use of Bt corn as an animal feed. The initial design of Bt corn was to make the corn withstand depredation by insect pests, initially corn borer and, later, other target pests such as rootworm. When ultimately commercialized, the modified corn also showed a dramatic reduction in fungal pathogen infections, with a consequently dramatic reduction in mycotoxin content, especially fumonisins (Wu, 2008). While the obvious benefit of this unintended effect was noted for human consumption, especially in poorer countries lacking regulatory filters of the ubiquitous mycotoxins, animals also benefit from feed corn carrying fewer of these toxins. For example, fumonisins promote leukoencephalomalacia in horses and porcine pulmonary edema in pigs (Wu, 2008). In addition to modifying plants, animals benefit from various modified microbes, for example as vaccines or as supplements to enhance silage and other feeds. Some microbial products increase lactate and reduce pH to reduce diarrhea. However, an analysis of microbial supplements, adjuvants, and vaccines is beyond the scope of this chapter.

CONTRIBUTIONS OF GENETICALLY MODIFIED ANIMALS TO SOCIETY

Rabbits, sheep, pigs, and mice were among the first animals genetically engineered a quarter century ago (Hammer et al., 1985a,b), but unlike transgenic crop plants being developed contemporaneously, transgenic animals have not successfully secured major segments of the food and agriculture market. To date, the only transgenic vertebrate animals in commerce are companion animals, including the green fluorescent protein (GFP)-producing zebrafish (GloFish™), and Alba, a transgenic GFP-producing bunny acquired by Chicago artist Eduardo Kac under suspicious circumstances from French researchers and displayed as modern high tech "art" (http://www.ekac.org/gfpbunny.html#gfpbunnyanchor). Alba's death in 2002 was similarly controversial (http://www.wired.com/medtech/health/news/2002/08/54399), as was the intensity of the rabbit's green glow in a popular online photograph. Other mammals (such as mice and pigs) have been transformed with this same GFP construct, but do not seem to have reached celebrity status or even the marketplace.

Curiously, such genetic alterations to the genomes of these animals are hardly adaptive or usefully pragmatic, and certainly do not appear to benefit the animals in question. One must wonder how long a small glowing fish would last in a natural environment shared with predators. At the same time, the GFP does not appear to harm the animal, although there is scant research into this question. Perhaps even more curiously, there is no record of any regulatory approvals or even regulatory oversight on these transgenic animals, apart from FDA disavowing them (see later discussion under "GloFish"). Regarding animal welfare, one might question the validity of using a controversial technology, rDNA, solely to create a living scientific curiosity to satisfy the artistic expression of an individual human's ego (as in the case of Alba). One might argue that companion animals are already genetically manipulated by humans, using classical animal breeding strategies, to serve human esthetic purposes (i.e., breeding out aggressive or other natural features unpleasant or undesirable from the human perspective, while breeding in features that serve no benefit to the animal other than make them more esthetically appealing to the ephemeral and fluctuating whim of human fashion and style). Such breeding is often benign (e.g., fur color patterns in fully domesticated companion mammals), but some are often detrimental to the welfare of the animal (e.g., those resulting in maladaptive skeletal or physiological features in companion canines). Nevertheless, here is a sample of some representative transgenic animals in or near commerce, starting with GloFish.

GLOFISH™

The glowing zebrafish (GloFish™) was transformed with fluorescence genes similar to those inserted into Alba's genome, and is certainly a commercial success. Since its release in the United States in 2003, the domestic aquaria stars have become available in three glows (red, orange, and green), each derived from bioluminescence genes isolated from marine organisms. At the time of GloFish's initial commercial release, FDA released a statement concerning the regulation of the glowing companion animals:

> Because tropical aquarium fish are not used for food purposes, they pose no threat to the food supply. There is no evidence that these genetically engineered zebra danio fish pose any more threat to the environment than their unmodified counterparts which have long been widely sold in the United States. In the absence of a clear risk to the public health, the FDA finds no reason to regulate these particular fish. (http://www.fda.gov/AnimalVeterinary/DevelopmentApprovalProcess/GeneticEngineering/GeneticallyEngineeredAnimals/ucm161437.htm).

ENVIROPIG™

The Enviropig™ is a transgenic line of Yorkshire pigs expressing phytase, giving them the ability to digest plant phosphorus in phytate more efficiently than conventional swine. Enviropigs produce the phytase in the salivary glands and it is then secreted in the saliva (Golovan et al., 2001). When cereal and soy grains are consumed, the phytase mixes with the feed as the pig chews. The phytase is active in the acidic environment of the stomach, thus degrading virtually all of the phytate in the feed (Forsberg et al., 2003). With this transgenic animal, feeds will not require special mixing, treatment, or supplementation with phytase, and the environmental management of manure would be substantially simplified. Adding a phytase gene to facilitate phosphorus availability and reduce pollution is being pursued in other animals, such as fish (Hostetler et al., 2005) and chickens (Cho et al., 2006). The Enviropig is now under regulatory review prior to commercial release.

HARVARD MOUSE

The Harvard mouse, or "OncoMouse™", is a commercially available transgenic mouse engineered to readily develop cancer and thus serve as a research tool for oncologists and others investigating the onset and progress of tumors in mammalian models. The Harvard mouse cannot be considered a companion animal, and certainly not a human food source. However, as a research tool, it is on the market. From the perspective of animal welfare, it is unlikely that a genetic predisposition to generate neoplasms could be considered in any way beneficial to the mouse.

AQUABOUNTY SALMON

In September 2010, FDA announced its approval (after over 10 years of "deliberation") of the transgenic Atlantic salmon (*Salmo salar* L.), which grows to market size faster than non-transgenic salmon. Thus, this fish is likely to become the first transgenic food animal in commercial production and on consumer's dinner plates. The Atlantic salmon, designated AquAdvantage™, carries a growth hormone gene from its relative Chinook salmon, driven by a promoter from the cold-water inhabiting ocean pout. The intent of the engineered salmon was to have it grow to market size faster than its non-transgenic brethren; hence, the growth hormone and, importantly, the cold-water functional promoter. In the wild, the production of growth hormone in normal salmon diminishes as the water temperature drops, so the growth rate also slows. With the addition of the ocean pout promoter to the hormone gene, growth hormone synthesis continues apace even in colder water (the natural home of the ocean pout) and so the salmon continue to grow. They do not grow to monster size, as feared by some, but instead stop growing at normal size; they just reach that size sooner.

The main concern with these fish is the potential for harm to natural salmon in the event of them escaping into the wild environment. The stated fear is that the transgenic salmon would outcompete their non-GE relatives for food or mates and come to dominate the population to the detriment of non-GE salmon. Although studies suggest the GE fish would not succeed in the wild, the developers have acceded to skeptics' concerns and arranged to farm the transgenic fish in landlocked pens, and to grow only triploid females, incapable of reproducing. The FDA briefing packet concerning the details of AquAdvantage salmon, including the FDA regulatory procedures outline and findings, are posted at: http://www.fda.gov/downloads/AdvisoryCommittees/CommitteesMeetingMaterials/VeterinaryMedicineAdvisoryCommittee/UCM224762.pdf.

The FDAs environmental assessment for the AquAdvantage salmon is at: http://www.fda.gov/downloads/AdvisoryCommittees/CommitteesMeetingMaterials/VeterinaryMedicineAdvisoryCommittee/UCM224760.pdf.

Various other fish and mammals have been transformed; the ability to insert any given gene construct to introduce a given trait into a species and have it grow into a whole, live, fertile adult is not a technological obstacle. The objectives of these various transgenic animal developments vary from simple research experiments using simple marker genes to using the animals as bioreactors to manufacture pharmaceuticals (usually expressed in the milk of mammals, to facilitate isolation and purification of the pharmaceutical), to enhancing the growth of the animal for use as feed or food, to improving the environmental footprint of the animal; for example, in making the feces less environmentally damaging. Few of these transformations are directed at improving the animal's well-being directly, although some clearly do so as a secondary benefit; for example, those facilitating feed digestion or those producing bacteriostatic compounds in the milk that benefit the animal. Transgenic goats producing lysozyme with bacteriostatic properties against mastitis-causing bacteria, for example, incidentally benefit from the technology (Maga et al., 2006).

Early reports on transgenic mammals producing lactoferrin, known to have broad-spectrum antimicrobial activity, gave hope that the transgenic protein would provide some resistance to infections responsible for mastitis in cattle or diarrhea in pigs (see http://www.agnet.org/library/ac/1999d/). Such outcomes would clearly have an animal welfare benefit. Unfortunately, transgenic animals expressing exogenous lactoferrin do not appear to be sufficiently protected for that technology to have commercial value (Hyvönen et al., 2006). However, transforming mammals to biosynthesize lactoferrin is being pursued, not for the benefit of the transgenic animal itself, but as a means to generate large quantities of lactoferrin for other commercial uses (Yang et al., 2008) and to serve as an antimicrobial lactoferrin supplement to feed (Lin et al., 2010). Moreover, transgenic animals are being created to synthesize various proteins and metabolites, especially in their milk, but most of these are to facilitate production, isolation, and purification for commercial use and not for the direct benefit of the transgenic animal (Sabikhi, 2007). One exception, where the objective is to benefit the animal directly, involves transgenic cattle producing lysostaphin in their milk, which kills *Staphylococcus aureus*, a major pathogen giving rise to mastitis (Wall, Powell, and Paape, 2005).

There are many other examples of potentially useful transgenic animals, including insects, mice, cattle, etc., but those are either not intended for marketing or they have not yet completed the regulatory review processes required prior to commercialization. In addition, a range of insects, mosquitoes, worms and other lower animals have been genetically transformed for various purposes over the past several years (usually with marker genes and used purely for research purposes, or else attempting to attenuate disease vectoring capacity). A discussion of those organisms is beyond the scope of this chapter. As for determining the impact of transgenesis on the welfare of these animals, considering that some are pests or pathogens, it is hindered by the lack of objective measures for the well-being of these species.

SUMMARY AND CONCLUSIONS

Biotechnology has many dimensions that, in the context of animal welfare, can be used to improve traits for production (meat and milk production), health and well-being (resistance to disease and parasites), esthetic value (various breeds of dogs and other companion animals), and novelty (GloFish™). These technologies may capitalize on the enormous biological variation among animals within a species for various phenotypes that may allow identification of genetic markers to allow one to select for desirable traits or against undesirable traits. Having identified desirable genes or genotypes, one can then use classical animal breeding, transgenics, or cloning to gain significant numbers of animals with desirable genotypes and phenotypes. There are genetic outcomes that benefit the animal in the case of resistance to disease and parasites; however, most outcomes are realized and exploited only if they benefit production animal agriculture, medicine (Harvard mouse), the environment (Enviropig™), pharmaceutical industry (biopharming pharmaceuticals from milk), human desire for novel animals (GloFish™), or unique companion animals (many breeds of dogs and cats). It has also been noted that plants and microbes have been genetically engineered to enhance qualities that benefit both animals and producers of livestock species.

Society is most uncomfortable with technologies such as transgenesis and cloning of animals and the scientific society recognizes that much needs to be learned to overcome problems with these technologies before either the public at large or the scientific community can accept them. Nevertheless, we have noted that results of numerous studies have indicated that products of cloned or transgenic livestock should be recognized as GRAS (generally recognized as safe), which is a criterion used to assess the safety of plants and plant products that each of us consume on a daily basis.

The evidence is that our global population is rapidly moving from 6 billion to 9 billion, increasing affluence is accompanied by increased demand for animal protein, and wild fish populations are severely over-harvested. Therefore, it seems prudent that we continue to use genomics, transgenesis, and cloning as "tools" in modern animal breeding to realize the most output from the fewest numbers of animals and with the fewest inputs of feedstuffs for which humans and animals compete.

We aspire to capitalize on energy from the sun to produce foods and there is no doubt that textured vegetable proteins will become increasingly important in diets of future generations. Until then, we must employ best practices in production animal agriculture and aquaculture to meet demands of our global societies for adequate, safe, and affordable sources of high-quality animal protein.

REFERENCES

Allan, M.F., Thallman, R.M., Cushman, R.A. et al. 2007. Association of a single nucleotide polymorphism in SPP1 with growth traits and twinning in a cattle population selected for twinning rate. *J Anim Sci* 85: 341–347.

Campbell, K.H., McWhin, J., Ritchie, W.A., and Wilmut, I. 1996. Sheep cloned by nuclear transfer from a cultured cell line. *Nature* 380: 64–66.

Cho, J.K., Choi, T., Darden, P.R. et al. 2006. Avian multiple inositol polyphosphate phosphatase is an active phytase that can be engineered to help ameliorate the planet's "phosphate crisis." *J Biotechnol* 126: 248–259.

Cibelli, J.B., Campbell, K.H., Seidel, G.E. et al. 2002. The health profile of cloned animals. *Nature Biotechnol* 20: 13–14.

Conrad, D.F., Pinto, D., Redon, R. et al. 2010. Origins and functional impact of copy number variation in the human genome. *Nature* 464: 704–712.

Council for Agricultural Science and Technology (CAST). 2010. Ethical implications of animal biotechnology: Considerations for animal welfare decision making: Animal agriculture's future through biotechnology. Issue Paper 46. Ames, IA: CAST.

Draghia-Akli, R., Malone, P.B., Hill, L.A. et al. 2002. Enhanced animal growth via ligand-regulated GHRH mygenic-injectable vectors. *Fed Am Soc Exp Biol J* 16: 426–428.

Federation of Animal Science Societies (FASS). 2010. *Guide for the Care and Use of Agricultural Animals in Research and Teaching*, 3rd ed. Champaign, IL: FASS (http://www.fass.org).

Forsberg, C.W., Phillips, J.P., Golovan, S.P. et al. 2003. The Enviropig physiology, performance, and contribution to nutrient management advances in a regulated environment: The leading edge of change in the pork industry. *J Anim Sci* 81: E68–E77.

Golding, M.C., Long, C.R., Carmell, M.A. et al. 2006. Suppression of prion protein in livestock by RNA interference. *Proc Natl Acad Sci USA* 103: 5285–5290.

Golovan, S.P., Meidinger, R.G., Ajakaiye, A. et al. 2001. Pigs expressing salivary phytase produce low-phosphorus manure. *Nature Biotech* 19:741–745.

Hall, J., Ali, S., Surani, M.A. et al. 1993 Manipulation of the repertoire of digestive enzymes secreted into the gastrointestinal tract of transgenic mice. *Nature Biotechnol* 11: 376–379.

Hammer, R.E., Brinster, R.L, Rosenfeld, M.G. et al. 1985a. Expression of human growth hormone-releasing factor in transgenic mice results in increased somatic growth. *Nature* 315: 413–416.

Hammer, R.E., Pursel, V.G., Rexroad, C.E. et al. 1985b. Production of transgenic rabbits, sheep and pigs by microinjection. *Nature* 315: 680–683.

Hepeng, J. 2010. Chinese green light for GM rice and maize prompts outcry. *Nature Biotech* 28: 390–391.

Hill, J.R. Chavatte-Palmer, P., Cibelli, J.B. et al. 2002. Pregnancy and neonatal care of cloned animals. In: *Principles of Cloning,* Vol.1. J. Cibelli, Ed. San Diego, CA: Academic Press, pp. 247–266.

Hofmann, A., Kessler, B., Ewerling, S. et al. 2004. Efficient transgenesis in farm animals by lentiviral vectors. *EMBO Rep* 4: 1054–1060.

Hostetler, H.A., Collodi, P., Devlin, R.H. et al. 2005. Improved phytate phosphorus utilization by Japanese medaka transgenic for the *Aspergillus niger* phytase gene. *Zebrafish* 2: 19–31.

Hyvönen, P., Suojala, L., Orro, T. et al. 2006. Transgenic cows that produce recombinant human lactoferrin in milk are not protected from experimental *Escherichia coli* intramammary infection. *Infect Immun* 74: 6206–6212.

International Embryo Transfer Society Health and Sanitary Advisory Committee (IETS). 2008. Health Assessment and Care for Animals Involved in the Cloning Process. Savoy, IL: International Embryo Transfer Society.

Johnson, W.H., Luskutoff, N.M., Plante, Y. et al. 1995. Production of four identical calves by the separation of blastomeres from an in vitro derived four-cell embryo. *Vet Rec* 137: 15–16.

Khan, A.S., Fiorotto, M.S., Hill, L.A. et al. 2002. Maternal GHRH plasmid administration changes pituitary cell lineage and improves progeny growth of pigs. *Endocrinology* 143: 3561–3567.

King, A.H., Jiang, Z., Gibson, J.P. et al. 2003. Mapping quantitative trait loci affecting female reproductive traits on porcine chromosome 8. *Biol Reprod* 68: 2172–2179.

Lacham-Kaplan, O., Diamente, M., Pushett, D. et al. 2000. Developmental competence of nuclear transfer cow oocytes after direct injection of fetal fibroblast nuclei. *Cloning* 2: 55–62.

Lai, L., Kolber-Simonds, D., Park, K.W. et al. 2002. Production of α-1, 3-galactosyltransferase-knockout inbred miniature swine by nuclear transfer cloning. *Science* 295: 1089–1092.

Laible, G. 2009. Enhancing livestock through genetic engineering — Recent advances and future prospects. *Comp Immunol Microbiol Infect Disease* 32: 123–137.

Lavitrano, M., Busnelli, M., Cerrito, M.G. et al. 2006. Sperm-mediated gene transfer. *Reprod Fertil Develop* 18: 19–23.

Lin, C.Y., Yang, P.H., Kao, C.L. et al. 2010. Transgenic zebrafish eggs containing bactericidal peptide is a novel food supplement enhancing resistance to pathogenic infection of fish. *Fish Shellfish Immunol* 28: 419–427.

Lois, C., Hong, E.J., Pease, S., Brown, E.J., and Baltimore, D. 2002. Germline transmission and tissue-specific expression of transgenes delivered by lentiviral vectors. *Science* 295: 868–872.

Long, C.R. 2010. Application of RNA interference-based gene silencing in animal agriculture. *Reprod Fertil Dev* 22: 47–58.

Maga, E.A., Cullor, J.S., Smith, W. et al. 2006. Human lysozyme expressed in the mammary gland of transgenic dairy goats can inhibit the growth of bacteria that cause mastitis and the cold-spoilage of milk. *Foodborne Pathog Dis* 3: 384–392.

Mortazavi1, A., Williams, B.S., McCue, K. et al. 2008. Mapping and quantifying mammalian transcriptomes by RNA-Seq. *Nature Methods* 5: 621–628.

National Research Council. 2002. *Animal Biotechnology: Science-Based Concerns*. Washington, DC: National Academy Press.

National Research Council. 2004. *Safety of Genetically Engineered Foods; Approaches to Assessing Unintended Health Effects*. Washington, DC: National Academy Press.

Norman, H.D., Lawlor, T.J, Wright, J.R. et al. 2004. Performance of Holstein clones in the United States. *J Dairy Sci* 87: 729–738.

Panarace, M., Aguero, J.I., Garrote, M. et al. 2007. How healthy are clones and their progeny: 5 years of field experience. *Theriogenology* 67: 142–151.

Perry, A.C., Wakayama, T., Kishikawa, H. et al. 1999. Mammalian transgenesis by intracytoplasmic sperm injection. *Science* 284: 1180–1183.

Pfeifer, A., Eigenbrod, S., Al-Khadra, S. et al. 2006. Lentivector-mediated RNAi efficiently suppresses prior protein and prolongs survival of scrapie-infected mice. *J Clin Invest* 116: 3204–3210.

Pinstrup-Andersen, P. and Pandya-Lorch, R. 1999. Securing and sustaining adequate world food production for the third millennium. In: *World Food Security and Sustainability: The Impacts of Biotechnology and Industrial Consolidation*. D.P. Weeks, J.B. Segelken, and R.W.F. Hardy, Eds. NABC Report 11. Ithaca, NY: National Agricultural Biotechnology Council, pp. 27–48.

Pursel, V.G., Pinkert, C.A., Miller, K.F. et al. 1989. Genetic engineering of livestock. *Science* 244: 1281–1288.

Rana, T.M. 2007. Illuminating the silence: Understanding the structure and function of small RNAs. *Nature Rev Mol Cell Biol* 8: 23–26.

Richt, J.A., Kasinathan, P., Hamir, A.N. et al. 2007. Production of cattle lacking prion protein. *Nature Biotech* 25: 132–138.

Robl, J.M., Wang, Z., Kasinathan, P.L. et al. 2007. Transgenic animal production and animal biotechnology. *Theriogenology* 67: 127–133.

Sabikhi, L. 2007. Designer milk. *Adv Food Nutr Res* 53: 161–198.

Sutter, N.B., Bustamante, C.D., Chase, K. et al. 2007. A single IGF1 allele is a major determinant of small size in dogs. *Science* 316: 112–115.

Thompson, P.B., Bazer, F.W., Einsiedel, E.F., and Riley, M.F. 2010. Ethical implications of animal biotechnology: Considerations for animal welfare decision making. In: CAST Issue Paper 26. Ames, IA: CAST.

U.S. Department of Agriculture (USDA). 2008a. *USDA Agricultural Projections to 2017*. http://www.ers.usda.gov/publications/oce081/oce20081.pdf (25 January 2010).

U.S. Department of Agriculture. 2008b. Statement by Bruce Knight, Under Secretary for Marketing and Regulatory Programs of FDA Risk Assessment on Animal Clones. Office of Communications Release No. 0012.08.

U.S. Department of Health and Human Services Food and Drug Administration Center for Veterinary Medicine. 2008. Guidance for Industry 179: Use of animal clones and clone progeny for human food and animal feed. USD HHS/FDA/CM.

U.S. Food and Drug Administration (USFDA). 2008a. Animal Cloning: A Risk Assessment. http://www.fda. gov/animalveterinary/safetyhealth/animalcloning/ucm055489.htm

U.S. Food and Drug Administration (USFDA). 2008b. FDA approves first human biologic produced by G E Animals. *FDA Veterinarian Newsletter* XXIII(VI).

U.S. Food and Drug Administration, Center for Veterinary Medicine (US FDA-CVM). 2009. Guidance for Industry 187; regulation of genetically engineered animals containing heritable recombinant DNA constructs. Communications Staff, Rockville, MD.

Vajta, G., Lewis, I.M., Hyttel, P. et al. 2001. Somatic cell cloning without micromanipulators. *Cloning* 3: 89–95.

Wall, R., Laible, G., Maga, E. et al. 2009. Animal productivity and genetic diversity: Cloned and transgenic animals. In: CAST Issue Paper 43. Ames, IA: CAST.

Wall, R.J., Powell M.J., and Paape, D.A. 2005. Genetically enhanced cows resist intramammary *Staphylococcus aureus* infection. *Nature Biotech* 23: 445–451.

Wang, J., Wang, W., Li, R. et al. 2008. The diploid genome sequence of an Asian individual. *Nature* 456: 60–66.

Wells, D.N., Misica, P.M., Tervit, H.R. et al. 1998. Adult somatic cell nuclear transfer is used to preserve the last surviving cow of the Ederby Island cattle breed. *Reprod Fertil Dev* 10: 369–378.

Westhusin, M.E., Shin, T., Templeton, J.W. et al. 2007. Rescuing valuable genomes by animal cloning: A case for natural disease resistance in cattle. *J Anim Sci* 85: 138–142.

Willadsen, S.M., 1986. Nuclear transplantation in sheep embryos. *Nature* 320: 63–65.

Wilmut, I., Schnieke, A.E., McWhin, J., Kind, A.J., and Campbell, K.H. 1997. Viable offspring derived from fetal and adult mammalian cells. *Nature* 385: 810–813.

Wilmut, I., Beaujean, N., DeSousa, P.A. et al. 2002. Somatic cell nuclear transfer. *Nature* 19: 583–586.

Wu, F. 2008. Field evidence: Bt corn and mycotoxin reduction. ISB News Report, February.

Yang, P., Wang, J., Gong, G. et al. 2008. Cattle mammary bioreactor generated by a novel procedure of transgenic cloning for large-scale production of functional human lactoferrin. *PLoS ONE* 3: e3453.

Zhang, J.X., Meidinger, R., Forsberg, C.W. et al. 1999. Expression and processing of a bacterial endoglucanase in transgenic mice. *Arch Biochem Biophys* 367: 317–321.

Index

For Product Safety Concerns and Information please contact our EU
representative GPSR@taylorandfrancis.com
Taylor & Francis Verlag GmbH, Kaufingerstraße 24, 80331 München, Germany

www.ingramcontent.com/pod-product-compliance
Ingram Content Group UK Ltd.
Pitfield, Milton Keynes, MK11 3LW, UK
UKHW051828180425
457613UK00007B/252